Nanomaterials Ecotoxicity Evaluation

Nanomaterials Ecotoxicity Evaluation

Editors

Xiaoshan Zhu
Jian Zhao

MDPI

Basel • Beijing • Wuhan • Barcelona • Belgrade • Novi Sad • Cluj • Manchester

Editors

Xiaoshan Zhu
School of Ecology and Environment
Hainan University
Haikou
China

Jian Zhao
College of Environmental Science and Engineering
Ocean University of China
Qingdao
China

Editorial Office
MDPI
St. Alban-Anlage 66
4052 Basel, Switzerland

This is a reprint of articles from the Special Issue published online in the open access journal *Nanomaterials* (ISSN 2079-4991) (available at: www.mdpi.com/journal/nanomaterials/special_issues/nano_ecotoxicity).

For citation purposes, cite each article independently as indicated on the article page online and as indicated below:

Lastname, A.A.; Lastname, B.B. Article Title. *Journal Name* **Year**, *Volume Number*, Page Range.

ISBN 978-3-0365-9978-6 (Hbk)
ISBN 978-3-0365-9977-9 (PDF)
doi.org/10.3390/books978-3-0365-9977-9

Contents

About the Editors

Xiaoshan Zhu

Xiaoshan Zhu is a professor at Hainan University. Prof. Zhu has authored >120 peer-reviewed publications on subjects related to ecotoxicology and marine environment science. His work is widely cited (>5,200 times), with a current H-index of 31 (Google Scholar). He received the first prize in Natural Science from the Ministry of Education in China in 2019 and the second prize in Natural Science from the Shenzhen Government in 2019 for his works on marine ecotoxicology and trophic transfer of manufactured nanomaterials. Currently, he has extensive experience in the area of marine environmental science and ecotoxicology, with an emphasis on marine ecotoxicology of emerging chemicals, especially the fate, transport, transformation, bioaccumulation, and toxicity of manufactured nanomaterials, plastic particles, POPs, heavy metals, etc. in aquatic environments. His interesting research fields also include understanding the responses of marine ecosystems to the single or combined impacts of climate change and human activities.

Jian Zhao

Jian Zhao is a professor at the Ocean University of China. Prof. Zhao is mainly engaged in research on environmental geochemistry and environmental toxicology. He has published more than 50 *SCI* papers, with an H-index of 26 and more than 2700 citations (Web of Science). In January 2014, he was introduced to the Ocean University of China under the "Young Talent Project". He received the first prize in Natural Science from the Ministry of Education in China in 2019. His interesting research fields include understanding the biological effects and biochemical transformation of nanoparticles and microplastics and the environmental applications of nanotechnology.

nanomaterials

Editorial

Editorial for the Special Issue "Nanomaterials Ecotoxicity Evaluation"

Xiaoshan Zhu [1,*] and Jian Zhao [2]

1 College of Ecology and Environment, Hainan University, Haikou 570208, China
2 College of Environmental Science and Engineering, Ocean University of China, Qingdao 266100, China; jzhao@ouc.edu.cn
* Correspondence: zhuxs@hainanu.edu.cn

Nanotechnology has made enormous progress over the last few decades, and the current use of nanomaterials is rapidly increasing. As a result, the continuous release of engineered nanomaterials (ENPs) into air, water, and soil has raised concerns about the possible adverse consequences for environmental and human health.

In this Special Issue, we aim to present a multidisciplinary overview of recent scientific articles that delve into various aspects of nanomaterials' ecotoxicity evaluation associated with their occurrences, behavior, fate, and bioavailability. Such an evaluation is critical for scientists, legislators, business leaders, and the public to understand and develop effective solutions to the potential impacts of nanomaterials. Two reviews are also noteworthy. To address the risks of nanomaterials in the aquatic environment, Yuan et al. reviewed the environmental impact of nanomaterials with inorganic sunscreens [1]. For the risks of nanomaterials in the soil environment, Suazo-Hernández et al. reviewed the impact of ENPs on the physical and chemical properties of soils [2].

The toxicity of nanomaterials remains a significant issue. Lu et al. studied how low concentrations of silver nanoparticles induced novel toxic effects on aquatic ferns [3]. Environmental factors can also affect the environmental behavior of nanomaterials and change their toxic effects, for example, humic acid affects the adsorption and suspension behavior of carbon nanotubes [4].

The co-effect of nanomaterials and heavy metal pollutants remains a concern. On the one hand, the combined effect of nanomaterials and heavy metals leads to higher ecological risks. Zhou et al. studied CuO and Fe_3O_4 nanomaterials to enhance the bioaccumulation and toxicity of arsenic in marine mussels [5]. Similarly, Wang et al. found that CuO NPs promoted the phytotoxicity and accumulation of cadmium in vegetables [6]. On the other hand, the application of nanomaterials will reduce the accumulation of heavy metals in crops. Nano-Fe may modify the microbial community and decrease the soil-available Cd and As contents, inhibit the absorption of Cd and As by the roots, and decrease the transport of Cd and As in rice grains and the risk of intake in humans [7].

The effects of nanoplastics on organisms have attracted much attention. Wang et al. identified multiple mechanisms of toxicity on microalgae induced by nanoplastics [8]. The synergistic impact of microplastics and organic pollutants remains poorly understood in the marine environment. Zhou et al. implied that MPs in synergy with organic pollutants can be more harmful to marine organisms [9].

Nanopesticides are increasingly considered an emerging alternative due to their higher efficiency and lower environmental impacts. Although the public is not familiar with nanopesticides, they have positive attitudes toward their future development and support labeling nanoscale ingredients on products [10]. Ganilho et al. also assessed the environmental risks of lipid nanoparticles loaded with lambda-cyhalothrin [11].

To conclude, this Special Issue presents several examples of the latest advancements in assessing nanomaterials' ecotoxicity. We hope readers will enjoy reading these articles and find them useful in their research.

check for updates

Citation: Zhu, X.; Zhao, J. Editorial for the Special Issue "Nanomaterials Ecotoxicity Evaluation". *Nanomaterials* **2023**, *13*, 2878. https://doi.org/10.3390/nano13212878

Received: 17 October 2023
Accepted: 25 October 2023
Published: 30 October 2023

Acknowledgments: We wish to thank all the editors in the Editorial Office of Nanomaterials, and all the reviewers for their valuable support on this Special Issue, and thank Ciara Chun Chen (Shantou University) for her outstanding coordination works, and Wei Qian (Hainan University) for his help in preparing this Editorial. The work was also supported by the National Natural Science Foundation of China (42077227), the Hainan Key Research and Development Program (ZDYF2022SHFZ317), and the Guangdong Basic and Applied Basic Research Foundation (2021A1515010158).

Conflicts of Interest: The author declares no conflict of interest.

References

1. Yuan, S.; Huang, J.; Jiang, X.; Huang, Y.; Zhu, X.; Cai, Z. Environmental Fate and Toxicity of Sunscreen-Derived Inorganic Ultraviolet Filters in Aquatic Environments: A Review. *Nanomaterials* **2022**, *12*, 699. [CrossRef] [PubMed]
2. Suazo-Hernández, J.; Arancibia-Miranda, N.; Mlih, R.; Cáceres-Jensen, L.; Bolan, N.; Mora, M.d.l.L. Impact on Some Soil Physical and Chemical Properties Caused by Metal and Metallic Oxide Engineered Nanoparticles: A Review. *Nanomaterials* **2023**, *13*, 572. [CrossRef] [PubMed]
3. Lu, Z.; Yin, L.; Li, W.; Jiang, H.-S. Low Concentrations of Silver Nanoparticles Inhibit Spore Germination and Disturb Gender Differentiation of *Ceratopteris thalictroides* (L.) Brongn. *Nanomaterials* **2022**, *12*, 1730. [PubMed]
4. Fang, M.; Zhao, T.; Zhao, X.; Tang, Z.; Liu, S.; Wang, J.; Niu, L.; Wu, F. Effect of Tube Diameters and Functional Groups on Adsorption and Suspension Behaviors of Carbon Nanotubes in Presence of Humic Acid. *Nanomaterials* **2022**, *12*, 1592. [PubMed]
5. Zhou, S.; Qian, W.; Ning, Z.; Zhu, X. Enhanced Bioaccumulation and Toxicity of Arsenic in Marine Mussel *Perna viridis* in the Presence of CuO/Fe_3O_4 Nanoparticles. *Nanomaterials* **2021**, *11*, 2769. [CrossRef] [PubMed]
6. Wang, S.; Fu, Y.; Zheng, S.; Xu, Y.; Sun, Y. Phytotoxicity and Accumulation of Copper-Based Nanoparticles in Brassica under Cadmium Stress. *Nanomaterials* **2022**, *12*, 1497. [CrossRef] [PubMed]
7. Li, T.; Li, J.; Zhan, X.; Wang, X.; He, B.; Cao, F.; Liao, C.; Yu, Y.; Zhang, Z.; Zhang, J.; et al. Application of Exogenous Iron Alters the Microbial Community Structure and Reduces the Accumulation of Cadmium and Arsenic in Rice (*Oryza sativa* L.). *Nanomaterials* **2022**, *12*, 1311. [CrossRef] [PubMed]
8. Wang, S.-C.; Liu, F.-F.; Huang, T.-Y.; Fan, J.-L.; Gao, Z.-Y.; Liu, G.-Z. Effects of Nanoplastics on the Dinoflagellate Amphidinium carterae Hulburt from the Perspectives of Algal Growth, Oxidative Stress and Hemolysin Production. *Nanomaterials* **2021**, *11*, 2471. [CrossRef] [PubMed]
9. Zhou, Y.; Li, Y.; Lan, W.; Jiang, H.; Pan, K. Short-Term Exposure to MPs and DEHP Disrupted Gill Functions in Marine Bivalves. *Nanomaterials* **2022**, *12*, 4077. [PubMed]
10. Liu, P.; Zheng, X.; Shangguan, S.; Zhao, L.; Fang, X.; Huang, Y.; Hermanowicz, S.W. Public Perceptions and Willingness-to-Pay for Nanopesticides. *Nanomaterials* **2022**, *12*, 1292. [PubMed]
11. Ganilho, C.; da Silva, M.B.; Paiva, C.; de Menezes, T.I.; dos Santos, M.R.; Pereira, C.M.; Pereira, R.; Andreani, T. Environmental Safety Assessments of Lipid Nanoparticles Loaded with Lambda-Cyhalothrin. *Nanomaterials* **2022**, *12*, 2576. [CrossRef] [PubMed]

nanomaterials

Review

Impact on Some Soil Physical and Chemical Properties Caused by Metal and Metallic Oxide Engineered Nanoparticles: A Review

Jonathan Suazo-Hernández [1,2,*], Nicolás Arancibia-Miranda [3,4], Rawan Mlih [5], Lizethly Cáceres-Jensen [6], Nanthi Bolan [7,8] and María de la Luz Mora [1,2,*]

1 Center of Plant, Soil Interaction and Natural Resources Biotechnology, Scientific and Biotechnological Bioresource Nucleus (BIOREN-UFRO), Universidad de La Frontera, Avenida Francisco Salazar 01145, Temuco 4780000, Chile
2 Department of Chemical Sciences and Natural Resources, Universidad de La Frontera, Avenida Francisco Salazar 01145, Temuco 4811230, Chile
3 Faculty of Chemistry and Biology, University of Santiago of Chile (USACH), Santiago 8320000, Chile
4 Center for the Development of Nanoscience and Nanotechnology, CEDENNA, Santiago 9170124, Chile
5 Institute of Bio- and Geosciences, Agrosphere (IBG-3), Forschungszentrum Juelich (FZJ), 52425 Juelich, Germany
6 Physical & Analytical Chemistry Laboratory (PachemLab), Nucleus of Computational Thinking and Education for Sustainable Development (NuCES), Center for Research in Education (CIE-UMCE), Department of Chemistry, Metropolitan University of Educational Sciences, Santiago 776019, Chile
7 School of Agriculture and Environment, The University of Western Australia, Perth, WA 6009, Australia
8 The UWA Institute of Agriculture, The University of Western Australia, Perth, WA 6009, Australia
* Correspondence: jonathan.suazo@ufrontera.cl (J.S.-H.); mariluz.mora@ufrontera.cl (M.d.l.L.M.); Tel.: +56-45-2734191 (M.d.l.L.M.)

Abstract: In recent years, the release of metal and metallic oxide engineered nanoparticles (ENPs) into the environment has generated an increase in their accumulation in agricultural soils, which is a serious risk to the ecosystem and soil health. Here, we show the impact of ENPs on the physical and chemical properties of soils. A literature search was performed in the Scopus database using the keywords ENPs, plus soil physical properties or soil chemical properties, and elements availability. In general, we found that the presence of metal and metallic oxide ENPs in soils can increase hydraulic conductivity and soil porosity and reduce the distance between soil particles, as well as causing a variation in pH, cation exchange capacity (CEC), electrical conductivity (EC), redox potential (Eh), and soil organic matter (SOM) content. Furthermore, ENPs or the metal cations released from them in soils can interact with nutrients like phosphorus (P) forming complexes or precipitates, decreasing their bioavailability in the soil solution. The results depend on the soil properties and the doses, exposure duration, concentrations, and type of ENPs. Therefore, we suggest that particular attention should be paid to every kind of metal and metallic oxide ENPs deposited into the soil.

Keywords: nanoparticles; soil properties; environment; emerging pollutants

Citation: Suazo-Hernández, J.; Arancibia-Miranda, N.; Mlih, R.; Cáceres-Jensen, L.; Bolan, N.; Mora, M.d.l.L. Impact on Some Soil Physical and Chemical Properties Caused by Metal and Metallic Oxide Engineered Nanoparticles: A Review. *Nanomaterials* **2023**, *13*, 572. https://doi.org/10.3390/nano13030572

Academic Editors: Xiaoshan Zhu and Jian Zhao

Received: 13 December 2022
Revised: 17 January 2023
Accepted: 21 January 2023
Published: 31 January 2023

1. Introduction

Engineered nanoparticles (ENPs) are materials intentionally produced with a particle size between 1 and 100 nm in at least one dimension, which are present in the form of a nanowire, spherical, nanotubes, and nanorods [1]. ENPs are divided into five classes; based on carbon, zero valence metal, metallic oxide, quantum points, and dendrimers [2]. These nanoparticles possess i) novel physicochemical characteristics such as a high surface area for reactions and interactions, and ii) exceptional optical, magnetic, and electrical properties compared to their bulk counterparts [3–5]. As a consequence of those advantages, in the last decade, the production and subsequent incorporation of ENPs in products such as cosmetics, clothes, pigments, industrial coatings, plastic additives, semiconductors, textiles, and antibacterial agents have increased considerably [6,7]. Currently, there are more than 1800 products containing ENPs in the market [8], and worldwide production of ENPs is expected to reach $125 billion by 2024 [9]. Therefore, nanotechnology is a science that has had and will continue to have great importance in improving the quality of life for humans [10]. However, this also means that the type and volume of ENPs released into the environment will increase [11].

Among the different ENPs that exist in the market, metal (e.g., Au, Al, Ag, Fe, and Cu) and metallic oxide (e.g., TiO_2, ZnO, Al_2O_3, Fe_3O_4, Fe_2O_3, NiO, CuO, Cu_2O, and CeO_2) ENPs are those with the greatest probability of being deposited in soils and in particular for agricultural use [12–14]. This is because they have antimicrobial properties, or the elements released from ENPs are nutrients for plants. Consequently, they can be incorporated and/or used as pesticides, insecticides, herbicides, fungicides, and fertilizers [15–18]. Various studies have shown that applying metal and metallic oxide ENPs are a promising alternative to treat infections in plants and increase plant development without impairing productivity by reducing the number of agrochemicals added to the soil. In this sense, it is estimated that the concentration of metal and metallic oxide ENPs deposited in agricultural soils could increase from 30 pg kg^{-1} in 2017 to 10 g kg^{-1} in 2050 [19]. Therefore, monitoring their presence in this non-renewable natural system is essential.

Once the metal and metallic oxide ENPs are in contact with the soil system, they can leach into groundwater or suffer biological, chemical, and photochemical transformations (e.g., homo/heteroaggregation, oxidation, dissolution, and precipitation) [20–22]. As a result, ENPs can cause changes in the biological (e.g., mesofauna, macrofauna, and microbiota), physical (e.g., hydraulic conductivity, porosity, texture, bulk density, aggregation), and chemical (e.g., cation exchange capacity (CEC), electricity conductivity (EC), redox potential (Eh), pH, dissolved organic matter (DOM), and organic matter (OM) content) properties of the soil [16,20,23–25]. In fact, due to their small particle size, ENPs can interact with plant nutrients, like phosphorus (P), affecting their availability in the soil solution [26–28].

Various reviews have been published about the transport, fate, and transformations of ENPs in soils and their effect on the abundance and diversity of microorganisms and on plant growth to date [21,29–31]. By contrast, there are only a few reviews about the effect of different metal and metallic oxide ENPs on the soil abiotic properties. One of the most recent reviews was published by Sun et al. [13], who concluded that an increase in the concentration of ENPs in soils might affect soil biochemical properties.

Therefore, there is a need to link the information about the effect of metal and metallic oxide ENPs on soil systems through a review. This review aimed to present the effects that different types of metal and metallic oxide ENPs cause on the physical and chemical properties of the soil. This review will help to understand the impact of ENPs on health and the balance in the soil system.

2. Soil Health and Quality

The physical and chemical properties of soil significantly influence soil health and quality. Therefore, the possibility that those factors decrease in the face of the effect of the millions of tons of ENPs accumulated annually is very high [19]. It is known that ENPs can suffer different processes and transformations in the soil system. In contrast, the impact on abiotic properties of soil and biogeochemical cycles has gone practically unnoticed due to the limited and scattered evidence [20]. However, more recently, studies have increased, revealing that physical and chemical properties are affected by the deposition of ENPs [13]. A list of studies investigating the impacts of ENPs on various abiotic soil properties is provided in Table 1 and a summary is provided in Figure 1.

Table 1. Effect of metal and metallic oxide engineered nanoparticles (ENPs) on some soil physical and chemical properties.

ENPs Type	Concentration	Type or Place of Soil	Duration	Remarks	Reference
			Physical properties		
Pt	0.1–1000 μg g^{-1}	Clay-free organic sapric histosol	5 weeks	Increase structural rigidity of SOM and aliphatic crystallites content; decrease in the enthalpy of evaporation of water in the SOM	[24]
γ-Al_2O_3 and CuO	0.05–0.3% γ-Al_2O_3, 0.15–0.7% CuO	Selangor, Malaysia	10 days	Reduction of the swelling stress and the shrinkage stress of the soil; decrease in hydraulic conductivity and density	[32]
Fe_3O_4 and MgO	1, 3, 5% (*w/w*)	Agricultural land in Hamedan, Iran	100 days	The bulk density of the soil increases with the dose of Fe_3O_4 ENPs but decreases with MgO ENPs	[33]
SiO_2 and Zn	50 mg L^{-1} of Zn or 2.5 mg L^{-1} of SiO_2	El-Serw Agricultural Research Station	7 weeks	Increase in the hydraulic conductivity of the soil; decrease in bulk density	[34]
Fe	1, 4, 7, 10 g L^{-1}	Oxisol	-	Concentrations <4 g L^{-1} do not affect the natural hydraulic conductivity of the soil. However, higher concentrations reduced the hydraulic conductivity value	[35]
Ag coated with polyvinylpyrrolidone (PVP) and citrate	2.5, 5.0, 10 mg L^{-1}	Red Soil	-	The surface coatings of Ag ENPs block the solid phase sites promoting the transport of the ENPs	[36]
			Chemical Properties		
Ag	20, 25, 50 and 100 mg kg^{-1}	Alluvial soil of Tezpur, India	60 days	Increase in the CEC, pH soil, and N and P bioavailability	[10]
CuO and Fe_3O_4	1 or 5% (*w/w*)	Red Sandy clay loam Mediterranean soil and Rendzina soil	24 h	Fe_3O_4 ENPs catalyze the oxidation of organic pollutants in aqueous suspensions, inducing changes in SOM	[23]
TiO_2, ZnO and CuO	50, 100 and 500 mg kg^{-1}	Paddy soils	90 days	Increase of soil pH, Eh, and EC in flooding-drying process	[28]
Fe	0.1, 1, 10 mg g^{-1}	Hangzhou, Taizhou, Haikou, Kunming, Honghe, Chifeng, Puer, and Yingtan	90 days	ENPs promote aromatic carbon sequestration and decrease the Eh of the soil. The impact of ENPs on soil pH, EC, ζ potential, dissolved organic carbon (DOC), and enzyme activity is dependent on the soil type and soil moisture content	[37]
Fe	10 mg g^{-1}	Silt loam soil	14 days	Decrease in Eh and increase in soil pH	[38]
Fe	28–36 mg g^{-1}	Acidic and calcareous	30 days	Modification of pH values depending on the buffering capacity of the soil; increased EC and water retention capacity of soils	[39]
Fe	1, 5, and 10% (*w/w*)	From El Terronal and Asturias.	72 h	No effect on soil pH and EC	[40]

Table 1. *Cont.*

ENPs Type	Concentration	Type or Place of Soil	Duration	Remarks	Reference
TiO_2	1 and 500 mg kg^{-1}	Sandy-loam, loam and silty-clay	90 days	Low doses of ENPs decrease the mineralization of C in a clay-silty soil	[41]
Fe_3O_4	1000 and 2000 mg L^{-1}	Loamy	30 days	Increase the CEC and total P content and P extractable	[42]
Fe	0.10–2.0 g L^{-1}	-	48 h	The residual DOM has a higher reduction capacity, % mineralization and photodegradation after the adsorption of ENPs	[43]
CuO	50,100, 500 and 1000 mg kg^{-1}	Hangzhou	88 days	High concentrations of ENPs decrease the Eh but improve EC; increased soil pH; increased phyto-availability of Cu in the soil	[44]
TiO_2	50 and 100 mg kg^{-1}	Seoul	40 days	Increase in EC and decrease in pH of the rhizosphere; improves P dissolution	[45]
CuO	10, 100, and 1000 mg kg^{-1}	Paddy soils	90 days	Increased degradation and mineralization of OM; increased in soil pH	[46]
CeO_2 and TiO_2	500 mg kg^{-1}	Southern Australian soils	260 days	Both ENPs alter the mineralization of organic N and/or the nitrification rates of the soil due to the catalytic and/or antimicrobial properties of the ENPs; increase in the phyto-availability of P and Zn in soils	[47]
ZnO	2.5 mg kg^{-1}	Inceptisol	60 days	Decrease in soil pH and SOC; increased EC and P available	[48]
SiO_2	4.5 mg L^{-1}	Wuhan, Chongqing, and Qianjiang	24 h	Decreased mobility of pesticides in soils, although this effect varies with the composition of the soil	[49]
ZnO	100 and 1000 mg kg^{-1}	Agronomy farm of Faisalabad	64 days	Increase in soil pH and C mineralization	[50]
Fe, Fe_3O_4, and Fe_2O_3	2 to 6 g kg^{-1}	Udic Ferrosols and Anthrosol	60 days	Fe ENPs increase in DOC and available NH_4^+-N but decrease available phosphorus (AP), while Fe_3O_4 and Fe_2O_3 ENPs slightly reduce soil pH and decrease available NH_4^+-N and AP	[51]
CuO	10 and 100 mg kg^{-1}	Sandy soil	31 days	Increase soil pH	[52]
ZnO	1.0 and 20.0 mg	Agricultural-clay soil and peaty soil	4 weeks	Decrease in the content of Al, Ca, Cu and Mg in the soil	[53]

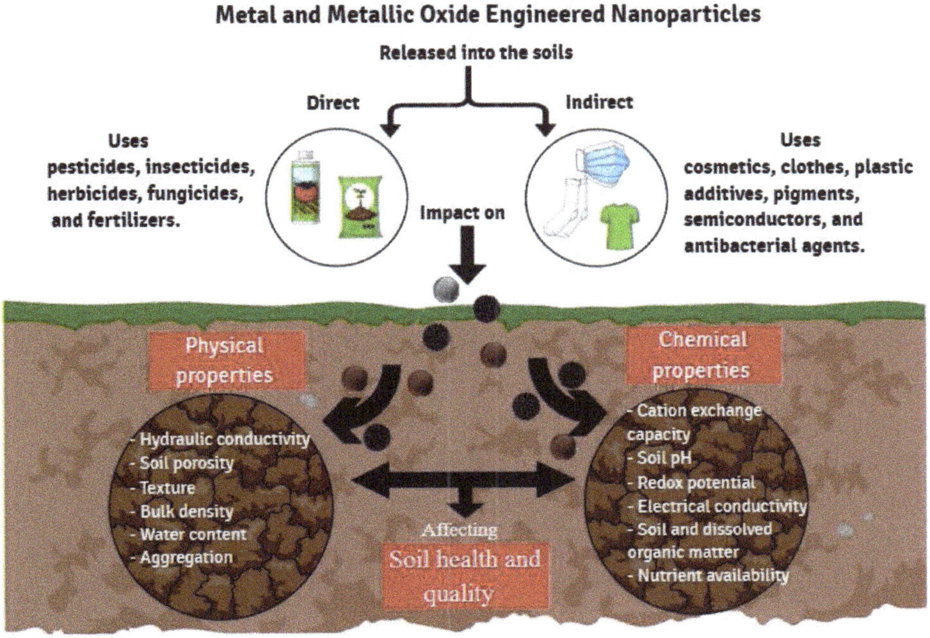

Figure 1. Summary of the impact of metal and metallic oxide engineered nanoparticles on physical and chemical soil properties.

3. Effect of ENPs on Soil Properties

3.1. Physical Properties

The physical and mechanical soil properties, including structure, bulk density, porosity, permeability, texture, temperature, moisture, and others, are relevant because they are correlated with the productivity of different plants and root growth [33]. Studies on physical properties have shown that ENPs such as Fe_3O_4, ZnO, MgO, SiO_2, and TiO_2 [33,34,54] can increase hydraulic conductivity and soil porosity and reduce the distance between soil particles (Table 1). As a consequence of this, ENPs aid in forming a more rigid matrix, favoring the increase in agricultural productivity and producing a safer environment and a healthier life. Concerning this, Bayat et al. [33] investigated the effects of the application of Fe_3O_4 and MgO ENPs (three doses of 1, 3, and 5% *w/w*) on total porosity, mean weight diameter aggregate, volumetric water content, penetration resistance, and saturated hydraulic conductivity during incubation periods of 40 and 100 days. They concluded that only MgO ENPs improved the soil's physical and mechanical properties due to their excellent adhesiveness, specific surface, activity, and reaction capacity.

Similarly, Bayat et al. [55] added MgO and Fe_3O_4 ENPs (<100 nm) into calcareous loamy soil after being subjected to various stresses. They found that MgO ENPs caused a decrease in soil bulk density compared to the effect produced by Fe_3O_4 ENPs. The reduction in density provides better aeration and penetration of roots in the soil. The difference was related to the smaller particle size of MgO ENPs (however, the specific size for both ENPs was not shown) compared to Fe_3O_4 ENPs. It was also found that MgO NPs improved soil structure, increased porosity, and reduced bulk density, whereas Fe_3O_4 ENPs only increased the tensile strength of the aggregates by strengthening the bonds between Fe and soil particles [55]. In addition, the physical properties of soil can be affected by the concentrations and particle sizes of ENPs. Komendová et al. [24] observed an increase in the strength of the water molecule bridges and the structural rigidity of the soil after using Pt ENPs of 3 nm in concentrations of 0.1, 1, and 10 μg by 300 mg^{-1} soil. However, at concentrations between 100 and 1000 μg by 300 mg^{-1} soil, they decreased the water content retained SOM. In the same way, Fe ENPs, with a smaller particle size than the pores of clay soil, managed to leach through it, but over time the soil pores could become clogged and consequently reduce the hydraulic conductivity due to the formation of aggregates [35].

3.2. Chemical Properties

3.2.1. Cation Exchange Capacity

The CEC of soil depends on the surface charge and surface area. In agricultural soils, CEC is a relevant parameter because it is an indicator of the ability of the soil to adsorb nutrients, in other words, of its fertilization [10,23]. A few studies have shown information on CEC in the presence of ENPs. De Souza et al. [42] showed that adding 2000 mg L^{-1} of Fe_3O_4 ENPs into a clay-textured soil caused an increase a 17% in the CEC (49.2 meq 100 g^{-1}) compared to the control soil (42.2 meq 100 g^{-1}). Meanwhile, Baragaño et al. [56] treated technosol soil with Fe ENPs in a 97.5:2.5 soil–ENPs ratio and obtained null variations in the CEC values.

Several researchers have stabilized metal and metallic oxide ENPs with organic molecules or immobilized and blended them with substrates. In this regard, the type of material used is highly relevant to changes shown by CEC values. Das et al. [10] synthesized Ag ENPs through green synthesis using an extract plant leaf (*Thuja occidentalis*) (GSENPs) and conventionally synthesized silver engineered nanoparticles (CSENPs). Both MPs were added in doses of 20, 25, 50, and 100 mg kg^{-1} to an alluvial soil, causing an increase in CEC between 1.01 and 3.35 times for CSENPs and between 1.27 and 3.47 times for GSENPs compared to control soil. This was because both ENPs caused an increase in soil porosity between 1.12 and 1.26 times for CSENPs and between 1.07 and 1.31 times for GSENPs, generating an improvement in the rate of stabilization of OM in soil. In addition, Ag GSENPs generated a change in the soil ionization, increasing the reactive surface and the net negative charge. Likewise, an increase in CEC has been reported between 9.4% and 64.1% for plowed soil with a dose range between 0.05–1.60%, w/w of Fe_3O_4 ENPs-biochar compared to soil without ENPs [57].

Similarly, adding a blend of Fe ENPs-compost-biochar composite to the soil from northern Spain after 15 and 75 days increased the CEC between 7 and 6.8 times, respectively, compared to the control soil [58]. As a control treatment, the authors added a sand-Fe ENPs mixture to the soil, and the CEC values obtained were similar to the control soil. Thus, they concluded that changes in the CEC were not associated with ENPs but were caused by biochar.

3.2.2. Soil pH

Soil pH is a factor that is directly related to soil fertility and health [59]. The pH values obtained in soils with ENPs are diverse. Studies carried out on different soils with ENPs of Ag [10], phytogenic iron oxide [60], ZnO [61], CuO [52], Fe_3O_4 [62], and ZnO and CuO [28] have shown a slight increase in pH values (Table 1). Gao et al. [52] used 10 mg kg^{-1} of CuO ENPs in sandy soil and determined that the pH ranged from 4.9 to 5, which is similar to the pH value of the control soil. Meanwhile, with 40 mg kg^{-1} CuO ENPs the pH increased from 5.1 to 5.4. This increase was less than expected, suggesting that the soil buffer capacity limited the increase in soil pH. The mechanism involved in the slight increase pH was the hydrolysis of CuO ENPs (it can also be used for ZnO ENPs) caused by the water contained in the soil pores and represented by Equations (1) and (2).

$$CuO_{(S)} + H_2O_{(I)} \leftrightarrow Cu(OH)_{2(S)} \tag{1}$$

$$Cu(OH)_{2(S)} + 2H^+_{(aq)} \leftrightarrow Cu^{2+} + 2H_2O_{(I)} \tag{2}$$

Fe ENPs have been widely used in studies of nanoremediation [63,64]. Therefore, a high amount of those ENPs can be deposited into natural soil systems. In this context, adding 10 mg g^{-1} Fe ENPs to soils from Hangzhou increased the pH between

0.10–0.40 units [37]. These results were attributed to the oxidation process of Fe ENPs in the environment, represented by Equations (3) and (4) [65,66].

$$Fe^{0}_{(s)} + 4H^{+}_{(ac)} + O_{2(aq)} \rightarrow 2Fe^{2+}_{(aq)} + 2H_2O \quad (3)$$

$$Fe^{0}_{(s)} + 2H_2O \rightarrow 2Fe^{2+}_{(aq)} + H_{2(g)} + 2OH^{-}_{(aq)} \quad (4)$$

Subsequently, the Fe^{2+} released can be oxidized according to Equations (5) and (6) [67].

$$2Fe^{2+}_{(aq)} + 2H^{+}_{(ac)} + 1/2O_{(2)(aq)} \rightarrow 2Fe^{3+}_{(aq)} + H_2O_{(l)} \quad (5)$$

$$2Fe^{2+}_{(aq)} + 2H_2O_{(l)} \rightarrow 2Fe^{3+}_{(aq)} + H_{2(g)} + 2OH^{-}_{(aq)} \quad (6)$$

On the other hand, there are studies where the changes in soil pH have been due to an indirect action of ENPs. In the study carried out by Zhang et al. [68], 100 mg kg^{-1} of Ag ENPs were added to soils in the absence or presence of cucumber (*Cucumis sativa*) plants. After 60 days, it was determined that the pH increased from 5.28 to 5.33 and from 5.18 to 5.26 for soil with and without the plant, respectively, which was associated with the alteration of metabolites in the soil by exposure to Ag ENPs.

Some studies have reported a slight decrease in soil pH values after the incorporation of ENPs. Duncan and Owens [47] found that after adding 500 mg kg^{-1} of CeO and TiO_2 ENPs to Australian soils, the pH values decreased between 0.1 and 0.3. On the other hand, Zahra et al. [45] showed that adding 50 and 100 mg kg^{-1} of TiO_2 ENPs to soil decreased the rhizosphere pH from 7.3 to 7.1. These authors did not provide information on the mechanism involved in the pH decrease because it was not the objective of their study. In the presence of 2, 4, and 6 g kg^{-1} of Fe, Fe_2O_3, and Fe_3O_4 ENPs in red soil, the pH decreased between 0.4 and 0.8 units on day 7 and in the Wushan soil between 0.60 and 1.10 units on the day 2 compared to the control soils. The acidification of both soils was related to the hydrolysis of Fe^{3+} ions [51].

The variation in soil pH by the presence of ENPs depends on matrix properties and the type of ENPs. After adding Fe ENPs to acidic soil, Mar Gil-Díaz et al. [39] found that the pH increased from ≈5.30 to 7.60, while for calcareous soil, the pH value was nearly 8.0 with and without ENPs. In the calcareous soil, pH values showed no variation, which was explained by the high carbonate content, and their capacity to buffer soil pH variations ($CaCO_3$ = 5.6% for calcareous soil and 0.15% for acid soil) [39]. After the incorporation of CuO ENPs at 10, 100, and 1000 mg kg^{-1} into two soils from Huizhou, Shi et al. [46] found that ENPs caused a significant pH increase in soil with less OM content. In addition, Cu^{2+} ions released from ENPs into the solution progressed towards the formation of more stable species such as Cu_2S and $Cu(OH)_2$, which also increased the soil pH.

3.2.3. Redox Potential

The soil Eh represents the oxidation-reduction reactions and depends on the oxygen (O_2) concentration, precipitation, temperature, and OM content [28]. Eh in agronomy is an essential parameter due to influences in the functioning of the soil–plant–microorganism system and the solubility of nutrients and contaminants. Studies of Eh soils without and with ENPs have received scant attention due to the interdependence between pH and Eh and the difficulty of reproducing, comparing, and interpreting the results obtained [69]. In soils, most metal ENPs tend to oxidize. In other words, they lose electrons, which are captured by substances from the external environment and, as a result, change the Eh values. Fe ENPs, due to their reduction potential (E^0, −0.41 V), are easily oxidized by O_2 of the environment, forming Fe^{2+}/Fe^{3+} species [70]. Those cations can form a superficial shell-core in ENPs formed by different iron oxides [71]. As a consequence of the redox process, Fe ENPs has been widely used to degrade organic pollutants such as chlorinated methane, benzenes, organochlorine pesticides, chlorinated phenols, and to reduce inorganic pollutants such as As^{V}, Se^{VI}, Cr^{VI}, Pb^{2+}, Hg^{2+}, and Zn^{2+} [70]. Vítková et al. [72] investigated the effect of Fe ENPs application on Zn and As availability in the rhizosphere of contaminated soils and found that Eh for the control soil ranged between 310–410 mV. After incubating the soil with As (15.9 g kg^{-1}), and Fe ENPs at 1 wt%, the Eh increased after a week, but after 5 weeks, it decreased. By contrast, when they added Zn (4.1 g kg^{-1}) and Fe ENPs at 1 wt%, there was an increase in Eh from 400–460 mV. The difference was associated with the presence of redox-active elements such as As, Fe, Mn, O_2, and NO_3^-, and their rapid reaction with Fe ENPs. The authors concluded that the variation in Eh values was highly dependent on doses of ENPs and incubation time, which was associated with the amount of reactive mass of ENPs.

In the case of metallic oxide ENPs, although elements are oxidized, they can influence the modification of the soil Eh. For instance, mixed-valence of Fe_3O_4 ENPs, uncoated and with dimercaptosuccinic acid (DMSA) coating, were added in natural wetland organic-rich soil. The Eh values obtained with ENPs were between 350 and 440 mV, while for the control soil they fluctuated between 417 and 457 mV [62]. Environmental conditions, such as aerobic and anaerobic systems as well as flooding conditions, are determining factors in the variations of Eh values. Studies on rice growth have related the variations in Eh values with the changes and transformations of ENPs [44,73]. Peng et al. [44] determined that 1000 mg kg^{-1} of CuO ENPs during the maturation stage of the rice caused an Eh decrease of 202.75 mV compared to the control system. The reason was that ENPs have catalytic properties; therefore, they can accelerate the generation of organic reducing substances. On the other hand, Peng et al. [61] reported that the addition of treatments of 50, 100, and 500 mg kg^{-1} ZnO, CuO and CeO_2 ENPs, increased the Eh values from −222,67 mV (control soil) to −130 mV–−75 mV for all treatments. In particular, Eh values proved to be highly influenced by doses of ENPs. In addition, they evaluated flood conditions for 30 days, where the Eh value decreased due to the presence of ENPs. This behavior was occasioned by the depletion of O_2 in the soil due to microbial respiration and by producing organic reducing substances through OM decomposition [61]. Conversely, Zhang et al. [28] determined that CuO and ZnO ENPs in flooding conditions in a paddy soil increased Eh values by about 20~30 mV, which was explained because, in flooding conditions, ENPs can consume the reducing substance (H^+) [28].

Other factors that influence soil Eh values are related to the presence of stabilizing agents such as proteins, humic acid, and chloride [74], and toxic effects of ENPs on soil microorganisms, which have been analyzed in various reviews [9,20,75,76]. In the case of Ag ENPs, which have antimicrobial properties, it has been reported that the variations of Eh values have been a consequence of the decrease in soil microbiology [28].

3.2.4. Electrical Conductivity

Soil electrical conductivity (EC) is a measure of total soluble salts. Various studies using different soils have reported that ENPs such as ZnO, CuO, and CeO_2 [44], CuO, TiO_2, ZnO [61], CuO [28], and TiO_2 [45] have increased the EC values. In particular, Zahra et al. [45] found that with 50 and 100 mg kg^{-1} of TiO_2 ENPs, the EC values of the rhizosphere increased from ≈0.36 µS cm^{-1} to 0.60 µS cm^{-1} and 0.52 µS cm^{-1}, respectively. The explanation was associated with the dissolution process of ENPs, which caused an increase in the number of cations in the solution. In the experiment carried out by García-Gómez et al. [77], the biological effect of ZnO ENPs on earthworms in agricultural soils was evaluated. They found at day 0 that the EC values were 284 µS cm^{-1} and 216 µS cm^{-1} for the soil control and system ZnO ENPs + soil, respectively, while after 35 days, the EC value for the control soil was 314 µS cm^{-1}, and for the system ZnO ENPs + soil was 283 µS cm^{-1}. The increase in the EC values obtained with the exposure time was related to the solubilization of ENPs. There are also studies where a decrease has been reported in EC values of soil due to the presence of ENPs. For example, in a study conducted by García-Gómez et al. [78] in soils located near Madrid, the EC decreased by ZnO ENPs, which was associated with the capacity of ENPs and/or the cations released from ENPs to combine with cations or anions contained in the soil. Similarly, after applying ZnO and SiO_2 ENPs (2% and 6%, respectively) in saline soils, Kheir et al. [34] reported a slight decrease in EC values compared to the control soil. However, in this case, the reasons involved in the EC values obtained were not explained.

There are studies where the addition of Ag ENPs has generated null effects on EC values, which has been mainly related to the low doses of ENPs [79,80]. Ag ENPs, in particular, are highly stable, so during laboratory experiments, it is unlikely they undergo oxidation processes and release cations into the soil solution [81]. Fabrega et al. [82] found that in a concentration range between 2–2000 $\mu g\ L^{-1}$ of Ag ENPs, less than 2% of ENPs were solubilized. Likewise, the stability of ENPs was increased when stabilizers such as PVP [81] or citrate [83] were used.

3.2.5. Soil and Dissolved Organic Matter

The most productive agricultural soils contain a high percentage of OM. Several studies have evaluated the effect of OM (SOM and DOM) on the toxicity, transformations, and mobility of ENPs [25,84–86]. It is known that DOM can be adsorbed on the surface of ENPs, improving their stability and preventing the release of ions from them [87,88], which reduces the toxicity of ENPs [89]. This is due to the mechanism that exists between ENPs being electrostatic and/or steric repulsion, which decreases the aggregation rate and the residence time of ENPs in the soils, thereby increasing the possibility of moving towards other natural systems such as groundwater and rivers [90].

On the other hand, the effects caused by ENPs on DOM are diverse. For example, in a study conducted by Lin et al. [60] in soils taken near a mine in Hunan, which was treated with 9% phytogenic iron oxide nanoparticle (PION), it was found that the DOC increased between 1.54 and 2.81 times compared to the system without ENPs. These results were related to the nature of PION because *Excoecaria cochinchinensis*, which was used as a reducing agent for ENPs, contains a large number of organic biomolecules. These molecules can be easily decomposed/degraded by soil microorganisms [60]. On the other hand, Zahra et al. [45], after investigating the effect of concentrations of 50 mg kg^{-1} and 100 mg kg^{-1} of TiO_2 ENPs in soils from China, found a dual behavior since 50 mg kg^{-1} of TiO_2 ENPs reduced the DOC by 11.6%, but with 100 mg kg^{-1} of TiO_2 ENPs the DOC increased by 25.5%. Specifically, the increase in DOC was explained by two reasons: (i) root–microbe interactions can stimulate roots to secrete a greater amount of exudate, and (ii) roots with a high quantity of ENPs can cause stress to the plant, inducing the release of low molecular weight substances (LMWS) such as oxalate, acetate, and malate [45,91].

In the soil, microorganisms are responsible for regulating OM decomposition and nutrient mineralization. However, ENPs due to catalytic and/or antimicrobial properties or as a consequence of the decrease in soil pH can decrease SOM content [47,92]. Some metal and metallic oxide ENPs that have shown those properties are Ag, Fe, TiO_2, ZnO, and CuO [46,93]. Rashid et al. [93] investigated the effect of 1000 mg kg^{-1} of ZnO ENPs on carbon and nitrogen mineralization of *Phoenix dactylifera* leaf litter in sandy soil. They found that ENPs reduced carbon (130%) and nitrogen (122%) mineralization efficiency from date palm leaf litter in sandy soil. The reason was due to the soil with ENPs having a lower microbial biomass carbon and the number of colonies of heterotrophic cultivable fungi and bacteria. By contrast, Shi et al. [46], after flooding a paddy soil for 60 days with a concentration of 1000 mg kg^{-1} of CuO ENPs, found that the mineralization of OM was accelerated, as well as increasing the Fe reduction process by increasing the Fe^{2+} content by 293%. These results were associated with the catalyst properties of ENPs.

On the other hand, null changes in total OM content have been determined using concentrations of 10 and 100 mg kg^{-1} of CeO_2, Fe_3O_4 and SnO_2 NPs [94], 1000 mg kg^{-1} of ZnO ENPs [93], 10 and 100 mg kg^{-1} of Ag ENPs [95], and 1% (w/w) Fe_3O_4 and CuO ENPs: soil [23]. The reason was related to the low amount of added ENPs. Specifically, Ben-Moshe et al. [23] added Fe_3O_4 and CuO ENPs to a Red Sandy clay loam and Rendzina soil.

3.2.6. Nutrients Availability

All plants require macronutrients like P, nitrogen (N), and potassium (K) for growth. In particular, P in the environment exists as H_3PO_4, $H_2PO_4^-$, HPO_4^{2-}, and PO_4^{3-}, the dissociation constants of which are: $pK_1 = 2.21$, $pK_2 = 7.21$, and $pK_3 = 12.67$, respectively [96]. In agricultural soil, phosphate in $H_2PO_4^-$ and HPO_4^{2-} helps plant growth and microorganisms, whose bioavailability may be affected by the deposition of pollutants, including ENPs [97]. Various studies conducted in aqueous systems have reported that phosphate can be adsorbed on ENPs like CeO_2 [98], magnetic iron oxide [99], Fe [100,101], ZnO [13], Fe/Cu [102], and TiO_2 [103]. It has been established that there is a chemical interaction between phosphate and active sites of different ENPs; the bonding is irreversible. In addition, those studies suggest that cations released from ENPs can form complexes and/or precipitates with phosphate. Although those investigations were not carried out in soils, they could be an approach to what could happen in the soil matrix. In fact, in the study carried out by Moharami and Jalali [104], they found that Al_2O_3 and Fe_3O_4 ENPs increased phosphate adsorption in calcareous soil. In addition, the presence of ENPs favored the transfer of phosphate from the HCl-P fraction to the Res-P and NaOH-P. Based on this, they concluded that the bioavailability of phosphate decreases due to the addition of ENPs [104]. In the same way, Koopmans et al. [105], using ferrihydrite of a size between 2–3 nm and a surface area of about 5.4 $m^2\ g^{-1}$, determined that the phosphate concentration in the 0.01 M $CaCl_2$ soil extracts decreased. Recently, Suazo-Hernández et al. [26,27] determined that L-ascorbic acid-coated Cu or Ag ENPs increased phosphate adsorption in an Andisol and its fractions. Particularly, in Suazo-Hernández et al. [26], using the Langmuir model, they concluded that by increasing Ag or Cu NPs content from 0 to 5%, the q_{max} values of Pi for the Andisol increased by 46% and 54% following the addition of Cu or Ag ENPs, respectively. These results were attributed to a decrease in soil solution, which is due to the coating of ENPs with L-ascorbic acid and probably some dissolved L-ascorbic acid. This study is relevant because Cu or Ag ENPs are being used as nano-pesticides, so large amount of ENPs can be deposited in soils.

One of the ENPs most likely to reach agricultural soils is ZnO because Zn is a necessary micronutrient for plants, and can therefore be incorporated into soils through agrochemicals [106]. The interaction between PO_4^{3-} and ZnO ENPs is related to the release of ions from ENPs. Subsequently, they can form a micrometer scale crystalline zinc phosphate and a nanoscale amorphous $Zn_3(PO_4)_2$ shell [107]. Likewise, Zn is one of the essential structural components of the enzymes phytase and phosphatase which participates in the mobilization of native P. Studies carried out by Verma et al. [48] found that ZnO ENPs increased the secretion of P mobilizing enzymes and consequently increased the concentration of phosphate bioavailable in the soil. Thus, the effect of ENPs on the bioavailability of P can be regulated by both direct and indirect factors. When TiO_2 and Fe_3O_4 ENPs were added to a sandy-loam soil, the phytoavailability of the P bound to the rhizosphere increased. The results were due to the acidification produced by the exudation of organic acids of *Lactuca sativa* roots Zahra et al. [108]. A similar mechanism was proposed for an increase in the concentration of available phosphate in the presence of Fe ENPs [99] and CeO_2 ENPs [47]. According to the report by Feng et al. [109], composites of CeO_2 ENPs-functionalized maize straw biochar (CeO_2-MSB) decreased the total phosphorus (TP) concentration of surface water by 27.33% and increased the TP content of the upper soil layer by 7.22%. Although this indeed caused an increase in P adsorption, it could be interesting to establish that the interaction between P and soil caused an increase in the height of the rice plant and the foliar area. Therefore, CeO_2-MSB could be used to reduce the risk of P loss from the surface of rice fields.

4. Conclusions and Perspectives

Metal and metallic oxide ENPs deposition/accumulation in soils will increase over time. In general, we have determined that ENPs can compact the particles, helping to improve their rigidity, as well as causing changes in pH, EC, Eh, and SOM. These results depended on the soil properties and the doses, concentrations, and types of ENPs. Furthermore, the presence of ENPs or the cations released from them in soils can interact with nutrients, forming complexes or precipitates and modifying their availability in the soil solution. Research into the impacts of ENPs on physical and chemical soil properties is still in its initial stage. For this reason, future studies should investigate not only the advantages of applications of metal and metallic oxide ENPs in agricultural systems but also their risks and disadvantages, like their impact on soil health and quality, considering abiotic properties as well as microorganisms and plants in the short and long term.

Author Contributions: J.S.-H., N.A.-M., R.M., L.C.-J., N.B. and M.d.l.L.M. wrote, coordinated, reviewed, and contributed to the scientific aspects of the article. All authors have read and agreed to the published version of the manuscript.

Funding: This work was partially funded by the Research Directorate of Universidad de La Frontera.

Data Availability Statement: Not applicable.

Acknowledgments: Jonathan Suazo-Hernández acknowledges the Technological Bioresource Nucleus (BIOREN-UFRO) and the Soil and Plant Laboratory. Lizethly Cáceres-Jensen acknowledges to FONDECYT REGULAR N° 1221634.

Conflicts of Interest: The authors declare no conflict of interest.

References

1. Peng, C.; Zhang, W.; Gao, H.; Li, Y.; Tong, X.; Li, K.; Zhu, X.; Wang, Y.; Chen, Y. Behavior and Potential Impacts of Metal-Based Engineered Nanoparticles in Aquatic Environments. *Nanomaterials* **2017**, *7*, 21. [CrossRef]
2. Shrivastava, M.; Srivastav, A.; Gandhi, S.; Rao, S.; Roychoudhury, A.; Kumar, A.; Singhal, R.K.; Jha, S.K.; Singh, S.D. Monitoring of Engineered Nanoparticles in Soil-Plant System: A Review. *Environ. Nanotechnol. Monit. Manag.* **2019**, *11*, 100218. [CrossRef]
3. Auffan, M.; Rose, J.; Bottero, J.-Y.; Lowry, G.V.; Jolivet, J.-P.; Wiesner, M.R. Towards a Definition of Inorganic Nanoparticles from an Environmental, Health and Safety Perspective. *Nat. Nanotechnol.* **2009**, *4*, 634–641. [CrossRef]
4. Simonin, M.; Colman, B.P.; Anderson, S.M.; King, R.S.; Ruis, M.T.; Avellan, A.; Bergemann, C.M.; Perrotta, B.G.; Geitner, N.K.; Ho, M.; et al. Engineered Nanoparticles Interact with Nutrients to Intensify Eutrophication in a Wetland Ecosystem Experiment. *Ecol. Appl.* **2018**, *28*, 1435–1449. [CrossRef]
5. Wang, Y.; Cai, R.; Chen, C. The Nano-Bio Interactions of Nanomedicines: Understanding the Biochemical Driving Forces and Redox Reactions. *Acc. Chem. Res.* **2019**, *52*, 1507–1518. [CrossRef]
6. Wigger, H.; Hackmann, S.; Zimmermann, T.; Köser, J.; Thöming, J.; von Gleich, A. Influences of Use Activities and Waste Management on Environmental Releases of Engineered Nanomaterials. *Sci. Total Environ.* **2015**, *535*, 160–171. [CrossRef]
7. Zhao, J.; Lin, M.; Wang, Z.; Cao, X.; Xing, B. Engineered Nanomaterials in the Environment: Are They Safe ? *Crit. Rev. Environ. Sci. Technol.* **2021**, *51*, 1443–1478. [CrossRef]
8. Nie, X.; Zhu, K.; Zhao, S.; Dai, Y.; Tian, H.; Sharma, V.K.; Jia, H. Interaction of Ag^+ with Soil Organic Matter: Elucidating the Formation of Silver Nanoparticles. *Chemosphere* **2020**, *243*, 125413. [CrossRef]
9. Abbas, Q.; Yousaf, B.; Ullah, H.; Ali, M.U.; Ok, Y.S.; Rinklebe, J. Environmental Transformation and Nano-Toxicity of Engineered Nano-Particles (ENPs) in Aquatic and Terrestrial Organisms. *Crit. Rev. Environ. Sci. Technol.* **2020**, *3389*, 2523–2581. [CrossRef]
10. Das, P.; Barua, S.; Sarkar, S.; Karak, N.; Bhattacharyya, P.; Raza, N.; Kim, K.H.; Bhattacharya, S.S. Plant Extract–Mediated Green Silver Nanoparticles: Efficacy as Soil Conditioner and Plant Growth Promoter. *J. Hazard. Mater.* **2018**, *346*, 62–72. [CrossRef]
11. Nowack, B.; Boldrin, A.; Caballero, A.; Hansen, S.F.; Gottschalk, F.; Heggelund, L.; Hennig, M.; Mackevica, A.; Maes, H.; Navratilova, J.; et al. Meeting the Needs for Released Nanomaterials Required for Further Testing—The SUN Approach. *Environ. Sci. Technol.* **2016**, *50*, 2747–2753. [CrossRef]
12. Chen, J.; Wu, L.; Lu, M.; Lu, S.; Li, Z.; Ding, W. Comparative Study on the Fungicidal Activity of Metallic MgO Nanoparticles and Macroscale MgO Against Soilborne Fungal Phytopathogens. *Front. Microbiol.* **2020**, *11*, 365. [CrossRef]
13. Sun, W.; Dou, F.; Li, C.; Ma, X.; Ma, L.Q. Impacts of Metallic Nanoparticles and Transformed Products on Soil Health. *Crit. Rev. Environ. Sci. Technol.* **2021**, *51*, 973–1002. [CrossRef]
14. Wang, X.; Liu, L.; Zhang, W.; Ma, X. Prediction of Plant Uptake and Translocation of Engineered Metallic Nanoparticles by Machine Learning. *Environ. Sci. Technol.* **2021**, *55*, 7491–7500. [CrossRef]
15. Aziz, Y.; Shah, G.A.; Rashid, M.I. ZnO Nanoparticles and Zeolite Influence Soil Nutrient Availability but Do Not Affect Herbage Nitrogen Uptake from Biogas Slurry. *Chemosphere* **2019**, *216*, 564–575. [CrossRef]
16. Gomez, A.; Peralta-Videaa, R.; Narayan, M.; Zhao, L.; Jia, X.; Bernal, R.A.; Lopez-Moreno, M.L.; Peralta-Videa, J.R. Effects of Nano-Enabled Agricultural Strategies on Food Quality: Current Knowledge and Future Research Needs. *Hazard. Mater.* **2021**, *401*, 123385. [CrossRef]

17. Landa, P. Positive Effects of Metallic Nanoparticles on Plants: Overview of Involved Mechanisms. *Plant Physiol. Biochem.* **2021**, *161*, 12–24. [CrossRef]
18. Liu, W.; Zeb, A.; Lian, J.; Wu, J.; Xiong, H.; Tang, J.; Zheng, S. Interactions of Metal-Based Nanoparticles (Mbnps) and Metal-Oxide Nanoparticles (Monps) with Crop Plants: A Critical Review of Research Progress and Prospects. *Environ. Rev.* **2020**, *28*, 294–310. [CrossRef]
19. Giese, B.; Klaessig, F.; Park, B.; Kaegi, R.; Steinfeldt, M.; Wigger, H.; Von Gleich, A.; Gottschalk, F. Risks, Release and Concentrations of Engineered Nanomaterial in the Environment. *Sci. Rep.* **2018**, *8*, 1565. [CrossRef]
20. Pérez-Hernández, H.; Fernández-Luqueño, F.; Huerta-Lwanga, E.; Mendoza-Vega, J.; Álvarez-Solís José, D. Effect of Engineered Nanoparticles on Soil Biota: Do They Improve the Soil Quality and Crop Production or Jeopardize Them? *L. Degrad. Dev.* **2020**, *31*, 2213–2230. [CrossRef]
21. Abbas, Q.; Yousaf, B.; Amina; Ali, M.U.; Munir, M.A.M.; El-Naggar, A.; Rinklebe, J.; Naushad, M. Transformation Pathways and Fate of Engineered Nanoparticles (ENPs) in Distinct Interactive Environmental Compartments: A Review. *Environ. Int.* **2020**, *138*, 105646. [CrossRef]
22. Shevlin, D.; O'Brien, N.; Cummins, E. Silver Engineered Nanoparticles in Freshwater Systems—Likely Fate and Behaviour through Natural Attenuation Processes. *Sci. Total Environ.* **2018**, *621*, 1033–1046. [CrossRef]
23. Ben-Moshe, T.; Frenk, S.; Dror, I.; Minz, D.; Berkowitz, B. Effects of Metal Oxide Nanoparticles on Soil Properties. *Chemosphere* **2013**, *90*, 640–646. [CrossRef]
24. Komendová, R.; Žídek, J.; Berka, M.; Jemelková, M.; Řezáčová, V.; Conte, P.; Kučerík, J. Small-Sized Platinum Nanoparticles in Soil Organic Matter: Influence on Water Holding Capacity, Evaporation and Structural Rigidity. *Sci. Total Environ.* **2019**, *694*, 133822. [CrossRef]
25. Van Koetsem, F.; Woldetsadik, G.S.; Folens, K.; Rinklebe, J.; Du Laing, G. Partitioning of Ag and CeO_2 Nanoparticles versus Ag and Ce Ions in Soil Suspensions and Effect of Natural Organic Matter on CeO_2 Nanoparticles Stability. *Chemosphere* **2018**, *200*, 471–480. [CrossRef]
26. Suazo-Hernández, J.; Klumpp, E.; Arancibia-miranda, N.; Poblete-grant, P.; Jara, A.; Bol, R.; Mora, M.D.L.L. Describing Phosphorus Sorption Processes on Volcanic Soil in the Presence of Copper or Silver Engineered Nanoparticles. *Minerals* **2021**, *11*, 373–380. [CrossRef]
27. Suazo-Hernández, J.; Klumpp, E.; Arancibia-Miranda, N.; Jara, A.; Poblete-Grant, P.; Sepúlveda, P.; Bol, R.; de la Mora, M.L. Combined Effect of Soil Particle Size Fractions and Engineered Nanoparticles on Phosphate Sorption Processes in Volcanic Soils Evaluated by Elovich and Langmuir-Freundlich Models. *J. Soil Sci. Plant Nutr.* **2022**, *22*, 3685–3696. [CrossRef]
28. Zhang, W.; Long, J.; Li, J.; Zhang, M.; Ye, X.; Chang, W.; Zeng, H. Effect of Metal Oxide Nanoparticles on the Chemical Speciation of Heavy Metals and Micronutrient Bioavailability in Paddy Soil. *Int. J. Environ. Res. Public Health* **2020**, *17*, 2482. [CrossRef]
29. Gladkova, M.M.; Terekhova, V.A. Engineered Nanomaterials in Soil: Sources of Entry and Migration Pathways. *Moscow Univ. Soil Sci. Bull.* **2013**, *68*, 129–134. [CrossRef]
30. Liu, G.; Zhong, H.; Ahmad, Z.; Yang, X.; Huo, L. Transport of Engineered Nanoparticles in Porous Media and Its Enhancement for Remediation of Contaminated Groundwater. *Crit. Rev. Environ. Sci. Technol.* **2020**, *50*, 2301–2378. [CrossRef]
31. Pan, B.; Xing, B. Applications and Implications of Manufactured Nanoparticles in Soils: A Review. *Eur. J. Soil Sci.* **2012**, *63*, 437–456. [CrossRef]
32. Taha, M.R.; Taha, O.M.E. Influence of Nano-Material on the Expansive and Shrinkage Soil Behavior. *J. Nanoparticle Res.* **2012**, *14*, 1190. [CrossRef]
33. Bayat, H.; Kolahchi, Z.; Valaey, S.; Rastgou, M.; Mahdavi, S. Iron and Magnesium Nano-Oxide Effects on Some Physical and Mechanical Properties of a Loamy Hypocalcic Cambisol. *Geoderma* **2019**, *335*, 57–68. [CrossRef]
34. Kheir, A.M.S.; Abouelsoud, H.M.; Hafez, E.M.; Ali, O.A.M. Integrated Effect of Nano-Zn, Nano-Si, and Drainage Using Crop Straw–Filled Ditches on Saline Sodic Soil Properties and Rice Productivity. *Arab. J. Geosci.* **2019**, *12*, 471. [CrossRef]
35. Reginatto, C.; Cecchin, I.; Salvagni Heineck, K.; Thomé, A.; Reddy, K.R. Influence of Nanoscale Zero-Valent Iron on Hydraulic Conductivity of a Residual Clayey Soil and Modeling of the Filtration Parameter. *Environ. Sci. Pollut. Res.* **2020**, *27*, 9288–9296. [CrossRef] [PubMed]
36. He, J.; Wang, D.; Zhou, D. Transport and Retention of Silver Nanoparticles in Soil: Effects of Input Concentration, Particle Size and Surface Coating. *Sci. Total Environ.* **2019**, *648*, 102–108. [CrossRef] [PubMed]
37. Wang, Y.; Liu, Y.; Su, G.; Yang, K.; Lin, D. Transformation and Implication of Nanoparticulate Zero Valent Iron in Soils. *J. Hazard. Mater.* **2021**, *412*, 125207. [CrossRef] [PubMed]
38. Cullen, L.G.; Tilston, E.L.; Mitchell, G.R.; Collins, C.D.; Shaw, L.J. Assessing the Impact of Nano- and Micro-Scale Zerovalent Iron Particles on Soil Microbial Activities: Particle Reactivity Interferes with Assay Conditions and Interpretation of Genuine Microbial Effects. *Chemosphere* **2011**, *82*, 1675–1682. [CrossRef]
39. Mar Gil-Díaz, M.; Pérez-Sanz, A.; Ángeles Vicente, M.; Carmen Lobo, M. Immobilisation of Pb and Zn in Soils Using Stabilised Zero-Valent Iron Nanoparticles: Effects on Soil Properties. *Clean-Soil Air Water* **2014**, *42*, 1776–1784. [CrossRef]
40. Gil-Díaz, M.; Alonso, J.; Rodríguez-Valdés, E.; Gallego, J.R.; Lobo, M.C. Comparing Different Commercial Zero Valent Iron Nanoparticles to Immobilize As and Hg in Brownfield Soil. *Sci. Total Environ.* **2017**, *584–585*, 1324–1332. [CrossRef]
41. Simonin, M.; Guyonnet, J.P.; Martins, J.M.F.; Ginot, M.; Richaume, A. Influence of Soil Properties on the Toxicity of TiO_2 Nanoparticles on Carbon Mineralization and Bacterial Abundance. *J. Hazard. Mater.* **2015**, *283*, 529–535. [CrossRef] [PubMed]

42. De Souza, A.; Govea-Alcaide, E.; Masunaga, S.H.; Fajardo-Rosabal, L.; Effenberger, F.; Rossi, L.M.; Jardim, R.F. Impact of Fe_3O_4 Nanoparticle on Nutrient Accumulation in Common Bean Plants Grown in Soil. *SN Appl. Sci.* **2019**, *1*, 308. [CrossRef]
43. Wang, Y.; Yang, K.; Lin, D. Nanoparticulate Zero Valent Iron Interaction with Dissolved Organic Matter Impacts Iron Transformation and Organic Carbon Stability. *Environ. Sci. Nano* **2020**, *7*, 1818–1830. [CrossRef]
44. Peng, C.; Xu, C.; Liu, Q.; Sun, L.; Luo, Y.; Shi, J. Fate and Transformation of CuO Nanoparticles in the Soil-Rice System during the Life Cycle of Rice Plants. *Environ. Sci. Technol.* **2017**, *51*, 4907–4917. [CrossRef]
45. Zahra, Z.; Maqbool, T.; Arshad, M.; Badshah, M.A.; Choi, H.K.; Hur, J. Changes in Fluorescent Dissolved Organic Matter and Their Association with Phytoavailable Phosphorus in Soil Amended with TiO_2 Nanoparticles. *Chemosphere* **2019**, *227*, 17–25. [CrossRef]
46. Shi, J.; Ye, J.; Fang, H.; Zhang, S.; Xu, C. Effects of Copper Oxide Nanoparticles on Paddy Soil Properties and Components. *Nanomaterials* **2018**, *8*, 839. [CrossRef]
47. Duncan, E.; Owens, G. Metal Oxide Nanomaterials Used to Remediate Heavy Metal Contaminated Soils Have Strong Effects on Nutrient and Trace Element Phytoavailability. *Sci. Total Environ.* **2019**, *678*, 430–437. [CrossRef]
48. Verma, Y.; Singh, S.K.; Jatav, H.S.; Rajput, V.D.; Minkina, T. Interaction of Zinc Oxide Nanoparticles with Soil: Insights into the Chemical and Biological Properties. *Environ. Geochem. Health* **2022**, *44*, 221–234. [CrossRef]
49. Huang, J.; Liang, C.; Zhang, X. Effects of Nano-SiO_2 on the Adsorption of Chiral Metalaxyl to Agricultural Soils. *Environ. Pollut. J.* **2017**, *225*, 201–210. [CrossRef]
50. Shemawar; Mahmood, A.; Hussain, S.; Mahmood, F.; Iqbal, M.; Shahid, M.; Ibrahim, M.; Ali, M.A.; Shahzad, T. Toxicity of Biogenic Zinc Oxide Nanoparticles to Soil Organic Matter Cycling and Their Interaction with Rice-Straw Derived Biochar. *Sci. Rep.* **2021**, *11*, 8429. [CrossRef]
51. Zhou, D.M.; Jin, S.Y.; Wang, Y.J.; Wang, P.; Weng, N.Y.; Wang, Y. Assessing the Impact of Iron-Based Nanoparticles on pH, Dissolved Organic Carbon, and Nutrient Availability in Soils. *Soil Sediment Contam.* **2012**, *21*, 101–114. [CrossRef]
52. Gao, X.; Spielman-Sun, E.; Rodrigues, S.M.; Casman, E.A.; Lowry, G.V. Time and Nanoparticle Concentration Affect the Extractability of Cu from CuO NP-Amended Soil. *Environ. Sci. Technol.* **2017**, *51*, 2226–2234. [CrossRef] [PubMed]
53. Baysal, A.; Saygın, H. Effect of Zinc Oxide Nanoparticles on the Trace Element Contents of Soils. *Chem. Ecol.* **2018**, *34*, 713–726. [CrossRef]
54. Luo, Z.; Wang, Z.; Wei, Q.S.; Yan, C.; Liu, F. Effects of Engineered Nano-Titanium Dioxide on Pore Surface Properties and Phosphorus Adsorption of Sediment: Its Environmental Implications. *J. Hazard. Mater.* **2011**, *192*, 1364–1369. [CrossRef]
55. Bayat, H.; Kolahchi, Z.; Valaey, S.; Rastgou, M.; Mahdavi, S. Novel Impacts of Nanoparticles on Soil Properties: Tensile Strength of Aggregates and Compression Characteristics of Soil. *Arch. Agron. Soil Sci.* **2018**, *64*, 776–789. [CrossRef]
56. Baragaño, D.; Gallego, J.L.R.; Baleriola, G.; Forján, R. Effects of Different In Situ Remediation Strategies for an As-Polluted Soil on Human Health Risk, Soil Properties, and Vegetation. *Agronomy* **2020**, *6*, 759–776. [CrossRef]
57. Zhang, J.Y.; Zhou, H.; Gu, J.F.; Huang, F.; Yang, W.J.; Wang, S.L.; Yuan, T.Y.; Liao, B.H. Effects of Nano-Fe_3O_4-Modified Biochar on Iron Plaque Formation and Cd Accumulation in Rice (*Oryza Sativa* L.). *Environ. Pollut.* **2020**, *260*, 11397. [CrossRef]
58. Baragaño, D.; Forján, R.; Fernández, B.; Ayala, J.; Afif, E.; Gallego, J.L.R. Application of Biochar, Compost and ZVI Nanoparticles for the Remediation of As, Cu, Pb and Zn Polluted Soil. *Environ. Sci. Pollut. Res.* **2020**, *27*, 33681–33691. [CrossRef]
59. Fajardo, C.; Gil-Díaz, M.; Costa, G.; Alonso, J.; Guerrero, A.M.; Nande, M.; Lobo, M.C.; Martín, M. Residual Impact of Aged nZVI on Heavy Metal-Polluted Soils. *Sci. Total Environ.* **2015**, *535*, 79–84. [CrossRef]
60. Lin, J.; Sun, M.; Su, B.; Owens, G.; Chen, Z. Immobilization of Cadmium in Polluted Soils by Phytogenic Iron Oxide Nanoparticles. *Sci. Total Environ.* **2019**, *659*, 491–498. [CrossRef]
61. Peng, C.; Tong, H.; Shen, C.; Sun, L.; Yuan, P.; He, M.; Shi, J. Bioavailability and Translocation of Metal Oxide Nanoparticles in the Soil-Rice Plant System. *Sci. Total Environ.* **2020**, *713*, 136662. [CrossRef]
62. Al-Sid-Cheikh, M.; Pédrot, M.; Dia, A.; Davranche, M.; Jeanneau, L.; Petitjean, P.; Bouhnik-Le Coz, M.; Cormier, M.A.; Grasset, F. Trace Element and Organic Matter Mobility Impacted by Fe_3O_4-Nanoparticle Surface Coating within Wetland Soil. *Environ. Sci. Nano* **2019**, *6*, 3049–3059. [CrossRef]
63. Gil-Díaz, M.; Diez-Pascual, S.; González, A.; Alonso, J.; Rodríguez-Valdés, E.; Gallego, J.R.; Lobo, M.C. A Nanoremediation Strategy for the Recovery of an As-Polluted Soil. *Chemosphere* **2016**, *149*, 137–145. [CrossRef]
64. Truskewycz, A.; Patil, S.; Ball, A.; Shukla, R. Iron Nanoparticles for Contaminated Site Remediation and Environmental Preservation. In *Nanobiotechnology*; CRC Press: Boca Raton, FL, USA, 2018; pp. 323–373. ISBN 9781402039959.
65. Filip, J.; Kolařík, J.; Petala, E.; Petr, M.; Šráček, O.; Zbořil, R. Nanoscale Zerovalent Iron Particles for Treatment of Metalloids. In *Nanoscale Zerovalent Iron Particles for Environmental Restoration: From Fundamental Science to Field Scale Engineering Applications*; Springer: Berlin, Germany, 2019; pp. 157–199. ISBN 9783319953403.
66. Ulucan-altuntas, K.; Debik, E. DDT Removal by Nano Zero Valent Iron Influence of pH on Removal Mechanism. *Eurasia Proc. Sci. Technol. Eng. Math.* **2017**, *1*, 339–346.
67. Kanel, S.R.; Manning, B.; Charlet, L.; Choi, H. Removal of Arsenic(III) from Groundwater by Nanoscale Zero-Valent Iron. *Environ. Sci. Technol.* **2005**, *39*, 1291–1298. [CrossRef] [PubMed]
68. Zhang, H.; Huang, M.; Zhang, W.; Gardea-Torresdey, J.L.; White, J.C.; Ji, R.; Zhao, L. Silver Nanoparticles Alter Soil Microbial Community Compositions and Metabolite Profiles in Unplanted and Cucumber-Planted Soils. *Environ. Sci. Technol.* **2020**, *54*, 3334–3342. [CrossRef] [PubMed]

69. Husson, O. Redox Potential (Eh) and pH as Drivers of Soil/Plant/Microorganism Systems: A Transdisciplinary Overview Pointing to Integrative Opportunities for Agronomy. *Plant Soil* **2013**, *362*, 389–417. [CrossRef]
70. Li, X.Q.; Zhang, W.X. Sequestration of Metal Cations with Zerovalent Iron Nanoparticles—A Study with High Resolution X-Ray Photoelectron Spectroscopy (HR-XPS). *J. Phys. Chem. C* **2007**, *111*, 6939–6946. [CrossRef]
71. Carroll, D.O.; Sleep, B.; Krol, M.; Boparai, H.; Kocur, C. Nanoscale Zero Valent Iron and Bimetallic Particles for Contaminated Site Remediation. *Adv. Water Resour.* **2013**, *51*, 104–122. [CrossRef]
72. Vítková, M.; Puschenreiter, M.; Komárek, M. Effect of Nano Zero-Valent Iron Application on As, Cd, Pb, and Zn Availability in the Rhizosphere of Metal(Loid) Contaminated Soils. *Chemosphere* **2018**, *200*, 217–226. [CrossRef]
73. Naja, G.; Apiratikul, R.; Pavasant, P.; Volesky, B.; Hawari, J. Dynamic and Equilibrium Studies of the RDX Removal from Soil Using CMC-Coated Zerovalent Iron Nanoparticles. *Environ. Pollut.* **2009**, *157*, 2405–2412. [CrossRef] [PubMed]
74. Köser, J.; Engelke, M.; Hoppe, M. Predictability of Silver Nanoparticle Speciation and Toxicity in Ecotoxicological Media. *Environ. Sci. Nano* **2017**, *4*, 1470–1483. [CrossRef]
75. Rajput, V.D.; Minkina, T.; Sushkova, S.; Tsitsuashvili, V.; Mandzhieva, S.; Gorovtsov, A.; Nevidomskyaya, D.; Gromakova, N. Effect of Nanoparticles on Crops and Soil Microbial Communities. *J. Soils Sediments* **2018**, *18*, 2179–2187. [CrossRef]
76. Chen, Y.; Zhang, X.; Liu, W. Effect of Metal and Metal Oxide Engineered Nano Particles on Nitrogen Bio-Conversion and Its Mechanism: A Review. *Chemosphere* **2022**, *287*, 132097. [CrossRef] [PubMed]
77. García-Gómez, C.; Babin, M.; Obrador, A.; Álvarez, J.M.; Fernández, M.D. Toxicity of ZnO Nanoparticles, ZnO Bulk, and $ZnCl_2$ on Earthworms in a Spiked Natural Soil and Toxicological Effects of Leachates on Aquatic Organisms. *Arch. Environ. Contam. Toxicol.* **2014**, *67*, 465–473. [CrossRef]
78. García-Gómez, C.; Babin, M.; Obrador, A.; Álvarez, J.M.; Fernández, M.D. Integrating Ecotoxicity and Chemical Approaches to Compare the Effects of ZnO Nanoparticles, ZnO Bulk, and $ZnCl_2$ on Plants and Microorganisms in a Natural Soil. *Environ. Sci. Pollut. Res.* **2015**, *22*, 16803–16813. [CrossRef] [PubMed]
79. Kumari, M.; Pandey, S.; Bhattacharya, A.; Mishra, A.; Nautiyal, C.S. Protective Role of Biosynthesized Silver Nanoparticles against Early Blight Disease in *Solanum lycopersicum*. *Plant Physiol. Biochem.* **2017**, *121*, 216–225. [CrossRef]
80. Montes de Oca-Vásquez, G.; Solano-Campos, F.; Vega-Baudrit, J.R.; López-Mondéjar, R.; Odriozola, I.; Vera, A.; Moreno, J.L.; Bastida, F. Environmentally Relevant Concentrations of Silver Nanoparticles Diminish Soil Microbial Biomass but Do Not Alter Enzyme Activities or Microbial Diversity. *J. Hazard. Mater.* **2020**, *391*, 122224. [CrossRef]
81. Cornelis, G.; DooletteMadeleine Thomas, C.; McLaughlin, M.J.; Kirby, J.K.; Beak, D.G.; Chittleborough, D. Retention and Dissolution of Engineered Silver Nanoparticles in Natural Soils. *Soil Sci. Soc. Am. J.* **2012**, *76*, 891–902. [CrossRef]
82. Fabrega, J.; Fawcett, S.R.; Renshaw, J.C.; Lead, J.R. Silver Nanoparticle Impact on Bacterial Growth: Effect of pH, Concentration, and Organic Matter. *Environ. Sci. Technol.* **2009**, *43*, 7285–7290. [CrossRef]
83. Coutris, C.; Joner, E.J.; Oughton, D.H. Aging and Soil Organic Matter Content Affect the Fate of Silver Nanoparticles in Soil. *Sci. Total Environ.* **2012**, *420*, 327–333. [CrossRef] [PubMed]
84. Chowdhury, I.; Walker, S.L.; Mylon, S.E. Aggregate Morphology of Nano-TiO_2: Role of Primary Particle Size, Solution Chemistry, and Organic Matter. *Environ. Sci. Process. Impacts* **2013**, *15*, 275–282. [CrossRef]
85. Quigg, A.; Chin, W.; Chen, C.; Zhang, S.-J.; Jiang, Y.; Miao, A.; Schwehr, K.; Xu, C.; Santschi, P.H. Direct and Indirect Toxic Effects of Engineered Nanoparticles on Algae: Role of Natural Organic Matter. *Sustain. Chem. Eng.* **2013**, *1*, 686–702. [CrossRef]
86. Collin, B.; Oostveen, E.; Tsyusko, O.V.; Unrine, J.M. Influence of Natural Organic Matter and Surface Charge on the Toxicity and Bioaccumulation of Functionalized Ceria Nanoparticles in *Caenorhabditis elegans*. *Environ. Sci. Technol.* **2014**, *48*, 1280–1289. [CrossRef]
87. Oldham, V.E.; Mucci, A.; Tebo, B.M.; Iii, G.W.L. Soluble Mn(III)-L Complexes Are Abundant in Oxygenated Waters and Stabilized by Humic Ligands. *Geochim. Cosmochim. Acta* **2017**, *199*, 238–246. [CrossRef]
88. Ren, M.; Horn, H.; Frimmel, F.H. Aggregation Behavior of TiO_2 Nanoparticles in Municipal Effluent: Influence of Ionic Strengthen and Organic Compounds. *Water Res.* **2017**, *123*, 678–686. [CrossRef]
89. Klitzke, S.; Metreveli, G.; Peters, A.; Schaumann, G.E.; Lang, F. The Fate of Silver Nanoparticles in Soil Solution—Sorption of Solutes and Aggregation. *Sci. Total Environ.* **2015**, *535*, 54–60. [CrossRef]
90. Van Hoecke, K.; De Schamphelaere, K.A.C.; Van Der Meeren, P.; Smagghe, G.; Janssen, C.R. Aggregation and Ecotoxicity of CeO_2 Nanoparticles in Synthetic and Natural Waters with Variable pH, Organic Matter Concentration and Ionic Strength. *Environ. Pollut.* **2011**, *159*, 970–976. [CrossRef] [PubMed]
91. Shang, H.; Guo, H.; Ma, C.; Li, C.; Chefetz, B.; Polubesova, T.; Xing, B. Maize (*Zea Mays* L.) Root Exudates Modify the Surface Chemistry of CuO Nanoparticles: Altered Aggregation, Dissolution and Toxicity. *Sci. Total Environ.* **2019**, *690*, 502–510. [CrossRef]
92. Gueye, M.T.; Di Palma, L.; Allahverdeyeva, G.; Bavasso, I.; Petrucci, E.; Stoller, M.; Vilardi, G. The Influence of Heavy Metals and Organic Matter on Hexavalent Chromium Reduction by Nano Zero Valent Iron in Soil. *Chem. Eng. Trans.* **2016**, *47*, 289–294. [CrossRef]
93. Rashid, M.I.; Shahzad, T.; Shahid, M.; Ismail, I.M.I.; Shah, G.M.; Almeelbi, T. Zinc Oxide Nanoparticles Affect Carbon and Nitrogen Mineralization of Phoenix Dactylifera Leaf Litter in a Sandy Soil. *J. Hazard. Mater.* **2017**, *324*, 298–305. [CrossRef]
94. Vittori Antisari, L.; Carbone, S.; Gatti, A.; Vianello, G.; Nannipieri, P. Toxicity of Metal Oxide (CeO_2, Fe_3O_4, SnO_2) Engineered Nanoparticles on Soil Microbial Biomass and Their Distribution in Soil. *Soil Biol. Biochem.* **2013**, *60*, 87–94. [CrossRef]

95. Carbone, S.; Vittori Antisari, L.; Gaggia, F.; Baffoni, L.; Di Gioia, D.; Vianello, G.; Nannipieri, P. Bioavailability and Biological Effect of Engineered Silver Nanoparticles in a Forest Soil. *J. Hazard. Mater.* **2014**, *280*, 89–96. [CrossRef]
96. Nagoya, S.; Nakamichi, S.; Kawase, Y. Mechanisms of Phosphate Removal from Aqueous Solution by Zero-Valent Iron: A Novel Kinetic Model for Electrostatic Adsorption, Surface Complexation and Precipitation of Phosphate under Oxic Conditions. *Sep. Purif. Technol.* **2019**, *218*, 120–129. [CrossRef]
97. Baysal, A.; Saygin, H.; Ustabasi, G.S. Influence of Al_2O_3 Nanoparticles on the Soil Elements. *Bull. Environ. Contam. Toxicol.* **2019**, *102*, 98–104. [CrossRef]
98. Dahle, J.T.; Livi, K.; Arai, Y. Effects of pH and Phosphate on CeO_2 Nanoparticle Dissolution. *Chemosphere* **2015**, *119*, 1365–1371. [CrossRef]
99. Yoon, S.Y.; Lee, C.G.; Park, J.A.; Kim, J.H.; Kim, S.B.; Lee, S.H.; Choi, J.W. Kinetic, Equilibrium and Thermodynamic Studies for Phosphate Adsorption to Magnetic Iron Oxide Nanoparticles. *Chem. Eng. J.* **2014**, *236*, 341–347. [CrossRef]
100. Eljamal, O.; Khalil, A.M.E.; Sugihara, Y.; Matsunaga, N. Phosphorus Removal from Aqueous Solution by Nanoscale Zero Valent Iron in the Presence of Copper Chloride. *Chem. Eng. J.* **2016**, *293*, 225–231. [CrossRef]
101. Wen, Z.; Zhang, Y.; Dai, C. Removal of Phosphate from Aqueous Solution Using Nanoscale Zerovalent Iron (nZVI). *Colloids Surf. A Physicochem. Eng. Asp.* **2014**, *457*, 433–440. [CrossRef]
102. Takami, S.; Eljamal, O.; Khalil, A.M.E.; Eljamal, R.; Matsunaga, N. Development of Continuous System Based on Nanoscale Zero Valent Iron Particles for Phosphorus Removal. *J. Japan Soc. Civ. Eng.* **2019**, *7*, 30–42. [CrossRef]
103. Bellani, L.; Siracusa, G.; Giorgetti, L.; Di Gregorio, S.; Ruffini Castiglione, M.; Spanò, C.; Muccifora, S.; Bottega, S.; Pini, R.; Tassi, E. TiO_2 Nanoparticles in a Biosolid-Amended Soil and Their Implication in Soil Nutrients, Microorganisms and Pisum Sativum Nutrition. *Ecotoxicol. Environ. Saf.* **2020**, *190*, 110095. [CrossRef]
104. Moharami, S.; Jalali, M. Effect of Time on the Sorption and Distribution of Phosphorus in Treated Soil with Minerals and Nanoparticles. *Environ. Earth Sci.* **2015**, *73*, 8599–8608. [CrossRef]
105. Koopmans, G.; Hiemstra, T.; Vaseur, C.; Chardon, W.; Voegelin, A.; Groenenberg, J. Use of Iron Oxide Nanoparticles for Immobilizing Phosphorus In-Situ: Increase in Soil Reactive Surface Area and Effect on Soluble Phosphorus. *Sci. Total Environ.* **2020**, *711*, 135220. [CrossRef] [PubMed]
106. Yusefi-Tanha, E.; Fallah, S.; Rostamnejadi, A.; Pokhrel, L.R. Zinc Oxide Nanoparticles (ZnONPs) as a Novel Nanofertilizer: Influence on Seed Yield and Antioxidant Defense System in Soil Grown Soybean (*Glycine Max* Cv. Kowsar). *Sci. Total Environ.* **2020**, *738*, 140240. [CrossRef] [PubMed]
107. Rathnayake, S.; Unrine, J.M.; Judy, J.; Miller, A.F.; Rao, W.; Bertsch, P.M. Multitechnique Investigation of the pH Dependence of Phosphate Induced Transformations of ZnO Nanoparticles. *Environ. Sci. Technol.* **2014**, *48*, 4757–4764. [CrossRef] [PubMed]
108. Zahra, Z.; Arshad, M.; Rafique, R.; Mahmood, A.; Habib, A.; Qazi, I.A.; Khan, S.A. Metallic Nanoparticle (TiO_2 and Fe_3O_4) Application Modifies Rhizosphere Phosphorus Availability and Uptake by *Lactuca sativa*. *J. Agric. Food Chem.* **2015**, *63*, 6876–6882. [CrossRef]
109. Feng, Y.; Lu, H.; Liu, Y.; Xue, L.; Dionysiou, D.D.; Yang, L.; Xing, B. Nano-Cerium Oxide Functionalized Biochar for Phosphate Retention: Preparation, Optimization and Rice Paddy Application. *Chemosphere* **2017**, *185*, 816–825. [CrossRef] [PubMed]

nanomaterials

Article

Short-Term Exposure to MPs and DEHP Disrupted Gill Functions in Marine Bivalves

Yanfei Zhou [1,2], Yanping Li [1], Wenlu Lan [3], Hao Jiang [4] and Ke Pan [1,*]

1 Shenzhen Key Laboratory of Marine Microbiome Engineering, Institute for Advanced Study, Shenzhen University, Shenzhen 518060, China
2 Key Laboratory of Optoelectronic Devices and Systems of Ministry of Education and Guangdong Province, College of Physics and Optoelectronic Engineering, Shenzhen University, Shenzhen 518060, China
3 Marine Environmental Monitoring Center of Guangxi, Beihai 536000, China
4 Key Laboratory of Aquatic Botany and Watershed Ecology, Wuhan Botanical Garden, Chinese Academy of Sciences, Wuhan 430074, China
* Correspondence: panke@szu.edu.cn; Tel.: +86-755-26940201

Abstract: The synergistic impact of microplastics (MPs) and organic pollutants remains poorly understood in the marine environment. This study aimed to assess the toxicity of polypropylene microplastics (PS) and/or di-(2-ethylhexyl) phthalate (DEHP) on marine clams. Both *Ruditapes philippinarum* and *Tegillarca granosa* were exposed to PS and DEHP individually and combined at environmentally relevant concentrations for 48 h. The filtration rate, antioxidant enzymes activity, lipid peroxidation, reactive oxygen species accumulation, and histological alterations were evaluated. Our results show that single or co-exposure to MPs and DEHP significantly decreases the filtration rate in both type of clams, but the latter exhibited stronger inhibition effect. Close examination of accumulation of reactive oxygen species and related biomarkers revealed that combined exposure exerts greater oxidative stress in the cells, which causes more serious histopathological damage in the gills of the bivalves. Our study implies that MPs, in synergy with organic pollutants, can be more harmful for marine organisms.

Keywords: microplastics; organic pollutants; bivalves; oxidative damage; synergic effects

check for updates

Citation: Zhou, Y.; Li, Y.; Lan, W.; Jiang, H.; Pan, K. Short-Term Exposure to MPs and DEHP Disrupted Gill Functions in Marine Bivalves. *Nanomaterials* **2022**, *12*, 4077. https://doi.org/10.3390/nano12224077

Academic Editor: Alexey Pestryakov

Received: 20 October 2022
Accepted: 16 November 2022
Published: 19 November 2022

1. Introduction

Improper use and disposal of plastics has made plastic pollution a major global environmental concern [1–3]. Large quantities of mismanaged plastic wastes enter the oceans worldwide every year [2]. The degradation of these wastes produces plastic fragments or particles less than 5 mm, known as primary microplastics (MPs) [4]. In addition, plastics deliberately manufactured in this size for use in cosmetics or as abrasives are another important source of MPs in the marine environment [5,6]. The ubiquitous distribution of these MPs presents a major threat to various marine organisms and has evolved into an overwhelming challenge for ocean health.

Marine bivalves have been considered particularly susceptible to microplastic pollution, as their strong filter-feeding activity make them readily accumulate MPs from the surrounding seawater [7–9]. Under sufficiently high concentrations, MPs result in altered feeding activities, slower growth, and impaired development in bivalves [7]. At the tissue and organ level, gills and the digestive glands are the most important target organs of MPs, in which histopathological alterations and inflammatory response can be observed [10–12]. At the cellular level, MPs can trigger a series of stress responses, such as oxidative injury, activation of antioxidant defenses, and destabilization of the lysosomal membrane [10,13]. Overall, these findings clearly indicate that MPs can produce a number of sublethal effects that could reduce the fitness of bivalves. This could impact survival of the individuals, recruitment of new members, and the sustainability of the bivalve population.

It should be noted that marine environments contain a great diversity of pollutants. MPs interact with co-existing pollutants via sorption and desorption. Because of their small size and large surface area, MPs can act as carriers for various pollutants, such as heavy metals, pharmaceuticals, and persistent organic pollutants [6,14,15]. Bivalves are believed to face simultaneous exposure to these chemicals and MPs [16]. While the toxic effects of MPs from single exposure are comparatively well known, the mechanisms and environmental relevance of combined toxicity resulting from the combined pollution deserve further study. Laboratory evidence has shown that MPs play a "carrier" role in increasing the bioaccumulation of sorbed organic pollutants [17–19]. Several studies have indicated that MPs exacerbate the adverse effects of organic pollutants [14,15,20,21]. Aquatic organisms suffer from greater histopathological damage or stresses when exposed to organic pollutants (PCBs, PAHs, brominated flame retardants, perfluorinated compounds, or methylmercury) together with MPs than from single pollutants alone [10,22]. For example, co-exposure to bisphenol A (BPA) and polystyrene (PS) nanoparticles generates greater neurotoxic effects in both the central nervous system and the dopaminergic system of zebrafish than to BPA alone [23]. Combined exposure of MPs with organic contaminants can also produce negative effects at genetic levels. Avio et al. [24] observed MP-associated organic contaminants caused irreversible DNA damage in the mussel *Mytilus galloprovincialis*. These studies have clearly indicated the synergistic or additive effects of MPs and organic pollutants. However, some studies contradictorily found that the presence of MPs decrease the toxicity of organic pollutants [25,26]. These inconsistent findings imply the complexity of interactions between MPs and organic pollutants and point to the need for further investigations of their combined toxicity. Currently, few studies have specifically studied the effect of combined exposure of MPs and endocrine disruptors on the gill function in terms of oxidative stress and histopathological changes in this important tissue.

To clarify the effects of exposure to MPs and/or organic pollutants on bivalves, accumulation of reactive oxygen species (ROS) was analyzed in the gills, the first organ. A set of biomarkers, including superoxide dismutase (SOD), peroxidase (POD), catalase (CAT), and peroxidase (POD), involved in oxidative injury and cellular antioxidant defense were studied to reveal the mechanisms counteracting the toxicity in the bivalve. Histopathological damage was observed via microscopic analysis after staining by hematoxylin and eosin. Polystyrene MPs were selected as model contaminants because they are one of the most common MPs in estuaries and coastal areas. Bis(2-ethylhexyl) phthalate (DEHP), one of the most widely used phthalic acid esters (PAEs), was chosen for the representative of endocrine disruptors in order to investigate the synergistic effects with MPs.

2. Materials and Methods

2.1. Preparations of Organisms and Materials

Healthy *R. philippinarum* (manila clam) and *T. granosa* (blood clam), with a mean shell length of 30.5 ± 1.5 mm and 28.5 ± 1.5 mm, were obtained from Shenzhen (Guangdong, China) and fed *Chlorella vulgaris* in the culture device (10 individuals in 10 L artificial seawater) for 48 h. Before the experiment, 80 *R. philippinarum* and *T. granosa* were acclimatized for one week in plexiglass containers (previously soaked with 10% HNO_3) with continued aerated and filtered artificial seawater (temperature, 20.5 ± 0.5 °C; pH, 8.1 ± 0.02; salinity, 30.5 ± 1.0) and fed *Chlorella vulgaris* (5.0×10^4 cell mL^{-1}) daily in the laboratory [8]. During the acclimatization and culture period, the properties of the artificial seawater were relatively constant. The seawater was changed daily to maintain quality, and no mortality was recorded during the acclimation and exposure period.

Polystyrene microplastics beads (approximately 1 μm in diameter, Figure 1, in the concentration of 3.0 mg L^{-1}) were used as typical MPs for this study, as PS is one of the most widely applied plastics worldwide and is a commonly encountered MP in marine environments [27,28]. According to previous studies, MPs in the shape of a bead are likely be easier to be ingest and translocate within invertebrates, leading to adverse biological impacts, hence PS microbeads were employed here [29,30]. In addition, the concentration of PS used in this work was considered as representative of the MPs polluted hotspot along coastal environments [31,32]. PS microplastics were synthesized with emulsion polymerization with styrene as a monomer, based on the methods described in Sun et al. [33] and Feng et al. [34]. Briefly, sodium dodecyl sulfate and ammonium persulfate were used as an emulsifier and an initiator, respectively. The synthesized product was transferred to a dialysis bag for removal of redundant styrene monomer, emulsifier, and initiator, and then freeze-dried before use.

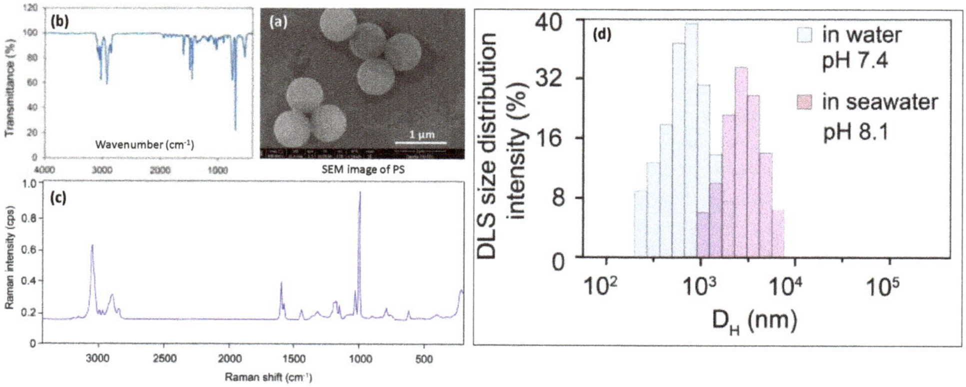

Figure 1. Characterization of PS particles used in this study. SEM image (**a**), infrared spectra (**b**), and Raman spectra (**c**), size distribution of the PS MPs measured by DLS (**d**). Aromatic C-H stretching vibration generated absorption peaks at the wave numbers of 3060 and 3026, and aromatic C=C stretching vibration generated three absorption peaks around the wave numbers of 1600, 1492, and 1452. The Raman spectrum of PS had a typical intense band at 1002 cm^{-1}, which was linked to the breathing mode of the aromatic carbon ring.

The surface morphologies and diameters of the MPs were observed and measured by field emission scanning electron microscopy (FEI-SEM, Quanta 250, FEI Company, Eindhoven, The Netherlands). The FTIR spectra were detected in the 4000–400 cm^{-1} region with a resolution of 4 cm^{-1} by a Nicolet iS10 (Thermo Fisher Scientific Inc., Waltham, WA, USA). The Raman microscope was set up as follows: number of sample scans 3, exposure times 15, background exposure times 512, laser 532 nm, laser energy 8 mW. Each suspected MP was manually located with 10× lens of DXR2 Raman. The average diameter of the PS was 735 ± 87 nm (n = 20). This diameter fell into the size category of nanoplastics (1–1000 nm, Hartmann et al., 2019) and the size range of MPs commonly observed in the coastal environment [35]. The SEM image, infrared spectra, and Raman spectra are shown in Figure 1. Dynamic light scattering (DLS, (Zetasizer, Malvern, UK)) analysis indicated that, in pure water (pH = 7,4), polystyrene microplastics were dispersed homogeneously, with a DLS size of 0.9 µm. In seawater (pH = 8.1), polystyrene MPs were gradually aggregated into approximately 1 to 7.8 µm, with average clusters of 4.9 µm.

2.2. Short-Term Exposure of Bivalves to MPs and DEHP

After one week of acclimation, forty healthy *R. philippinarum* and *T. granosa* were randomly assigned to four treatments (in triplicate for 48 h) respectively: Control (with only *R. philippinarum* or *T. granosa*); PS (with 3 mg L^{-1} PS); DEHP (with 50 µg L^{-1} DEHP); and PS + DEHP (with 3 mg L^{-1} PS and 50 µg L^{-1} DEHP). During exposure treatment, ten individuals were randomly translocated into a 12 L plexiglass container (containing 10 L artificial seawater). The concentration of 3 mg L^{-1} PS was approximately 2.6×10^{12} particles mL^{-1}, respectively. The cultivation conditions were kept the same as the acclimation period, and PS or DEHP were added after feeding *Chlorella vulgaris*.

2.3. Measurements of Filtration Rates

The filtration rates assessment was conducted on the basis of previous studies [7,36]. Approximately 150 mL of *Chlorella vulgaris* was added into each plexiglass container, providing a cell density of 5.0×10^4 cell mL^{-1} in artificial seawater. According to the function of microalgae quantity and time, the filtration rate can be calculated. Three repeated samples were collected from each container, in brief, 20 mL seawater was taken before feeding and 1 h after feeding. The concentrations of microalgae were determined by a flow cytometry system (VS-IV, Fluid Imaging Technology, Scarborough, ME, USA). The filtration rates were calculated as follows:

$$\text{Filtration rate} = V/(n{\cdot}t) \times \log(C_0/C_{60})$$

where n is the number of individuals in each container, t is the consumption time (1 h), and C_0 and C_{60} is the number of microalgae at t = 0 and t = 60, respectively. The results were given as L $clam^{-1}$ h^{-1}.

2.4. Analysis of SOD, CAT, POD, and MDA in the Gills

The visceral mass of the clams was dissected under low temperature conditions, with three biological samples for each replicate. The collected samples were homogenized in cold phosphate-buffered saline (three organisms per replicate). All pretreated samples were stored at −80 °C until analysis. The enzyme activity, including superoxide dismutase (SOD), catalase (CAT), and peroxidase (POD), were measured using the corresponding commercial assay kit (Nanjing Jiancheng Bioengineering Institute, Nanjing, China). SOD is an antioxidant enzyme, which can not only effectively remove free radicals from the body, but also specifically remove the damage of superoxide free radicals from cells [37]. Generally, the CAT–SOD system is considered to be the first line of defense against the toxicity of reactive oxygen species under adverse stress [38,39]. POD is one of the key enzymes in the enzymatic defense system under adverse conditions, and it cooperates with SOD and CAT to remove excess internal free radicals [14]. Malondialdehyde (MDA), a commonly used index of membrane lipid peroxidation, was determined using an MDA assay kit (Jiancheng, Nanjing, China).

2.5. Examining the Accumulation of Reactive Oxygen Species (ROS) in the Gills

In order to visualize O^{2-} and H_2O_2 in situ, fresh cut clam tissues were stained with 2′,7′-dichlorodihydrofluorescein diacetate acetyl ester (DCFH-DA; Sigma Aldrich, Burlington, MA, USA) and the ROS fluorescent probe dihydroethidium (DHE; Sigma-Aldrich), respectively [40,41](Waters et al., 2021; Yu et al., 2021). The staining operations were performed according to manufacturer protocols. After incubation, transverse sections were imaged using a fluorescence microscope, with the fluorescence intensity calculated by Image-Pro plus 6.0. In brief, firstly, taking images under 200× of visual field randomly from each treated earthworm slice, we ensured that the background light of each image was consistent. Subsequently, we converted green/red fluorescent monochrome photos into black-and-white images, and then selected the same black as the unified standard (positive control). We analyzed each image to obtain the positive cumulative optical density (IOD) and the pixel area (AREA) of samples, then calculated the average optical density value IOD/AREA (mean density) [42](Adler and Parmryd, 2010).

2.6. Histological Analysis of the Gills

To evaluate the histopathological alterations after stressor exposure, the gills of *R. philippinarum* and *T. granosa* were separated and fixed overnight in 10% (*v*/*v*) formaldehyde. Subsequently, each of the gill samples were cut into 3 μm thick sections and embedded in paraffin. The slices were then stained with hematoxylin and eosin (HE) before examination with a microscope, according to Sarasquete and Gutiérrez [43]. The histological changes were assessed using a microscope (3D HISTECH, Pannoramic MIDI, 3D-Histech Ltd., Budapest, Hungary) and the images were analyzed by CaseViewer.

2.7. Statistical Analysis

The data were managed using IBM SPSS statistics and OriginPro software. Unless stated otherwise, data in the graphs/plots and tables are displayed as the mean ± s.d. One-way analysis of the variance with an LSD post-hoc test was used for multiple comparisons. When p values were less than 0.05, the results were deemed statistically significantly different, but exact p values are not shown in the graphs/plots and tables.

3. Results and Discussion

3.1. Effects of MPs/DEHP Exposures on Filtration Rates of Bivalve Mollusks

After a short-term exposure treatment (48 h), the filtration rates (L Clam^{-1} h^{-1}) of both types of clams were affected. Exposure of manila and blood clams to the MPs/DEHP tested resulted in significantly higher reductions in the filtration rate (Figure 2), indicating a significant decline in food intake [7,44,45]. As observed in this study, exposure to MPs and DEHP significantly attenuated the filtration rates of both types of clams, and significant differences were observed among the different stressor treatments. These findings are in accordance with the results observed in previous studies [16,46].

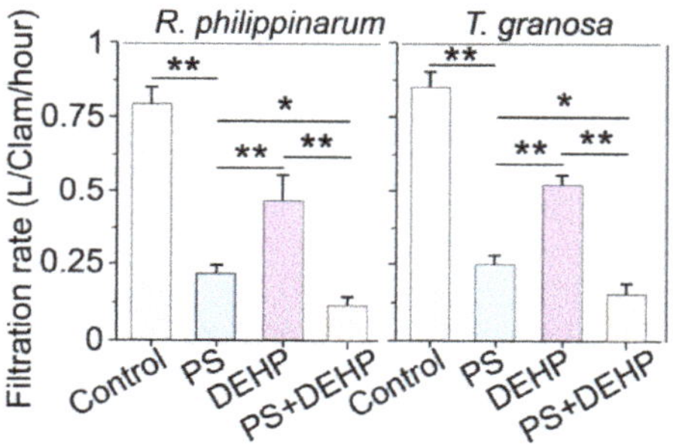

Figure 2. Filtration rates of manila clams (*R. philippinarum*) and blood clams (*T. granosa*) after 48 h of exposure to different PS microplastics and DEHP. Data are presented as mean ± s.d., n = 5. Asterisks above the data indicate significant differences between the two groups, * $p < 0.05$, ** $p < 0.01$.

Specifically, compared to that of control, both types of clams treated with MPs + DEHP had significantly lower filtration rates, which were significantly decreased by approximately 80.3% and 80.4% for manila and blood clams, respectively. Similarly, the filtration rates of manila clams exposed to single MPs (3 mg L^{-1}) or DEHP (50 μg L^{-1}) declined 70.2% and 38.6%, and for blood clams declined to 77.7% and 41.6% of that of the control, respectively. In this work, for manila clams, the filtration rate in MPs + DEHP group dropped to approximately 34% and 67.9% than that in the single MPs and DEHP group, respectively. For blood clams, compared with that of single treatment, the filtration rate in MPs + DEHP group declined to 12.2% and 66.4%, respectively.

It has been demonstrated that MPs may release endocrine-disrupting compounds to ambient environments such as DEHP, which was found to induce adverse effects in various organisms [47–49]. There were significantly lower filtration rates in clams treated with MPs + DEHP, suggesting that the negative impacts induced by single MPs or DEHP might be augmented by the synergic effects of MPs and DEHP [21,50]. Generally, the decrease in the filtration rate after exposure is largely probably owing to the tendency of the tested clams to shut off their valves to avoid continuous exposure [13,51]. As shown in previous studies, *Corbicula fluminea* and *Mytilus edulis L.* closed their valves after exposure to MPs and mercury/bisphenol A, which is an adaptation for bivalves [43,52].

3.2. Effects of MPs/DEHP Exposures on Enzymatic and Non-Enzymatic Antioxidant Defenses

In the context of MPs and organic pollutants, exposure is the factor that may induce various responses on the antioxidant markers of aquatic organisms [10,20]. In this study, antioxidant parameters, i.e., SOD, CAT, and POD, in the gill tissues were assessed, due to their critical roles in the uptake and elimination of pollutants, and in different physiological process [9,53]. Lipid peroxidation (MDA content), as the process of ROS (reactive oxygen species) oxidation of biofilm after enhanced oxygen stress [14], was separately evaluated in the gill tissues.

As an important component of the enzyme antioxidant system, SOD plays an irreplaceable role in the reaction process of homeostatic oxidative stress. The SOD results are shown in Figure 3A. The lowest values were detected in clams exposed to MPs (3 mg L^{-1}) + DEHP (50 µg L^{-1}), with 1.5 and 2 U mg^{-1} protein in manila and blood clams, respectively, implying that co-exposure to MPs and organic pollutants can impede the free radicals capability of bivalves [54]. Nevertheless, the change trend of SOD activity induced by pollutants exposure was inconsistent in manila and blood clams. For manila clams, both the MPs/DEHP group and the combined exposure group significantly decreased SOD activity. The SOD activity in the MPs + DEHP group was 45.5% and 33.5%, lower than that of MPs (2.8 U mg^{-1} protein) and DEHP (2.3 U mg^{-1} protein) alone. For blood clams, the SOD activity was significantly increased after single exposure, while significantly decreased in MPs + DEHP treatment, which was 40% lower than that of the control (3.4 U mg^{-1} protein).

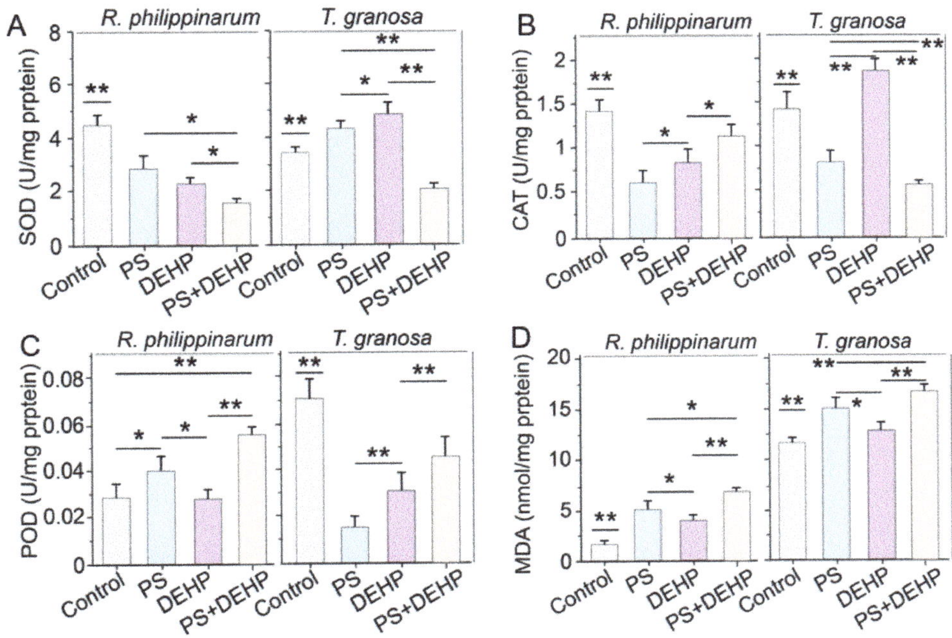

Figure 3. Effects of MPs/DEHP on the activity of antioxidant enzymes (**A**): SOD, superoxide dismutase; (**B**): CAT, catalase; (**C**): POD, peroxidase and lipid peroxidation level (**D**): indicated by malondialdehyde content in manila clams (*R. philippinarum*) and blood clams (*T. granosa*) after 48 h of exposure to different PS microplastics and DEHP. Data are presented as mean ± s.d., n = 5. Asterisks above data indicate significant differences between the two groups, * $p < 0.05$, ** $p < 0.01$.

The results from CAT are shown in Figure 3B. The exposure of clams to MPs or MPs + DEHP decreased the CAT antioxidant activity after 48 h compared with control. The reduction in CAT activity suggests that MPs alone or combined with organic pollutants might cause actual inhibitory effects on the CAT–SOD system after a short-term exposure [14], whereas increased activity was observed when blood clams were exposed to DEHP alone. For manila clams, exposure to MPs/DEHP alone or combined significantly decreased CAT activity, and co-exposure to MPs + DEHP led to less impact on CAT activity (1.1 U mg^{-1} protein), which was 89.6% and 36.9% higher than that of MPs (0.6 U mg^{-1} protein) and DEHP (0.8 U mg^{-1} protein) alone. For blood clams, MPs and MPs + DEHP exposure significantly decreased CAT activity, however, CAT activity was significantly promoted in the DEHP group (1.9 U mg^{-1} protein), which was 30% higher than that of the control (1.4 U mg^{-1} protein).

Total POD activities varied from 0.028 to 0.06 U mg^{-1} protein in manila clams and from 0.012 to 0.072 U mg^{-1} protein in blood clams (Figure 3C), with the lowest value being recorded in the DEHP and MPs group, respectively, after 48 h of exposure. For manila clams, MPs (0.04 U mg^{-1} protein) or MPs + DEHP (0.055 U mg^{-1} protein) treatment significantly increased POD activity, while DEHP exposure alone (0.029 U mg^{-1} protein) had no effect. For blood clams, POD activities were largely decreased by all treatments; additionally, the decrease of POD activity (0.055 U mg^{-1} protein) in the MPs + DEHP group was lower than that of the MPs (0.045 U mg^{-1} protein) or DEHP (0.031 U mg^{-1} protein) group.

The lipid peroxidation (indicated by MDA content) results display a trend to increase throughout the exposure tests for both types of clams (manila and blood clams) (Figure 3D). The lowest values were measured in control organisms (1.51 and 11.5 nM mg^{-1} protein, respectively) and the highest values (6.8 and 16.6 nM mg^{-1} protein, respectively) were observed in organisms exposed to MPs + DEHP after 48 h of exposure. For manila clams, the MDA content in clams exposed to MPs + DEHP was 36.1% and 79.9% higher than those exposed to MPs and DEHP alone, respectively. For blood clams, the content of MDA in the MPs + DEHP group was 11.1% and 28.9% higher than that of single group, respectively.

When bivalves are exposed to MPs or organic pollutants, ROS, i.e., O^{2-}, H_2O_2, or HO, will be over-accumulated, which can directly cause LPO damage [15,55]. The excessive ROS thus leads to an increase of MDA, which is one of the byproducts of LPO [15]. Generally, the present findings of growth inhibition, antioxidant enzyme activities (SOD, CAT and POD) alteration, and MDA increase in single/combined toxicity indicate that the presence of microplastics and organic pollutants could cause the gills of bivalves to accelerate the production of ROS [12].

3.3. *Effects of MPs/DEHP Exposures on ROS Accumulation*

To assess the accumulations of ROS in gills of manila clams (*R. philippinarum*) and blood clams (*T. granosa*), O^{2-} and H_2O_2 were detected using DHE and DCFH-DA. The ROS (O^{2-} and H_2O_2) staining observations revealed that the ROS concentration in gills of manila and blood clams were significantly enhanced by MPs/DEHP exposure for 48 h (Figure 4). This is consistent with the results of MDA increase, and is also in accordance with previous studies' findings [54,56]. For manila clams, regardless of whether MPs and DEHP exposure was alone or combined, ROS accumulation occurred in the gill tissues. The combined exposure showed higher accumulation levels, followed by MPs alone exposure. Compared to the control, the O^{2-} and H_2O_2 concentrations in MPs + DEHP were approximately 6 and 8.6 times higher, respectively (Figure 4B,C). For blood clams, similar results were observed, in which MPs and DEHP exposed alone or combined, obvious ROS accumulation was detected in the gill tissues. Co-exposure led to higher accumulation levels, followed by MPs alone exposure. Compared to the control, the O^{2-} and H_2O_2 concentrations in MPs + DEHP were approximately 5 and 12.2 times higher, respectively (Figure 3B,C).

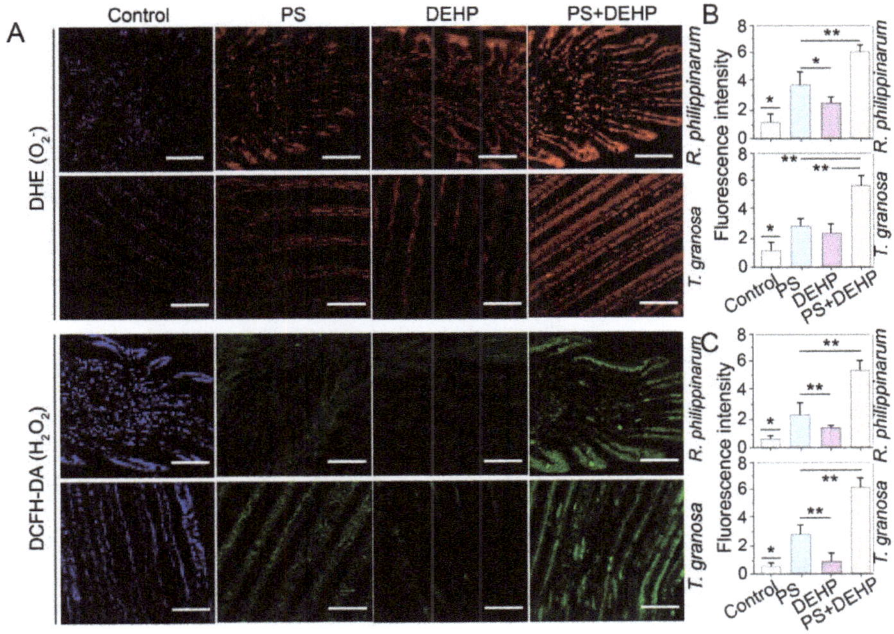

Figure 4. Effects of MPs/DEHP on the accumulation of ROS in gills of manila clams (*R. philippinarum*) and blood clams (*T. granosa*) after 48 h exposure to different PS microplastics and DEHP (**A**). Red and green fluorescence indicate O^{2-} and H_2O_2, respectively, representative pictures. Data are presented as mean ± s.d, n = 5. The fluorescence intensity of red (**B**) and green (**C**) values were calculated from six different pictures for each data, n = 6, scale bar 200 μm. Asterisks above data indicate significant differences between the two groups, * $p < 0.05$, ** $p < 0.01$.

In general, MPs exposure in organisms can disrupt the redox homeostasis and cause oxidative challenges by rapidly accumulating ROS concentrations in tissue [38]. In this work, ingestion of microplastics by bivalves induced oxidative stress that destabilized the homeostasis and produced free radical and leads to increases ROS in gill tissue of all treated clams. Polystyrene microplastics have been found to increase the ROS production level in various aquatic organisms [56–58]. This could possibly be due to the insufficient elimination of free radicals in clams, eventually causing oxidative damage and histopathological changes [35,58]. Over-accumulated ROS affect the antioxidant system response and cause oxidative stress [59]. As a result, short-term dietary exposure to polypropylene microplastics and/or organic pollutant (DEHP) influences the antioxidant defense system of both manila and blood clams.

3.4. Effects of MPs/DEHP Exposures on Histological Alterations

Obvious histopathological changes were observed in the gills of both types of clams, as a result of 48 h actual toxicity tests, after the microscopic examination of the gill tissues sections (Figure 5). Co-exposure to MPs and DEHP in particular caused discrete pathologies in the gills. No histopathological alterations were found in the gill tissues of the control groups, which had normal structure, displaying basic features of the primary and secondary lamellae with typical pillar, chloride cells in the epithelium, and hemocytes throughout the basal to frontal zones (especially in blood clams). MPs exposure has been reported to cause remarkable tissue pathologies in various aquatic organisms, i.e., *Crassostrea gigas*, *Mytilus* spp., *Pomacea paludosa*, and *Mytilus edulis*. [11,12,60]. Similar effects were also observed to be induced by combined MPs and DEHP. For example, polystyrene plastics particles and DEHP co-exposure caused significant histological damages in *Micropterus salmoides* [61].

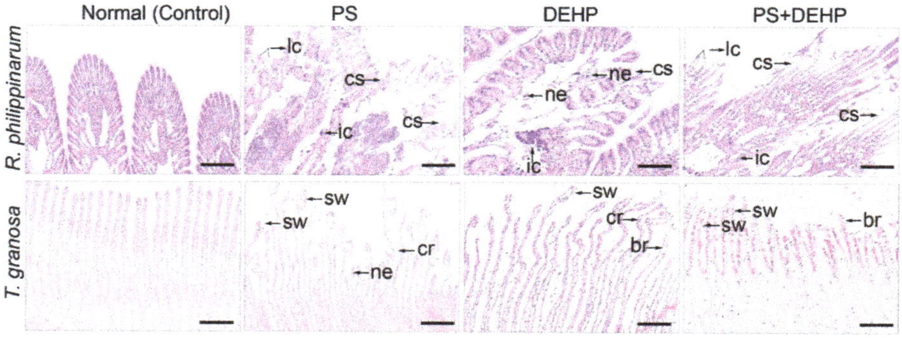

Figure 5. Representative images of haemotoxylin/eosin-stained sections of gill extracted from manila clams (*R. philippinarum*) and blood clams (*T. granosa*). Normal structure shown in the control group, histopathological changes occurred after single/co-exposure to polystyrene MPs/DEHP, Scale bar 200 μm. cs: ciliary structure destruction; lc: loss of contact between gill filaments; ic: inflammatory infiltration; ne: necrosis; cr: curly. sw: swollen; lh: rare in hemocytes; br: broken; ne: necrosis.

However, in comparison to single treatment of DEHP alone, the histopathological degree was more serious in the MPs group and the gill filaments were morphologically changed. For manila clams, pathological changes occurred in the tissues, such as ciliary structure destruction (cs), loss of contact between gill filaments (lc), inflammatory infiltration (ic), and necrosis (ne). For blood clams, there were curly (cr) and swollen (sw) gill filaments in the frontal lobe, and rare in hemocytes (lh). Co-exposure to MPs + DEHP led to abnormal gill structure, mainly manifested in the gill filament terminal being swollen (sw) or broken (br), and cell necrosis (ne). In addition, compared with single exposure to MPs/DEHP, more obvious morphological changes were found in the gills of the MPs + DEHP group, which completely lost their original features and became necrotic.

It has been demonstrated that MPs and DEHP exposure can interfere with carbohydrate metabolism and ROS elimination of tested organisms [62,63] (Brate et al., 2018; Romano et al., 2018). This leads to significant pathologies alterations, such as inflammation and cell necrosis, in both vertebrate and invertebrate species following MPs or organic pollutants exposure [7,61] (Sikdokur et al., 2020; Liao et al., 2022). The histopathological alterations that may occur in tested organisms can vary depending on biotic and abiotic factors, e.g., animal (species, life stage, tolerance capability etc.), pollutant (type and characteristics), and exposure condition (concentration, duration, and pathway) [3,64–66]. In this study, the pathological changes that occurred in the gills of both types of clams may be attributed as a cause of attenuated filtration rate.

4. Conclusions

The results of the current study indicate that both *R. philippinarum* and *T. granosa* are vulnerable to short-term microplastics and DEHP exposure. The filtration rate of *R. philippinarum* and *T. granosa* were significantly decreased as a result of the exposure to MPs and DEHP, especially co-exposure treatment, suggesting single/combined exposure causes significant feeding inhibition. In regard to antioxidant enzymes activities, like SOD, CAT, and POD, the gills of both types of clams were considerably changed as a result of the MPs and DEHP exposure. The lipid peroxidation level (MDA) occurred in single/combined toxicity, implying that the presence of MPs and DEHP could induce the gill of bivalves to accelerate the production of ROS. Further analysis confirmed the over-accumulation of ROS (O^{2-} and H_2O_2) in gill tissues of *R. philippinarum* and *T. granosa*, thus leading to oxidative damage and histopathological changes, which results in a sharp decrease of the filtration rate.

Considering the widespread pollution of marine ecosystems by MPs and organic pollutants, once in severely contaminated environments, bivalves are inevitably at risk of suffering adverse effects deriving from both MPs and organic pollutants. However, it should be noted that the size of MPs was around 700 nm in this study, which is larger than the typical nanoplastics (1 to 100 nm) [67]. The scope of this study is limited, as it only investigated one size of nanoplastic. Size is a critical parameter that determines the toxicity of nanoplastics [68,69]. To fully understand the ecological impacts of nanoplastics and the associated organic pollutants, it is imperative that more research is conducted to investigate the effect of size. Greater toxicity is expected for smaller particles because of their high specific surface area and their ability to cross cell membranes where they may cause oxidative stress within the cell.

Author Contributions: Conceptualization, Y.Z. and K.P.; methodology, Y.Z. and Y.L.; investigation, H.J.; resources, W.L.; data curation, Y.Z. and K.P.; writing—original draft preparation, Y.Z. and Y.L.; writing—review and editing, Y.Z., Y.L. and K.P.; visualization, Y.Z. and Y.L.; supervision, K.P.; project administration, K.P.; funding acquisition, K.P. All authors have read and agreed to the published version of the manuscript.

Funding: This work was supported by the Shenzhen Science and Technology Innovation Commission of China (JCYJ20180507182227257), the Guangxi Key R&D Program of China (GUIKE AB20297018), and the Guangdong Basic and Applied Basic Research Foundation (2019A1515011630), the Innovation Team Project of Universities in Guangdong Province (No. 2020KCXTD023).

Data Availability Statement: Not applicable.

Acknowledgments: We thank the three anonymous reviewers for their constructive comments.

Conflicts of Interest: The authors declare that they have no known competing financial interests or personal relationships that could have appeared to influence the work reported in this paper.

References

1. Schmaltz, E.; Melvin, E.C.; Diana, Z.; Gunady, E.F.; Rittschof, D.; Somarelli, J.A.; Virdin, J.; Dunphy-Daly, M.M. Plastic pollution solutions: Emerging technologies to prevent and collect marine plastic pollution. *Environ. Int.* **2020**, *144*, 106067. [CrossRef]
2. MacLeo, M.; Arp, H.P.H.; Tekman, M.B.; Jahnke, A. The global threat from plastic pollution. *Science* **2021**, *373*, 61–65. [CrossRef]
3. Wang, C.; Zhao, J.; Xing, B. Environmental source, fate, and toxicity of microplastics. *J. Hazard. Mater.* **2021**, *407*, 124357. [CrossRef]
4. Law, K.L.; Thompson, R.C. Microplastics in the seas. *Science* **2014**, *345*, 144–145. [CrossRef] [PubMed]
5. Turner, A. Foamed Polystyrene in the Marine Environment: Sources, Additives, Transport, Behavior, and Impacts. *Environ. Sci. Technol.* **2020**, *54*, 10411–10420. [CrossRef] [PubMed]
6. Wang, Y.; Yang, Y.; Liu, X.; Zhao, J.; Liu, R.; Xing, B. Interaction of Microplastics with Antibiotics in Aquatic Environment: Distribution, Adsorption, and Toxicity. *Environ. Sci. Technol.* **2021**, *55*, 15579–15595. [CrossRef]
7. Sikdokur, E.; Belivermis, M.; Sezer, N.; Pekmez, M.; Bulan, O.K.; Kilic, O. Effects of microplastics and mercury on manila clam Ruditapes philippinarum: Feeding rate, immunomodulation, histopathology and oxidative stress. *Environ. Pollut.* **2020**, *262*, 114247. [CrossRef] [PubMed]
8. Jiang, W.W.; Fang, J.H.; Du, M.R.; Gao, Y.P.; Fang, J.G.; Jiang, Z.J. Microplastics influence physiological processes, growth and reproduction in the Manila clam, Ruditapes philippinarum. *Environ. Pollut.* **2022**, *293*, 118502. [CrossRef]
9. Zhou, Z.; Ni, X.; Chen, S.; Wu, Z.; Tang, J.; Su, Y.; Wang, X.; Wang, L. Ingested microplastics impair the metabolic relationship between the giant clam Tridacna crocea and its symbionts. *Aquat. Toxicol.* **2022**, *243*, 106075. [CrossRef] [PubMed]
10. Paul-Pont, I.; Lacroix, C.; Gonzalez Fernandez, C.; Hegaret, H.; Lambert, C.; Le Goic, N.; Frere, L.; Cassone, A.-L.; Sussarellu, R.; Fabioux, C.; et al. Exposure of marine mussels Mytilus spp. to polystyrene microplastics: Toxicity and influence on fluoranthene bioaccumulation. *Environ. Pollut.* **2016**, *216*, 724–737. [CrossRef] [PubMed]
11. Teng, J.; Zhao, J.; Zhu, X.; Shan, E.; Zhang, C.; Zhang, W.; Wang, Q. Toxic effects of exposure to microplastics with environmentally relevant shapes and concentrations: Accumulation, energy metabolism and tissue damage in oyster Crassostrea gigas. *Environ. Pollut.* **2021**, *269*, 116169. [CrossRef] [PubMed]
12. Jeyavani, J.; Sibiya, A.; Gopi, N.; Mahboob, S.; Riaz, M.N.; Vaseeharan, B. Dietary consumption of polypropylene microplastics alter the biochemical parameters and histological response in freshwater benthic mollusc Pomacea paludosa. *Environ. Res.* **2022**, *212*, 113370. [CrossRef] [PubMed]
13. Monteiro, R.; Costa, S.; Coppola, F.; Freitas, R.; Vale, C.; Pereira, E. Toxicity beyond accumulation of Titanium after exposure of Mytilus galloprovincialis to spiked seawater. *Environ. Pollut.* **2019**, *244*, 845–854. [CrossRef] [PubMed]

14. Yang, W.; Gao, X.; Wu, Y.; Wan, L.; Tan, L.; Yuan, S.; Ding, H.; Zhang, W. The combined toxicity influence of microplastics and nonylphenol on microalgae Chlorella pyrenoidosa. *Ecotoxicol. Environ. Saf.* **2020**, *195*, 110484. [CrossRef] [PubMed]
15. Zhang, J.; Meng, H.; Kong, X.; Cheng, X.; Ma, T.; He, H.; Du, W.; Yang, S.; Li, S.; Zhang, L. Combined effects of polyethylene and organic contaminant on zebrafish (Danio rerio): Accumulation of 9-Nitroanthracene, biomarkers and intestinal microbiota. *Environ. Pollut.* **2021**, *277*, 116767. [CrossRef] [PubMed]
16. Guo, X.; Cai, Y.; Ma, C.; Han, L.; Yang, Z. Combined toxicity of micro/nano scale polystyrene plastics and ciprofloxacin to Corbicula fluminea in freshwater sediments. *Sci. Total Environ.* **2021**, *789*, 147887. [CrossRef]
17. Besseling, E.; Wegner, A.; Foekema, E.M.; Van Den Heuvel-Greve, M.J.; Koelmans, A.A. Effects of microplastic on fitness and PCB bioaccumulation by the lugworm Arenicola marina (L.). *Environ. Sci. Technol.* **2013**, *47*, 593–600. [CrossRef]
18. Wardrop, P.; Shimeta, J.; Nugegoda, D.; Morrison, P.D.; Miranda, A.; Tang, M.; Clarke, B.O. Chemical pollutants sorbed to ingested microbeads from personal care products accumulate in fish. *Environ. Sci. Technol.* **2016**, *50*, 4037–4044. [CrossRef]
19. Zhao, H.J.; Xu, J.K.; Yan, Z.H.; Ren, H.Q.; Zhang, Y. Microplastics enhance the developmental toxicity of synthetic phenolic antioxidants by disturbing the thyroid function and metabolism in developing zebrafish. *Environ. Int.* **2020**, *140*, 105750. [CrossRef]
20. O'Donovan, S.; Mestre, N.C.; Abel, S.; Fonseca, T.G.; Carteny, C.C.; Cormier, B.; Keiter, S.H.; Bebianno, M.J. Ecotoxicological Effects of Chemical Contaminants Adsorbed to Microplastics in the Clam Scrobicularia plana. *Front. Mar. Sci.* **2018**, *5*, 143. [CrossRef]
21. Shi, Q.; Tang, J.; Wang, L.; Liu, R.; Giesy, J.P. Combined cytotoxicity of polystyrene nanoplastics and phthalate esters on human lung epithelial A549 cells and its mechanism. *Ecotoxicol. Environ. Saf.* **2021**, *213*, 112041. [CrossRef]
22. Rainieri, S.; Conlledo, N.; Larsen, B.K.; Granby, K.; Barranco, A. Combined effects of microplastics and chemical contaminants on the organ toxicity of zebrafish (Danio rerio). *Environ. Res.* **2018**, *162*, 135–143. [CrossRef]
23. Chen, Q.; Yin, D.; Jia, Y.; Schiwy, S.; Legradi, J.; Yang, S.; Hollert, H. Enhanced uptake of BPA in the presence of nanoplastics can lead to neurotoxic effects in adult zebrafish. *Sci. Total Environ.* **2017**, *609*, 1312–1321. [CrossRef]
24. Avio, C.G.; Gorbi, S.; Milan, M.; Benedetti, M.; Fattorini, D.; d'Errico, G.; Regoli, F. Pollutants bioavailability and toxicological risk from microplastics to marine mussels. *Environ. Pollut.* **2015**, *198*, 211–222. [CrossRef] [PubMed]
25. Oliveira, M.; Ribeiro, A.; Hylland, K.; Guilhermino, L. Single and combined effects of microplastics and pyrene on juveniles (0+ group) of the common goby Pomatoschistus microps (Teleostei, Gobiidae). *Ecol. Indic.* **2013**, *34*, 641–647. [CrossRef]
26. Rehse, S.; Kloas, W.; Zarfl, C. Microplastics reduce short-term effects of environmental contaminants. Part I: Effects of bisphenol A on freshwater zooplankton are lower in presence of polyamide particles. *Int. J. Environ. Res. Public Health* **2018**, *15*, 280. [CrossRef] [PubMed]
27. Courtene-Jones, W.; Quinn, B.; Ewins, C.; Gary, S.F.; Narayanaswamy, B.E. Consistent microplastic ingestion by deep-sea invertebrates over the last four decades (1976–2015), a study from the North East Atlantic. *Environ. Pollut.* **2019**, *244*, 503–512. [CrossRef] [PubMed]
28. Kutralam-Muniasamy, G.; Perez-Guevara, F.; Elizalde-Martinez, I.; Shruti, V.C. Review of current trends, advances and analytical challenges for microplastics contamination in Latin America. *Environ. Pollut.* **2020**, *267*, 115463. [CrossRef]
29. Hanachi, P.; Karbalaei, S.; Yu, S. Combined polystyrene microplastics and chlorpyrifos decrease levels of nutritional parameters in muscle of rainbow trout (Oncorhynchus mykiss). *Environ. Sci. Pollut. Res.* **2021**, *28*, 64908–64920. [CrossRef]
30. Huang, W.; Wang, X.; Chen, D.; Xu, E.G.; Luo, X.; Zeng, J.; Huan, T.; Li, L.; Wang, Y. Toxicity mechanisms of polystyrene microplastics in marine mussels revealed by high-coverage quantitative metabolomics using chemical isotope labeling liquid chromatography mass spectrometry. *J. Hazard. Mater.* **2021**, *417*, 126003. [CrossRef]
31. Goldstein, M.C.; Rosenberg, M.; Cheng, L. Increased oceanic microplastic debris enhances oviposition in an endemic pelagic insect. *Biol. Lett.* **2012**, *8*, 817–820. [CrossRef] [PubMed]
32. Ivar do Sul, J.A.; Costa, M.F.; Fillmann, G. Microplastics in the pelagic environment around oceanic islands of the Western Tropical Atlantic Ocean. *Water Air Soil Pollut.* **2014**, *225*, 2004. [CrossRef]
33. Sun, X.D.; Yuan, X.Z.; Jia, Y.; Feng, L.J.; Zhu, F.P.; Dong, S.; Liu, J.J.; Kong, X.P.; Tian, H.Y.; Duan, J.L.; et al. Differentially charged nanoplastics demonstrate distinct accumulation in Arabidopsis thaliana. *Nat. Nanotechnol.* **2020**, *15*, 755–760. [CrossRef] [PubMed]
34. Feng, L.J.; Wang, J.J.; Liu, S.C.; Sun, X.D.; Yuan, X.Z.; Wang, S.G. Role of extracellular polymeric substances in the acute inhibition of activated sludge by polystyrene nanoparticles. *Environ. Pollut.* **2018**, *238*, 859–865. [CrossRef]
35. Chai, B.; Li, Y.; Wang, L.; Zhang, X.T.; Wan, Y.P.; Chen, F.; Pan, K. Microplastic contamination on the beaches of South China. *Front. Mar. Sci.* **2022**, *9*, 863652. [CrossRef]
36. Riisgard, H.U.; Pleissner, D.; Lundgreen, K.; Larsen, P.S. Growth of mussels Mytilus edulis at algal (Rhodomonas salina) concentrations below and above saturation levels for reduced filtration rate. *Mar. Biol. Res.* **2013**, *9*, 1005–1017. [CrossRef]
37. Vale, G.; Mehennaoui, K.; Cambier, S.; Libralato, G.; Jomini, S.; Domingos, R.F. Manufactured nanoparticles in the aquatic environment-biochemical responses on freshwater organisms: A critical overview. *Aquat. Toxicol.* **2016**, *170*, 162–174. [CrossRef]
38. Winston, G.W.; Digiulio, R.T. Prooxidant and antioxidant mechanisms in aquatic organisms. *Aquat. Toxicol.* **1991**, *19*, 137–161. [CrossRef]

39. Britto, R.S.; Nascimento, J.P.; Serodre, T.; Santos, A.P.; Soares, A.M.V.M.; Furtado, C.; Ventura-Lima, J.; Monserrat, J.M.; Freitas, R. Oxidative stress in Ruditapes philippinarum after exposure to different graphene oxide concentrations in the presence and absence of sediment. *Comp. Biochem. Physiol. C-Toxicol. Pharmacol.* **2021**, *240*, 108922. [CrossRef]
40. Waters, E.C.T.; Baark, F.; Yu, Z.; Mota, F.; Eykyn, T.R.; Yan, R.; Southworth, R. Detecting Validated Intracellular ROS Generation with F-18-dihydroethidine-Based PET. *Mol. Imaging Biol.* **2022**, *24*, 377–383. [CrossRef]
41. Yu, D.; Zha, Y.; Zhong, Z.; Ruan, Y.; Li, Z.; Sun, L.; Hou, S. Improved detection of reactive oxygen species by DCFH-DA: New insight into self-amplification of fluorescence signal by light irradiation. *Sens. Actuators B-Chem.* **2021**, *339*, 129878. [CrossRef]
42. Adler, J.; Parmryd, I. Quantifying colocalization by correlation: The pearson correlation coefficient is cuperior to the mander's overlap coefficient. *Cytom. Part A* **2010**, *77*, 733–742. [CrossRef] [PubMed]
43. Sarasquete, C.; Gutierrez, M. New tetrachromic VOF stain (Type III-G.S) for normal and pathological fish tissues. *Eur. J. Histochem.* **2005**, *49*, 211–220. [PubMed]
44. Esperanza, M.; Seoane, M.; Servia, M.J.; Cid, A. Effects of Bisphenol A on the microalga Chlamydomonas reinhardtii and the clam Corbicula fluminea. *Ecotoxicol. Environ. Saf.* **2020**, *197*, 110609. [CrossRef] [PubMed]
45. Zha, S.; Tang, Y.; Shi, W.; Liu, H.; Sun, C.; Bao, Y.; Liu, G. Impacts of four commonly used nanoparticles on the metabolism of a marine bivalve species, Tegillarca granosa. *Chemosphere* **2022**, *296*, 134079. [CrossRef]
46. Oliveira, P.; Antao Barboza, L.G.; Branco, V.; Figueiredo, N.; Carvalho, C.; Guilhermino, L. Effects of microplastics and mercury in the freshwater bivalve Corbicula fluminea (Muller, 1774): Filtration rate, biochemical biomarkers and mercury bioconcentration. *Ecotoxicol. Environ. Saf.* **2018**, *164*, 155–163. [CrossRef]
47. Coffin, S.; Lee, I.; Gan, J.; Schlenk, D. Simulated digestion of polystyrene foam enhances desorption of diethylhexyl phthalate (DEHP) and In vitro estrogenic activity in a size-dependent manner. *Environ. Pollut.* **2019**, *246*, 452–462. [CrossRef]
48. Molino, C.; Filippi, S.; Stoppiello, G.A.; Meschini, R.; Angeletti, D. In Vitro evaluation of cytotoxic and genotoxic effects of Di(2-ethylhexyl)-phthalate (DEHP) on European sea bass (Dicentrarchus labrax) embryonic cell line. *Toxicol. Vitr.* **2019**, *56*, 118–125. [CrossRef]
49. Rios-Fuster, B.; Alomar, C.; Capo, X.; Gonzalez, G.P.; Martinez, R.M.G.; Rojas, D.L.S.; Silva, M.; Hernando, P.F.; Sole, M.; Freitas, R.; et al. Assessment of the impact of aquaculture facilities on transplanted mussels (Mytilus galloprovincialis): Integrating plasticizers and physiological analyses as a biomonitoring strategy. *J. Hazard. Mater.* **2022**, *424*, 127264. [CrossRef] [PubMed]
50. Heinder, F.M.; Alajmi, F.; Huerlimann, R.; Zeng, C.; Newman, S.J.; Vamvounis, G.; van Herwerden, L. Toxic effects of polyethylene terephthalate microparticles and Di(2-ethylhexyl)phthalate on the calanoid copepod, Parvocalanus crassirostris. *Ecotoxicol. Environ. Saf.* **2017**, *141*, 298–305. [CrossRef]
51. Araujo, C.V.M.; Gomez, L.; Silva, D.C.V.R.; Pintado-Herrera, M.G.; Lara-Martin, P.A.; Hampel, M.; Blasco, J. Risk of triclosan based on avoidance by the shrimp Palaemon varians in a heterogeneous contamination scenario: How sensitive is this approach? *Chemosphere* **2019**, *235*, 126–135. [CrossRef] [PubMed]
52. Wegner, A.; Besseling, E.; Foekema, E.M.; Kamermans, P.; Koelmans, A.A. Effects of nanopolystyrene on the feeding behavior of the blue mussel (*Mytilus edulis* L.). *Environ. Toxicol. Chem.* **2012**, *31*, 2490–2497. [CrossRef] [PubMed]
53. Ozdilek, S.Y.; Demir, N.; Gurkan, S.E. Assessment of pollution biomarker and stable isotope data in Mytilus galloprovincialis tissues. *Environ. Monit. Assess.* **2019**, *191*, 60. [CrossRef] [PubMed]
54. Guilhermino, L.; Vieira, L.R.; Ribeiro, D.; Tavares, A.S.; Cardoso, V.; Alves, A.; Almeida, J.M. Uptake and effects of the antimicrobial florfenicol, microplastics and their mixtures on freshwater exotic invasive bivalve Corbicula fluminea. *Sci. Total Environ.* **2018**, *622*, 1131–1142. [CrossRef]
55. Piddington, D.L.; Fang, F.C.; Laessig, T.; Cooper, A.M.; Orme, I.M.; Buchmeier, N.A. Cu,Zn superoxide dismutase of Mycobacterium tuberculosis contributes to survival in activated macrophages that are generating an oxidative burst. *Infect. Immun.* **2001**, *69*, 4980–4987. [CrossRef]
56. Han, Y.; Zhou, W.; Tang, Y.; Shi, W.; Shao, Y.; Ren, P.; Zhang, J.; Xiao, G.; Sun, H.; Liu, G. Microplastics aggravate the bioaccumulation of three veterinary antibiotics in the thick shell mussel Mytilus coruscus and induce synergistic immunotoxic effects. *Sci. Total Environ.* **2021**, *770*, 145273. [CrossRef]
57. Mao, Y.; Ai, H.; Chen, Y.; Zhang, Z.; Zeng, P.; Kang, L.; Li, W.; Gu, W.; He, Q.; Li, H. Phytoplankton response to polystyrene microplastics: Perspective from an entire growth period. *Chemosphere* **2018**, *208*, 59–68. [CrossRef]
58. Zhao, T.; Tan, L.; Zhu, X.; Huang, W.; Wang, J. Size-dependent oxidative stress effect of nano/micro-scaled polystyrene on Karenia mikimotoi. *Mar. Pollut. Bull.* **2020**, *154*, 111074. [CrossRef]
59. Trestrail, C.; Nugegoda, D.; Shimeta, J. Invertebrate responses to microplastic ingestion: Reviewing the role of the antioxidant system. *Sci. Total Environ.* **2020**, *734*, 138559. [CrossRef]
60. Revel, M.; Lagarde, F.; Perrein-Ettajani, H.; Bruneau, M.; Akcha, F.; Sussarellu, R.; Rouxel, J.; Costil, K.; Decottignies, P.; Cognie, B.; et al. Tissue-Specific Biomarker Responses in the Blue Mussel Mytilus spp. Exposed to a Mixture of Microplastics at Environmentally Relevant Concentrations. *Front. Environ. Sci.* **2019**, *7*, 33. [CrossRef]
61. Liao, H.; Liu, S.; Junaid, M.; Gao, D.; Ai, W.; Chen, G.; Wang, J. Di-(2-ethylhexyl) phthalate exacerbated the toxicity of polystyrene nanoplastics through histological damage and intestinal microbiota dysbiosis in freshwater Micropterus salmoides. *Water Res.* **2022**, *219*, 118608. [CrossRef]
62. Brate, I.L.N.; Blazquez, M.; Brooks, S.J.; Thomas, K.V. Weathering impacts the uptake of polyethylene microparticles from toothpaste in Mediterranean mussels (*M. galloprovincialis*). *Sci. Total Environ.* **2018**, *626*, 1310–1318. [CrossRef] [PubMed]

63. Romano, N.; Ashikin, M.; Teh, J.C.; Syukri, F.; Karami, A. Effects of pristine polyvinyl chloride fragments on whole body histology and protease activity in silver barb Barbodes gonionotus fry. *Environ. Pollut.* **2018**, *237*, 1106–1111. [CrossRef] [PubMed]
64. Prata, J.C.; da Costa, J.P.; Lopes, I.; Duarte, A.C.; Rocha-Santos, T. Environmental exposure to microplastics: An overview on possible human health effects. *Sci. Total Environ.* **2020**, *702*, 134455. [CrossRef]
65. Kim, J.-H.; Yu, Y.-B.; Choi, J.-H. Toxic effects on bioaccumulation, hematological parameters, oxidative stress, immune responses and neurotoxicity in fish exposed to microplastics: A review. *J. Hazard. Mater.* **2021**, *413*, 125423. [CrossRef] [PubMed]
66. Rahman, A.; Sarkar, A.; Yadav, O.P.; Achari, G.; Slobodnik, J. Potential human health risks due to environmental exposure to nano-and microplastics and knowledge gaps: A scoping review. *Sci. Total Environ.* **2021**, *757*, 143872. [CrossRef] [PubMed]
67. Koelmans, A.A. Modeling the role of microplastics in bioaccumulation of organic chemicals to marine aquatic organisms. A critical review. *Mar. Anthropog. Litter* **2015**, 309–324.
68. Miao, L.; Hou, J.; You, G.; Liu, Z.; Liu, S.; Li, T.; Qu, H. Acute effects of nanoplastics and microplastics on periphytic biofilms depending on particle size, concentration and surface modification. *Environ. Pollut.* **2019**, *255*, 113300. [CrossRef] [PubMed]
69. Ning, Q.; Wang, D.; An, J.; Ding, Q.; Huang, Z.; Zou, Y.; You, J. Combined effects of nanosized polystyrene and erythromycin on bacterial growth and resistance mutations in Escherichia coli. *J. Hazard. Mater.* **2022**, *422*, 126858. [CrossRef] [PubMed]

nanomaterials

Article

Environmental Safety Assessments of Lipid Nanoparticles Loaded with Lambda-Cyhalothrin

Catarina Ganilho [1], Márcia Bessa da Silva [1], Cristiana Paiva [1], Thacilla Ingrid de Menezes [2], Mayara Roncaglia dos Santos [2], Carlos M. Pereira [2], Ruth Pereira [1,*] and Tatiana Andreani [1,2,3,*]

1 GreenUPorto, Sustainable Agrifood Production Research Centre & INOV4AGRO, Department of Biology, Faculty of Sciences, University of Porto, Rua Campo Alegre s/n, 4169-007 Porto, Portugal; up201707255@edu.fc.up.pt (C.G.); bessamiss@gmail.com (M.B.d.S.); cristiana.pgpaiva@fc.up.pt (C.P.)
2 Chemistry Research Centre (CIQUP) & Institute of Molecular Sciences (IMS), Department of Chemistry and Biochemistry, Faculty of Sciences, University of Porto, Rua do Campo Alegre s/n, 4169-007 Porto, Portugal; up201811986@edu.fc.up.pt (T.I.d.M.); mayara.santos@fc.up.pt (M.R.d.S.); cmpereir@fc.up.pt (C.M.P.)
3 Centre for Research and Technology of Agro-Environmental and Biological Sciences (CTAB) & INOV4AGRO, University of Trás-os-Montes e Alto Douro, UTAD, 5000-801 Vila Real, Portugal
* Correspondence: ruth.pereira@fc.up.pt (R.P.); tatiana.andreani@fc.up.pt (T.A.); Tel.: +351-220-402-000 (R.P. & T.A.)

Abstract: Lipid nanoparticles (LN) composed of biodegradable lipids and produced by green methods are candidates for the encapsulation of pesticides, potentially contributing to decreasing their release in the environment. From a safety-by-design concept, this work proposes LN for the encapsulation of insecticide active ingredients (AI). However, given the complexity of nanoparticles, ecotoxicological studies are often controversial, and a detailed investigation of their effects on the environment is required. Accordingly, this work aimed to produce and characterize LN containing the insecticide lambda-cyhalothrin (LC) and evaluate their safety to crops (*Solanum lycopersicum* and *Zea mays*), soil invertebrates (*Folsomia candida* and *Eisenia fetida*), and soil microbial parameters. The average particle size for LN-loaded with LC (LN–LC) was 165.4 ± 2.34 nm, with narrow size distribution and negative charge (−38.7 ± 0.954 mV). LN were able to encapsulate LC with an entrapment efficacy of 98.44 ± 0.04%, maintaining the stability for at least 4 months. The LN–LC showed no risk to the growth of crops and reproduction of the invertebrates. The effect on microbial parameters showed that the activity of certain soil microbial parameters can be inhibited or stimulated by the presence of LN at highest concentrations, probably by changing the pH of soil or by the intrinsic properties of LN.

Keywords: nanopesticides; insecticides; ecotoxicology; non-target terrestrial organisms; soil microbial parameters

Citation: Ganilho, C.; da Silva, M.B.; Paiva, C.; de Menezes, T.I.; dos Santos, M.R.; Pereira, C.M.; Pereira, R.; Andreani, T. Environmental Safety Assessments of Lipid Nanoparticles Loaded with Lambda-Cyhalothrin. *Nanomaterials* **2022**, *12*, 2576. https://doi.org/10.3390/nano12152576

Academic Editors: Xiaoshan Zhu and Jian Zhao

Received: 21 June 2022
Accepted: 21 July 2022
Published: 27 July 2022

1. Introduction

Modern agriculture practices have been suffering constant pressure to increase food production, due to the rapid expansion of global population density, as well as the need to respond to highly demanding markets and consumers. The intensification of the use of agrochemicals, especially insecticides for pest control, has been one of the major consequences [1] and is partly responsible for different environmental issues, such as soil, water, and food contamination; pest resistance; bioaccumulation of pesticide residues in food chains; and decrease of biodiversity [2]. During the field application, only a fraction of the active ingredient (AI) of pesticide formulations reaches the desired target. The largest amount is lost because of improper pulverization practices and precipitation washing, being decomposed and/or becoming mobile in the environment through volatilization, soil percolation, and runoffs. Variable environmental conditions also affect pesticides' stability [3], thus increasing the amounts that have to be applied [4] and the costs to the farmers.

In most cases, and depending on its chemical properties, the agricultural ecosystems are unable to completely neutralize the applied insecticides, resulting in the accumulation of its residues in the non-target compartments, such as biota [5].

In recent years, different strategies have been proposed for the sustainable control of agricultural pests and diseases. In this context, nanoscience and nanotechnology developments have been targeting the agri-food sector through the design of nanopesticides (NPest) formulations [6], in which, for example, pesticide's active ingredients (AI) are encapsulated in nanomaterials (NM) [7], with a size between 1 and 1000 nm [8]. The use of NPest has been announced as a promising strategy for pest management, since nanoformulations decrease the amount of pesticide AI needed by improving its stability (protecting AI from external agents, such as temperature, light, and moisture) and by controlling its targeted release in the environment with minimal environmental and human impact. In this context, lipid nanoparticles (LN) appeared as promising nanocarriers for insecticide AI loading, as they are composed of biocompatible and biodegradable lipids [3], have good physicochemical stability, promote sustainable release of AI, and their synthesis offers competitive costs and ease of manufacturing [9]. In addition, LN can be produced by ecofriendly methods, without the use of organic solvents [10]. However, only one study described the application of LN for the encapsulation of insecticide AI [11]. Furthermore, these studies only evaluate the efficacy of insecticide-loaded LN on target species, and, thus, there is still limited information about the safety of these systems to non-target organisms, specially to terrestrial species.

In this context, the present work aimed at encapsulating the insecticide lambda-cyhalothrin (LC) in LN, using physiological lipids and ecofriendly methods, as well as at evaluating their safety on agriculture crops (*Solanum lycopersicum* and *Zea mays*), on soil invertebrates (*Folsomia candida* and *Eisenia fetida*), and on soil microbial parameters.

Lambda-cyhalothrin (LC) is a type II pyrethroid insecticide [12] that mimics natural molecules extracted from chrysanthemum flowers (*Chrysanthemum cinerariaefolium* and *C. cineum*) with biocidal effect and therefore has been widely used to control insect pests in agriculture, veterinary, and domestic applications. However, after spraying products containing pyrethroids, their residues can be spilled on soil, where they can accumulate, as pyrethroids are highly hydrophobic compounds, binding to organic matter and other soil components [13]. Nevertheless, due to it risk to aquatic organisms and non-target arthropods, LC has been banned from the European market since 2021, based on the risk assessment of LC presented by The European Food and Safety Authority [14]. Thus, the encapsulation of LC in physiological and biodegradable lipids may be an excellent strategy to reduce its harmful effects on the environment and to recycle the use of the banned pesticides with a more sustainable perspective. In addition, the gathered data in the present study can be used in a safety-by-design (SbD) approach to select the concentrations of the encapsulated AI that are less harmful to non-target species, and they can then be tested for their efficacy to target species. To the best of our knowledge, the environmental safety of LC encapsulated in LN has not been previously reported, and, therefore, these findings can facilitate the understanding of the impact of encapsulated LC on the environment.

2. Materials and Methods

2.1. Materials

The AI lambda-cyhalothrin (LC) was purchased from LGC Standards (Barcelona, Spain). Precirol® ATO 5 (Prec) (Glyceryl palmitostearate) (Gattefosse S.A., Saint-Priest, France) was used as a solid lipid, and Capryol 90® (Propylene glycol monocaprylate) was used as a liquid lipid (Gattefosse S.A., Saint-Priest, France). Soy lecithin, used as amphipathic surfactant, was obtained from VWR (Alfragide, Portugal). The surfactant TegoCare 450® (TG450) was supplied by Evonik Industries (Essen, Germany). Dimethyl sulfoxide (DMSO) was used as a dispersant agent to suspend LC and was obtained from Sigma-Aldrich (Lisbon, Portugal).

Preparation and Characterization of Lipid Nanoparticles (LN)

LN were prepared by a modified oil-in-water (O/W) emulsion method without the addition of organic solvents, as described previously [15]. Briefly, 0.1% (*w/w*) of LC was added to the oil phase composed of 3% (*w/w*) of glyceryl palmitostearate, 0.5% (*w/w*) of soy lecithin, 2% (*w/w*) of Capryol 90, and 1.5 % (*w/w*) of TG450 and heated at 5–10 °C above the melting point of the solid lipid (~65 °C). The aqueous phase was heated to the same temperature and then added to the oil phase, followed by high-speed homogenization at 15,000 rpm (Ultra-Turrax T25, IKA Labortechnik, Deutschland, Germany) for 10 min, with a probe (18 G). After homogenization, the dispersion was placed in an ice bath for 10 min for recrystallization of the lipid matrix and formation of LN.

LN were characterized for the average particle size, particle distribution expressed as polydispersity index (PI), and zeta potential (ZP). The average size and PI were determined by dynamic light scattering (DLS) with a Zeta Sizer NanoZS equipment (Malvern Panalytical, Malvern, UK), using disposable polystyrene cells, 173° scattering angle, and refractive index of 1.6. The ZP analysis of LN was performed in the same equipment by electrophoretic light scattering, using disposable plain folded capillary zeta cells. All measurements were performed at 25 °C, and the LN dispersions were diluted in deionized water in a 1:100 (*v/v*) ratio and analyzed in triplicate. The values were expressed as mean values ± standard deviation (SD). After production, unloaded and loaded LN were stored in the dark at 25 °C, and the variation of average particle size, PI, and ZP were recorded at predetermined time intervals (day 0 and 30, 60, and 120 days after LN synthesis).

Glyceryl palmitostearate in LN was indirectly measured by the filtration/centrifugation method, followed by the determination of the free amount of LC (non-encapsulated) by high-performance liquid chromatography (HPLC). A volume of 250 µL of LC-loaded LN dissolved in ethanol (1:1) was centrifuged for 10 min at 13,500 rpm, at 25 °C, in a centrifugal filter device (Amicon Ultra 0.5, NMWL 30 K, Merck Millipore Ltd., Dublin, Ireland). Then 100 µL of filtrate was diluted with ethanol at 1:4, and the concentration of LC was analyzed by HPLC. The chromatographic analysis was performed by Shimadzu Nexera-I-LC2040C-3D (Shimadzu, Kioto, Japan) equipped with Inertsil ODS 3V (150 mm × 4.6 mm, 3 µm) column (GL Science Inc., Tokio, Japan) at 30 °C and a UV–Vis detector set at 240 nm. The mobile phase consisted of acetonitrile:water (85:15, *v/v*). The flowrate was set to 0.1 mL/min, and the injection volume was 10 µL. The area under the curve was measured for the calculation of the LC concentration based on the calibration curve. The EE% was calculated by using the following equation:

$$\%\,\mathrm{EE} = \frac{\text{Total amount of LC AI} - \text{Free AI}}{\text{Total amount of AI}} \times 100$$

2.2. *Environmental Safety Assessment of LN–LC on Terrestrial Organisms*

2.2.1. Test Soil

Natural soil used in the present study was collected from the 0–20 cm topsoil layer in the region of Vairão (Vila do Conde, Portugal), in the open fields of the GreenUPorto Research Centre, with no historical application of pesticides at least in the last three decades, sieved through a 4 mm mesh size and air dried. The soil physicochemical characterization presented in Table 1 was performed in the laboratory with soil sieved at 2 mm. Five replicates were used for all the following parameters: pH (H_2O and KCl (1 M)), conductivity, organic matter content (%OM), and water-holding capacity (% WHCmax). The pH and conductivity were determined in a 1:5 (*v/v*) soil suspension. After 60 ± 10 min of agitation, the suspension was allowed to stand for 60 min before measuring pH with a pre-calibrated multiparameter (Edge, HANNA Instruments, Povoa de Varzim, Portugal) [16]. Conductivity was recorded with the pre-calibrated meter on the H_2O suspension [17]. For OM content, soil samples were dried at 105 °C until constant weight and then obtained by loss on ignition of dried soil samples at 450 °C, for 8 h, and expressed in percentage [18]. For WHC_{max} (%), soil samples were introduced in polypropylene containers whose bottom was replaced with filter paper and immersed in water for 3 h. Subsequently, soil samples were drained with successive exchanges of absorbent paper for 2 h and dried at 105 °C [19].

Table 1. Physicochemical characterization (mean values ± standard deviation (SD)) of the natural soil.

Natural Soil	pH (H_2O)	pH (KCl)	Conductivity (mS cm^{-1})	% OM	% WHC_{max}
	6.46 ± 0.020	5.39 ± 0.32	0.77 ± 0.045	5.21 ± 0.15	43.00 ± 0.15

% OM, percentage of organic matter; % WHCmax, water-holding capacity in percentage of dry mass.

2.2.2. Assessment LN–LC Safety to Terrestrial Plants Growth

For the plant growth assay, two species were used: one dicotyledonous plant, *Solanum lycopersicum* (tomato); and one monocotyledonous plant, *Zea mays* (corn). Seeds were purchased from a local supplier in the city of Porto and used for the assay, following the standard protocol OECD 227 [20]. For this purpose, 200 g of natural soil was added to plastic pots moistened to the WHC_{max} of the natural soil (~43%). For *S. lycopersicum* and *Z. mays*, we added 20 and 10 seeds to each pot, respectively. The test started after the germination of 70% of the seeds. Then 10 seedlings for *S. lycopersicum* and 7 seedlings for *Z. mays* were kept in each pot, thus avoiding intraspecific competition. The plants germinated and grew in a growth chamber with a controlled temperature (20 ± 2 °C), a photoperiod (16 h^L:8 h^D), and photosynthetically active radiation (25,000 lux). After 15 days of growth and formation of the 2nd and 3rd true leaf, the plants were sprayed on the shoots with a volume of 5 mL of the nanoformulation containing LC at concentrations of 0, 7, 10, 14, 20, and 28 g LC ha^{-1} (based on the dose of LC recommended in the commercial formulation Karate+ (Syngenta)). The compounds tested were LC AI dispersed in DMSO at 5% (*v/v*) and LC-loaded in lipid nanoparticles (LN) in an aqueous dispersion (LN–LC). The effect of DMSO at 5% (*v/v*) was also tested in a DMSO control. The unloaded LN were not tested in this case because the LC-loaded LN did not cause significant effects. For the plants' growth assays, 5 replicates were tested for the control group and for each compound tested concentration. In the control, plants were only sprayed with deionized water. After 10 days of exposure, plants from each replicate were harvested, and their fresh and dry biomass (roots and shoots) and the length of the roots were evaluated.

2.2.3. Assessments of LN–LC Safety to Soil Invertebrates

Since the interaction between the AI encapsulated into LN and the biological membranes can be mediated by the release profile of AI from nanoparticles, the short- and long-term effects of nanoformulations were evaluated on the avoidance, survival, and reproduction of the earthworm *Eisenia fetida* (Oligochaeta: Lumbricidae) and the collembola *Folsomia candida* (Collembola: Isotomidae). All organisms were obtained from laboratorial cultures kept under controlled conditions (temperature, 20 ± 2 °C; photoperiod, 16 h^L:8 h^D). The organisms were fed weekly. Test organisms are collected from synchronized cultures with homogeneous age.

Reproduction Test with *F. candida*

The reproduction test with *F. candida* species was performed according to the OECD standard protocol 232 [21]. The assay was performed in small plastic containers containing 30 g of natural soil thoroughly mixed with the nanoformulations at the concentrations 0, 7, 10, 14, 20, and 28 g LC ha^{-1} (based on the doses of LC recommended in the commercial formulation). The formulations tested were LC AI solution (LC dispersed in DMSO at 5% (*v/v*)), unloaded LN, and LC-loaded LN (LN–LC). The concentrations of LN without LC were based on the amount of solid lipid (SL) used in the synthesis ranging from 0, 150, 210, 300, 480, and 660 g SL ha^{-1}. The concentrations of all LN constituents (lipids and surfactants) were the same for unloaded and loaded LN with LC. The effect of DMSO at 5% (*v/v*) was also tested in a DMSO control. The soil WHC_{max} was adjusted to 50%. The water used to adjust soil moisture was used to add the formulations at the concentrations mentioned. Thereafter, ten individuals (9–12 days old) were placed in the plastic containers, and dry yeast was added as food. Each concentration and the control groups were tested in five replicates. After 28 days of exposure, under the same conditions as described for the cultures, the plastic containers were filled with water, gently mixed with the soil, and then transferred to larger plastic pots. Afterward, few drops of China ink were added and carefully homogenized. The collembola floated on the water surface, which was photographed to allow us to count the number of the adults and juveniles in each pot, using a public domain software ImageJ version 1.53k (Wayne Rasband and contributors, National Institutes of Health, Maryland, USA – Java 1.8.0_172 (64-bit) – http://imagej.nih.gov/ij).

Avoidance and Reproduction Tests with *E. fetida*

Avoidance tests with *E. fetida* were carried out in plastic boxes and followed the standard protocol ISO 17512-1 [22]. The boxes were divided into two equal compartments with a paperboard divider. To each compartment, 200 g of natural soil were added with the soil WHC_{max} adjusted to 50%. The water used to adjust soil moisture was used to add the formulations at the concentrations mentioned. On the left side, we added non-contaminated soil (control), while on the right side, we placed soil spiked with different concentrations of unloaded LN and loaded LN with LC (LN–LC) dispersions, as well as the AI LC. The concentrations tested range included 0, 7, 10, 14, 20, and 28 g LC ha^{-1} (based on the dose of LC recommended in the commercial formulation Karate+, Syngenta), and the concentrations of unloaded LN were based on the amount of solid lipid (SL) used in the synthesis range that included 0, 150, 210, 300, 480, and 660 g SL ha^{-1}. The concentrations of all LN constituents (lipids and surfactants) were the same for unloaded and loaded LN with LC. The control group was moistened only with deionized water and tested in five replicates, while four replicates were tested per concentration for the contaminated soils. After preparing all replicates, 10 adult earthworms (300–600 mg) were placed between both compartments. Boxes were covered with a perforated cap and maintained for 48 h in the same conditions (Section 2.2.3), without adding food. After 48 h of exposure, the number of earthworms on each side (control and treatments) of the boxes was counted, and the average avoidance percentage was calculated.

The reproduction test with *E. fetida* followed the standard OECD 222 protocol [23]. Before the assay, clitellate adult oligochaetes (300–600 mg) were acclimatized in the natural soil used in this study, for 48 h, under the same culture maintenance conditions (Section 2.2.3). The test was conducted in plastic containers with 500 g of natural soil homogenized, with deionized water (control group) and with the nanoformulation at the concentrations 0, 7, 10, 14, 20, and 28 g LC ha^{-1}, based on the dose of LC recommended in the commercial formulation. The formulations tested were LC AI dispersed in DMSO at 5% (*v/v*), unloaded LN, and LC-loaded LN (LN–LC) dispersions. The concentrations of unloaded LN were based on the amount of solid lipid (SL) used in the synthesis range that included 0, 150, 210, 300, 480, and 660 g SL ha^{-1}. The concentrations of all LN constituents (lipids and surfactants) were the same for unloaded and loaded LN. The effect of DMSO at 5% (*v/v*) was also tested. The soil WHC_{max} was adjusted to 50%. The water used to adjust soil moisture was used to add the formulations at the concentrations mentioned. Ten oligochaetes with clitellum were added to each container and kept for 56 days. Each concentration was tested in quadruplicate, while the control group was tested in quintuplicate. After 28 days of exposure, the adults were removed and counted, and the test was extended for a further 28 days to allow the juveniles to grow. At the end of the test (56 days), the juveniles were extracted from the soil with heat, in a water bath (60 °C for 15 min), and the number of juveniles per replicate was counted.

2.2.4. Assessments of LN–LC Safety to Soil Microbial Parameters

The collected natural soil was sieved at 2 mm. The stock solution of the formulations was diluted at different concentrations in the amount of water required to adjust the WHCmax to 80% and maintained for 15 days, at 20 ± 2 °C, and for a photoperiod of

$16h^L$:$8h^D$. The moisture of the control group was adjusted only with deionized water. During this period, the soil was weighted every 3 days, and the moisture was adjusted with deionized water when necessary. After 15 days of exposure, for each replicate (both control and contaminated soils), 3 sub-replicates were prepared by transferring 1 g of soil to individual Falcon tubes, which were stored at −20 °C for a maximum of 1 month.

The pH variation in soil due to the presence of contaminants can affect the behavior of the soil microbial community and, consequently, its biological activity. Thus, after incubation, the pH of the control soil and of the soils treated with the highest concentrations tested was again checked in a suspension in KCl (1 M) [16].

For the assessment of soil microbial parameters, changes in the activity of the enzymes dehydrogenase, acid phosphatase, arylsulfatase, carboximetil cellulase (CM-cellulase), and urease were measured. In addition, the potential nitrification and nitrogen mineralization were also evaluated.

For the analysis of dehydrogenase activity [24], soil samples were incubated at 40 °C, for 24 h, with a solution of triphenyltetrazolium chloride (TTC) (10g L^{-1}) in Tris buffer 0.1 M at a pH of 7.6, while the blank tubes were incubated only with Tris buffer 0.1 M at a pH of 7.6. The triphenylformazan (TPF), which is produced by the reduction of TTC, was extracted with acetone, forming a pink-colored complex. The absorbance was then measured spectrophotometrically at 546 nm. The dehydrogenase enzymes' activity was determined from a standard curve obtained for TPF in acetone and expressed in µg TPF g^{-1}dm (% dry matter) h^{-1}.

For the acid phosphatase enzyme [25], soil samples were incubated at 37 °C for 1 h with p-nitrophenylphosphate solution (115 mM) in standard buffer at pH 6.5. The blank tubes were incubated only with standard buffer at a pH of 6.5. The p-nitrophenol (pNP) released by the phosphomonoesterase activity was extracted with sodium hydroxide (0.5 M), forming a yellow complex, which was measured at 405 nm. The acid phosphatase enzyme activity was determined from a standard curve and expressed in µg pNP g^{-1}dm h^{-1}.

For arylsulfatase enzyme analysis [26], soil samples were incubated at 37 °C for 1 h with a p-nitrophenylsulfate solution (0.02 M) in acetate buffer 0.5 M at a pH of 5.8. The blank tubes were incubated only with sodium acetate trihydrate buffer (0.5 M) at a pH of 5.8. The nitrophenol (pNP) released by the arylsulfatase activity was extracted by NaOH 0.5 M resulting in a yellow-colored complex, which was measured at 420 nm. The activity of the arylsulfatase enzyme was determined from a standard curve and expressed in µg pNP g^{-1}dm h^{-1}.

For the analysis of the carboximetil cellulase (CM-cellulase) [27], using carboxymethyl cellulose as the substrate (0.7% *w/v*), the soil samples were incubated for 24 h at 50 °C and a pH of 5.5. The blank tubes were incubated only with acetate buffer (2M) at a pH of 5.5. The reducing sugars produced during the incubation period caused the reduction of potassium hexacyanoferrate (III) to potassium hexacyanoferrate (II) in an alkaline solution. The latter compound reacted with ferric ammonium sulfate in an acid solution to form a blue-colored ferric hexacyanoferrate (II) complex, which was measured at 690 nm. The activity of the CM-cellulase enzyme was calculated from a standard curve obtained for defined concentrations of glucose, in aqueous solution, and expressed in µg glucose (GLU) g^{-1}dm 24 h^{-1}.

For the analysis of the urease enzyme activity [28], the soil samples were incubated at 37 °C, for 2 h, with a borate buffer 0.1 M at pH 10 and a solution of urea (720 mM) in the same buffer. The blank tubes were incubated only with borate buffer solution (0.1 M; pH 10). The ammonia released by the enzyme activity was extracted with KCl 2 M, forming a green-colored complex, which was measured at 690 nm. The activity of the urease enzyme was calculated from a standard curve, using defined concentrations of NH_4^+, in aqueous solution, and expressed in µg NH_4^+ g^{-1}dm 2h^{-1}.

For the evaluation of the nitrification potential [29], soil samples were incubated at 25 °C, under in an orbital shaker, and the blanks were incubated at −20 °C for 5 h, with ammonium sulfate $(NH_4)_2SO_4$ 1 mM as the substrate and NaCl 1.5 M. The nitrite released by the enzyme activity was extracted with KCl 2 M, forming a pink-colored complex, which was measured at 502 nm. The nitrification was determined from a standard curve, using defined concentrations of NO_2 in aqueous solution, and expressed in ng of nitrite NO_2^- g^{-1}dm $5h^{-1}$.

For the evaluation of nitrogen mineralization [30], soil samples and the blanks were incubated at 40 °C for 7 days, with deionized water. In this process, the organic forms of N were metabolized into inorganic N forms, such as NH_4^+, which were determined after the extraction with KCl 2 M. After the reaction of ammonia with sodium salicylate and sodium nitroprusside in the presence of sodium dichloroisocyanurate, a green-colored complex was formed and measured spectrophotometrically at 690 nm. The mineralization of nitrogen was determined from a standard curve, using defined concentrations of NH_4^+, in aqueous solution, and expressed as µg NH_4^+ g^{-1}dm d^{-1}.

2.3. Statistical Analysis

All statistical procedures were performed by using Prism version 8.0.2 (263), created by GraphPad Software (San Diego, California, USA). For all the endpoints, data were expressed as mean values ± standard deviation (SD). Regarding hypothesis testing, the data obtained for the average particle size, PI, and ZP of nanoformulations were tested for significant differences between storage periods (30, 90, and 120 days) and time 0 (day of the production) for LN or LN–LC and between unloaded LN and loaded LN–LC, within the same storage period by analysis of variance (two-way ANOVA), with subsequent post hoc comparison, using Tukey's test. The analysis of the effects on plants growth endpoints (*S. lycopersicum* and *Z. mays*), on reproduction of the soil invertebrates (*E. fetida* and *F. candida*), and on soil microbial parameters was performed by analysis of variance (ANOVA), followed by a Dunnett multi-comparison test to check for differences from the control group (CTL). Whenever ANOVA assumptions were not met, a non-parametric Kruskal–Wallis test, followed by a Dunn's multi-comparison test, was performed. For the avoidance test with *E. fetida*, Fisher's exact test was performed by using GraphPad online software (https://www.graphpad.com/quickcalcs/contingency1.cfm, accessed on: 20 June 2022) to test for no avoidance between soils. The variations in the pH value were calculated by analysis of variance (ANOVA), followed by a Dunnett multi-comparison test to check for differences with the control group (CTL). Statistically significant differences were considered for a significance level $\alpha = 0.05$.

3. Results and Discussions

3.1. Physicochemical Characterization of LN

Synthesized LN were evaluated with respect to their average particle size (nm), PI, and ZP (mV), which are essential physicochemical parameters that are necessary in order to obtain information concerning the appearance, consistency, and stability of colloidal systems [31]. The measurements of unloaded and loaded LN (LN–LC) were performed immediately after production and 30, 60, and 120 days of storage in the dark, at 25 °C (Figure 1).

Initially, the average particle size values of unloaded LN were 175.4 ± 2.6 nm and showed no significant variation over the period (120 days) ($p > 0.05$) (Figure 1A). The encapsulation of LC in LN resulted in a significant decrease ($p < 0.05$; F (9, 18) = 91.50) in the average particle size to165.4 ± 2.343 nm; this may indicate that LC, by having a high hydrophobic character, has great affinity with the lipid matrix, thus providing a greater homogenization of all constituents that compose the LN structure (Figure 1A). Similarly, to unloaded LN, the average particle size of LN–LC remains in the same range for 120 days, thus indicating no tendency to form aggregation and the high stability of LN after LC encapsulation when stored at 25 °C.

Regarding particle size distribution, immediately after production (day 0), unloaded LN formulations showed low PI values (0.293 ± 0.006) and were not significantly affected after LC encapsulation (PI = 0.264 ± 0.012) ($p > 0.05$; F (9, 18) = 10.19) (Figure 1B). Over the 120 days, the PI values decreased significantly for both formulations LN and LN–

LC, when compared with day 0 ($p < 0.05$; F (9, 20) = 10.12) (Figure 1B)). The PI values of formulations were all lower than 0.3, which indicates that LN formed homogeneous suspensions [32]. The ZP values represent the electrical charge at the nanoparticle surface, as well as the degree of repulsion between similarly charged particles which avoids the occurrence of particle aggregation [33]. After synthesis, the ZP values of unloaded LN were −50 ± 1.52 mV (Figure 1C). However, after LC encapsulation, the ZP values decreased to −38.7 ± 0.954 mV (Figure 1C), which indicates that an amount of the insecticide could be located on the LN surface, and thus influencing the surface charge of nanoparticles ($p < 0.05$; F (9, 20) = 70.99). Nevertheless, In the present study, all ZP values recorded were above | 30 mv |, which is an indicator of higher electrostatic repulsion between particles, providing a greater colloidal stability over time [34].

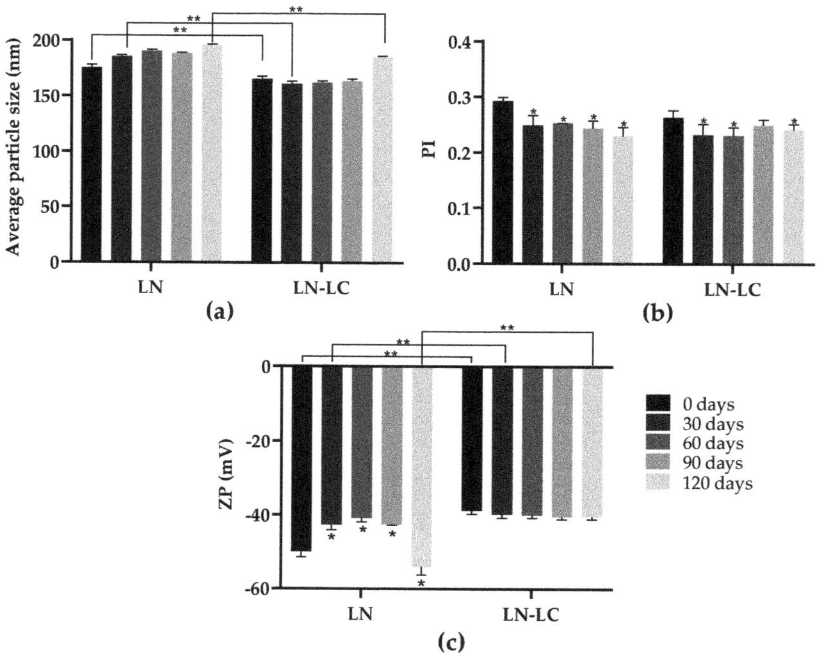

Figure 1. Average particle size (nm) (**A**), polydispersity index (PI) (**B**), and ZP (mV) (**C**) of synthesized LN, monitored at 25 °C immediately after synthesis (0 days) and after 30, 90, and 120 days of storage. Results are expressed as mean ± SD (n = 3). Significant differences between 0 days and other periods for the same LN are represented by (*), $p < 0.05$ (ANOVA followed by Tukey test). Significant differences between unloaded LN and LN loaded with LC for the same period are represented by (**), $p < 0.05$ (two-way ANOVA, followed by Tukey's test).

The interaction of LC and LN was investigated by using the evaluation of entrapment efficiency (EE). Since LC is considered a poorly aqueous soluble compound (solubility in water = 0.005 mg L^{-1} at 20 °C), LN are suitable nanocarriers for LC encapsulation, showing a %EE of 98.44 ± 0.04. The high %EE can be attributed to the reduced particle size of synthesized LN and to the presence of liquid lipid (Capryol 90) in the LN structure, as this can favor the solubility of LC in the melted lipid matrix, as well as the formation of LN with a more amorphous structure to improve the LC entrapment. Thus, the synthesized LN showed good properties regarding average particle size, homogeneity, and surface charge, demonstrating that LN are excellent systems for LC encapsulation.

3.2. Assessment of LN Environmental Safety

NPest have been widely developed in recent years. However, knowledge about their fate and environmental effects is very limited, and it is still unclear whether NPest will result in significant benefits over conventional products. From the point of view of study design and correct interpretation of results, the data relating to the safety of developed NPest is still unclear, and, thus, this present study tries to fill this gap, at least for LC-loaded LN.

The effect of nanoformulations loaded with LC was evaluated by using a monocotyledonous plant, *Z. mays*, and a dicotyledonous plant, *S. lycopersicum*. According to the results, no statistically significant differences were recorded on the fresh biomass of the shoots ($p > 0.05$; $F\ (12, 50) = 1.210$) and roots ($p > 0.05$); $F\ (12, 50) = 0.9060$) of *S. lycopersicum* (Figure 2a,b); moreover, no significant differences were recorded on the fresh biomass of the shoots ($p > 0.05$) and roots ($p > 0.05$; $F\ (12, 50) = 1.230$) of *Z. mays* (Figure 3a,b) after exposure to increasing concentrations of LC and LN–LC by foliar application when compared to the control group. Thus, LN–LC demonstrated itself to be safe for the plants selected in this study. However, LC negatively affected the dry biomass of the roots of *S. lycopersicum* (LOEC = 28 g LC ha^{-1}, NOEC = 20 g LC ha^{-1}; $p < 0.05$; Kruskal–Wallis statistics value = 21.44), as indicated in Figure 2d, while LC stimulated *Z. mays* root length at 20 g LC ha^{-1} ($p < 0.05$; $F\ (12, 302) = 5.01$) (Figure 3e). No statistically differences were recorded for both species after exposure to DMSO at 5% (*v/v*) in comparison with the control group (data not shown).

Plants' growth and development depend on their adaptation to a constantly changing abiotic environment, as well as their contact with pesticide residues present in soil [35]. Although there are several works about the effect of nanomaterials on plants, the most are related to metallic/inorganic nanomaterials [36,37], and, to the best of our knowledge, no data on LN phytotoxicity exist.

Concerning the LC effects on terrestrial plants, the data are also scarce and depend on the plant species. In our study, dicotyledonous species appeared to be more sensitive than monocotyledonous species were to LC exposure, and, in the same way, the development of roots was more affected than the that of the shoots. However, a comparative study conducted by Bragança et al. (2018) to evaluate the effects of pyrethroids (0 to 500 µg $kg^{-1}{}_{soil}$) on *Cucumis sativa* showed that LC was not toxic to plants, while cypermethrin decreased the shoot length (50 to 500 µg $kg^{-1}{}_{soil}$) compared to the control group [38]. This is in agreement with our results, since the maximum concentration tested for LC was 28 g LC ha^{-1} (which corresponded to 133 µg $kg^{-1}{}_{soil}$). On the contrary, Liu et al. (2009) showed that the application of the insecticide cypermethrin tested at concentrations ranging from 0 to 64 mg $kg^{-1}{}_{soil}$ (0–13 kg LC ha^{-1}) reduced the root elongation (mm) of Pakchoi (Chinese cabbage) [39] [However, the maximum concentration of LC tested by these authors (13 kg LC ha^{-1}) is much higher than those we tested and seemed to be well above recommended application doses.

Regarding the avoidance test with *E. fetida*, the results revealed that oligochaetes did not avoid LC contaminated soil (Figure 4a). Contrarily, the earthworms significantly avoided LN–LC contaminated soil at the highest doses (20 and 28 g LC ha^{-1}) ($p < 0.05$) (Figure 4c): LOEC 20 g LC ha^{-1} and NOEC 14 g LC ha^{-1}) and significantly avoided LN contaminated soil at 150 and 480 g SL ha^{-1} ($p < 0.05$) (Figure 4b) compared to control soil. However, a maximum avoidance percentage of 75% was only recorded for the dose 20 g LC ha^{-1} when the active ingredient was encapsulated (LN–LC), which means that the habitat function of these soils was not compromised (<80% avoidance) even for the highest concentration tested (Figure 4c). Although the avoidance behavior is usually interpreted as a negative response, the results obtained in our study allow to conclude that these invertebrates may be less exposed to LC when it is encapsulated in LN, as somehow LN may be promoting an avoidance behavior, letting them to escape from a higher exposure level.

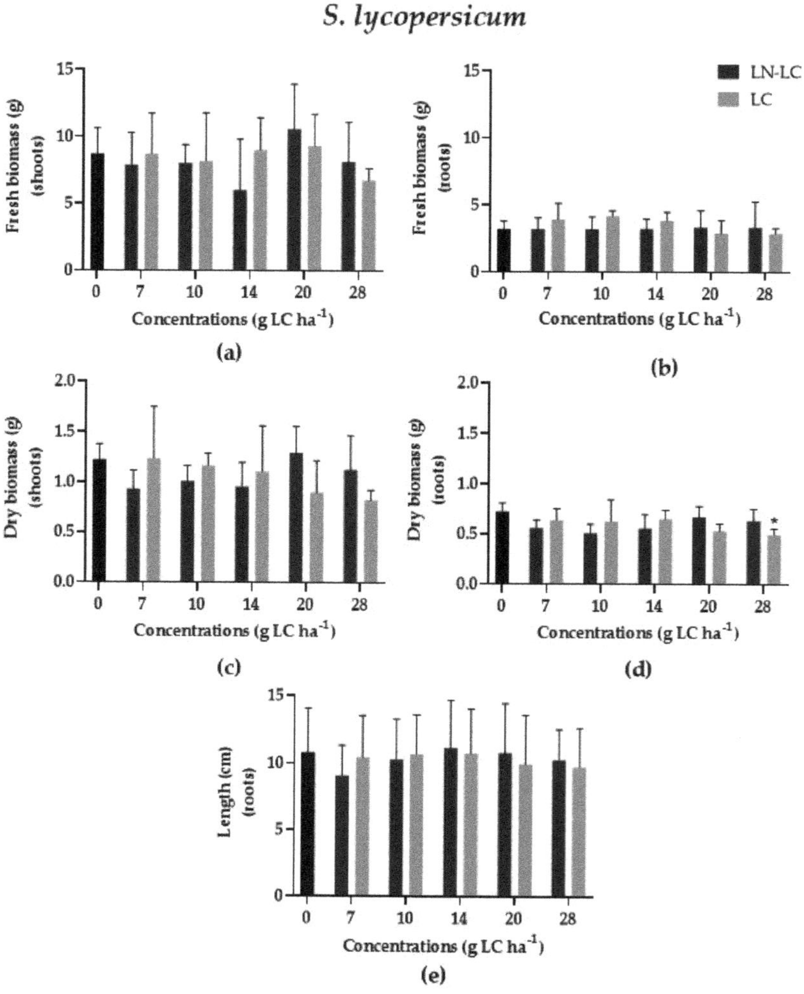

Figure 2. Variation of fresh biomass of shoots and roots (**a**,**b**), dry biomass of shoots, and roots (**c**,**d**) and length (**e**) of roots of *S. lycopersicum* after exposure to LC (light) and LN–LC (dark). The concentrations tested are based on the amount of LC. Results are represented by mean ± SD values. Asterisks mark the significant differences compared to the control group (0 g LC ha^{-1}) ((**a**,**b**) $p < 0.05$, Dunnett; (**c**–**e**) $p < 0.05$, Dunn's).

In addition, these findings may indicate that, despite detecting the presence of LC, the earthworms behavior did not appear to be affected by the concentrations tested or these concentrations may have inhibited the response ability of the organisms to the insecticide, leading to muscle paralysis, inconsistent movements, or the absence of sensorial detection [40].

Despite the avoidance results, the reproduction tests with *E. fetida* were not affected by the presence of LN ($p > 0.05$; F (5, 19) = 1.054) and LN–LC ($p > 0.05$; F (5, 19) = 1.643) after 56 days of exposure to increasing concentrations when compared to the control group (Figure 5b,c), and no statistical differences were recorded after exposure to DMSO at 5% (*v/v*) (data not shown). Although there are no data regarding the effect of LN on oligochaetes, for comparison, and despite being a work where a surfactant was tested, a study conducted by Gavina et al. (2016) showed that the earthworms *Eisenia andrei* significantly avoided the contaminated soil with nanovesicles made from the anionic surfactant sodium dodecyl sulphate and the cationic lipid didodecyl dimethylammonium bromide (SDS/DDAB), while the reproduction of the same invertebrates was not affected, probably by the rapid degradation of the vesicles in the soil [41]. Since the nanoparticles in the present study were produced by using lipids, it is possible that, initially, the earthworms were affected by their presence, but along the time, these LN can be (bio)degraded, and the soil habitat function can be restored. The reproduction of *E. fetida* was not also significantly affected by LC as shown in Figure 5a and the survival of adult earthworms was not affected when exposed to LN and LC (data not shown) at the concentrations tested. However, there are studies that state that LC is very toxic to *E. fetida* due to the high cutaneous absorption of these compounds by earthworms (LC_{50} from 1055 to 2570 g ha^{-1}) [42], but these concentrations are higher than those tested in our study, which are likely more ecologically relevant, as they are close to the application dose.

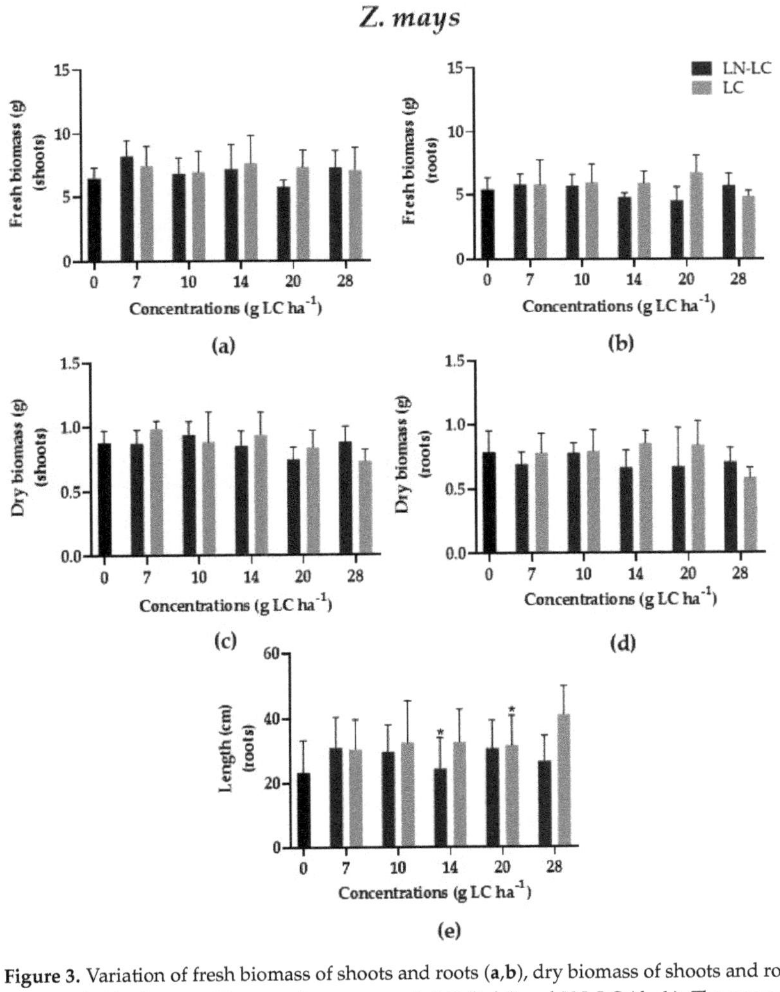

Figure 3. Variation of fresh biomass of shoots and roots (**a**,**b**), dry biomass of shoots and roots (**c**,**d**), and length (**e**) of roots of *Z. mays* after exposure to LC (light) and LN–LC (dark). The concentrations tested are based on the amount of LC. Results are represented by mean ± SD values. Asterisks mark the significant differences compared to the control group (0 g LC ha^{-1}): (**a**,**c**,**e**) $p < 0.05$, Dunnett; (**b**,**d**) $p < 0.05$, Dunn's.

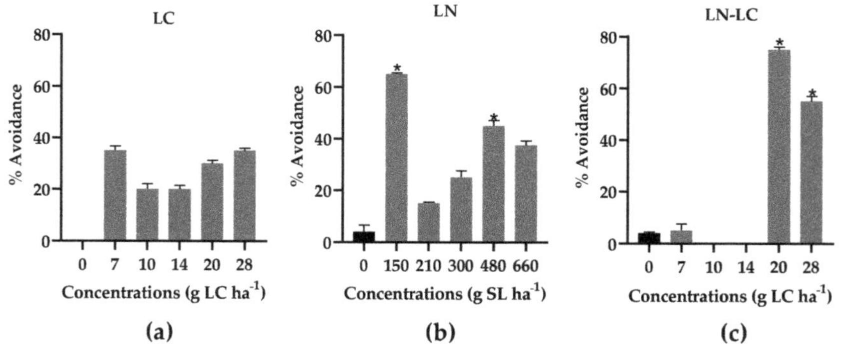

Figure 4. Percentage avoidance of *E. fetida* after exposure to different concentrations of LC (**a**), LN, (**b**) and LN–LC (**c**). The concentrations tested are based on the amount of LC. In the case of LN, the concentrations indicated are of solid lipid (SL) used in the synthesis of LN. The same concentrations of SL were tested for LN–LC. Results are represented by mean ± SD values. Asterisks mark significant differences compared to the control group (0 g LC or LS ha^{-1}) ($p < 0.05$, Fisher´s test).

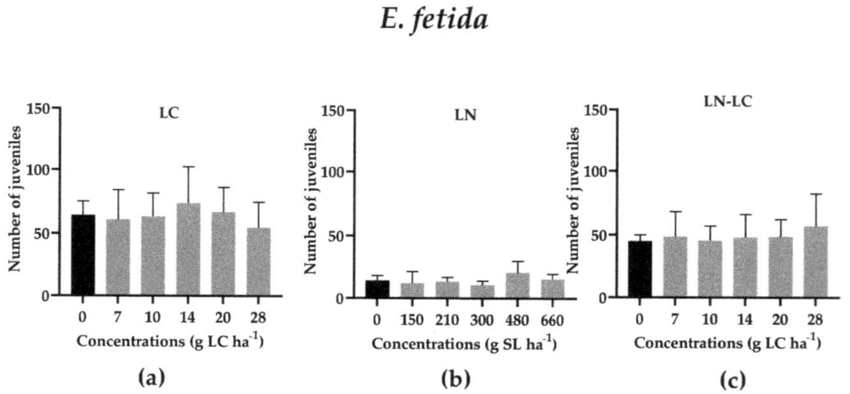

Figure 5. *E. fetida* juveniles after the exposure of adults to different concentrations of LC (**a**), LN (**b**) and LN–LC (**c**). The concentrations tested are based on the amount of LC. In the case of LN, the concentrations indicated are of solid lipid (SL) used in the synthesis of LN. The same concentrations of SL were tested for LN–LC. Results are represented by mean ± SD values ($p < 0.05$, Dunnett).

According to the results, the reproduction of *F. candida* was significantly affected only by the presence of LC ($p < 0.05$; $F(5, 24) = 5.903$) at all the concentrations tested, as illustrated in Figure 6a (NOEC < 7g LC ha^{-1}, LOEC ≤ 7 g LC ha^{-1}), while LN ($p > 0.05$; $F(5, 24) = 0.7850$) and LN–LC ($p > 0.05$; $F(5, 24) = 0.8134$) did not affect the reproduction of collembola (Figure 6b,c). In fact, the encapsulation of LC in LN decreased the toxicity of LC after 56 days of exposure to increasing concentrations of the insecticide (Figure 6c). The survival of adult springtails was not affected when exposed to LN and LC (data not shown), and as expected, the reproduction of *F. candida* was more sensitive than mortality to insecticide application. No statistical differences were recorded after the exposure of invertebrates to DMSO at 5% (*v/v*) (data not shown).

Soil microbiological and biochemical parameters are essential indicators of microorganism metabolism and function, nutrient cycling, and soil contamination [43]. Some studies have reported that the soil microbial community can be significantly compromised upon exposure to nanomaterials [44]. However, these studies were conducted to understand the impact on soil microbial community of metal/metal oxide nanomaterials [45] and polymeric nanoparticles [46].

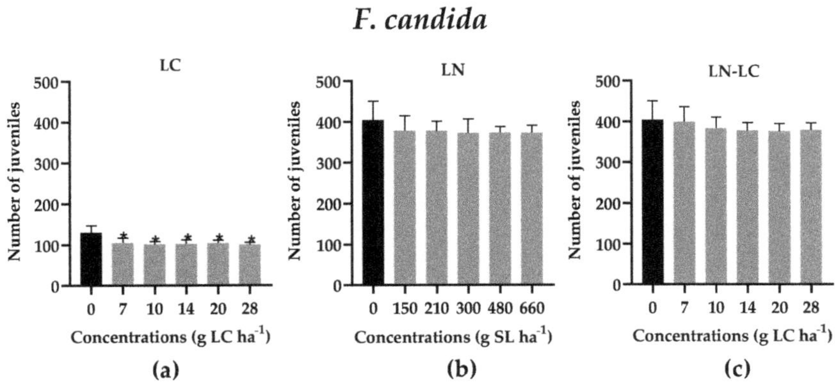

Figure 6. *F. candida* juveniles after the exposure of adults to different concentrations of LC (**a**), LN (**b**), and LN–LC (**c**). The concentrations tested are based on the concentration of LC in the commercial formulation. In the case of LN, the concentrations indicated are of solid lipid (SL) used in the synthesis of LN. The same concentrations of SL were tested for LN–LC. Results are represented by mean ± SD values. Asterisks mark the significant differences compared to the control group (0 g LC ha^{-1}) ($p < 0.05$, Dunnett).

In the present study, the effect of formulations was also evaluated in soil microbial activity through the analysis of enzymatic activities, N mineralization, and potential nitrification. As far as authors are aware, the present work is the first study gathering data concerning the effect of LC and LN on the soil microbial community. Since the soil physicochemical properties can influence the activity of soil enzymes, the pH was evaluated after the incubation of soil with different formulations, namely LC, LN, and LN–LC, and the results are presented in Table 2. The pH results demonstrated that the unloaded LN significantly increased the pH value of the soil ($p < 0.05$; F (2, 6) = 10.50), and the LC-loaded LN significantly decreased the pH of the soil ($p < 0.05$; F (2, 6) = 35.00).

Table 2. Average pH values determined for the natural soil containing the different formulations of lambda-cyhalothrin (LC) at the concentration of 28 g LC ha^{-1} after 15 days of incubation. Results are represented by mean values ± standard deviation (SD) (n = 3).

pH (KCl, 1M)	pH Values for the Natural Soil		
	Control soil	**Soil containing LC**	**Soil containing LN**
pH (KCl, 1M)	4.59 ± 0.006	4.61 ± 0.010	4.64 * ± 0.006
	Control soil	**Soil containing LN–LC**	
	4.69 ± 0.006	4.67 * ± 0.012	

* Significance α = 0.05, compared with respective control soil.

A study has shown that pesticides exhibit inhibitory effects on dehydrogenase enzymes' activity [46]. The dehydrogenases are involved in cellular respiration and are considered an indicator of overall microbial metabolic activity. Furthermore, the dehydrogenases have a relevant role in the oxidation of organic matter, with their activity being an excellent parameter for assessing the effect of NPest on the soil microbial community [47]. According to the obtained results, LC did not affect dehydrogenase activity for all concentrations tested ($p > 0.05$; F (10, 108) = 19.01) (Table 3 and Supplementary Figure S1a), while loaded LN with LC ($p < 0.05$; F (10, 24) = 16.34) significantly reduced the activity of dehydrogenase at highest concentrations compared to the control group (Table 3 and Supplementary Figure S1c). Unloaded LN significantly stimulated the activity at 210 and 300 g SL ha^{-1} and significantly inhibited the activity at the highest concentration tested ($p < 0.05$; F (10, 108) = 19.01) (Table 3 and Supplementary Figure S1b). Thus, our study suggests that, as nanoparticles are composed of lipids, at high concentrations, these lipids may promote the occlusion of soil pores [48]. The occlusion of soil pores can reduce soil oxygenation, decreasing the microbial population and, therefore, decreasing the activity of dehydrogenases at the highest concentrations tested (28 g LC ha^{-1}/660 g SL ha^{-1}) (Table 3 and Supplementary Figure S1b,c, respectively).

Table 3. Soil microbial parameters. Enzymatic activity of dehydrogenase, CM-cellulase, urease, arylsulfatase, and acid-phosphatase after exposure to different concentration of LC, LN, and LN–LC. The concentrations tested are based on the amount of LC. In the case of LN, the concentrations indicated are of solid lipid (SL) used in the synthesis of LN. Results are represented by mean ± SD values. The same concentrations of SL were tested for LN–LC. Asterisks mark the significant differences in relation to the control group (0 g LC ha^{-1}).

Soil Microbial Parameters						
		Dehydrogenase (μg TPF g^{-1}dm h^{-1})	CM-Cellulase (μg GLU g^{-1}dm 24 h^{-1})	Urease (μg NH_4^+ g^{-1}dm 2 h^{-1})	Arylsulfatase (μg pNP g^{-1}dm h^{-1})	Acid Phosphatase (μg pNP g^{-1}dm h^{-1})
LC (g LC ha^{-1})	0	0.71 ± 0.24	107.90 ± 33.28	0.99 ± 0.74	84.57 ± 10.89	234.19 ± 22.34
	7	0.62 ± 0.24	128.27 ± 64.39	1.08 ± 0.94	88.69 ± 14.70	225.16 ± 10. 99
	10	0.82 ± 0.28	116.05 ± 52.04	1.37 ± 0.80	87.04 ± 8.50	231.52 ± 17.59
	14	0.65 ± 0.35	105.57 ± 37.94	3.03 ± 1.54	82.87 ± 6.75	158.41 * ± 59.82
	20	0.73 ± 0.16	94.83 ± 106.75	1.93 ± 2.22	90.57 ± 20.78	179.46 * ± 47.73
	28	0.83 ± 0.24	137.56 ± 31.04	1.85 ± 1.42	87.86 ± 8.00	159.38 * ± 34.37
LN (g SL ha^{-1})	0	0.71 ± 0.24	107. 90 ± 33.28	0.99 ± 0.74	84.57 ± 10.90	234.19 ± 22.34
	150	0.38 ± 0.25	118.48 ± 38.01	1.81 ± 1.14	74.42 ± 7.00	145.39 * ± 38.61
	210	1.68 * ± 0.38	82.83 ± 28.80	1.25 ± 0.80	76.17 ± 8.76	142.83 * ± 44.58
	300	1.75 * ± 0.36	74.50 ± 35.41	1.62 ± 1.30	76.74 ± 9.60	147.80 * ± 36.68
	480	0.48 ± 0.22	103.75 ± 42.01	2.74 * ± 1.34	72.56* ± 11.33	138.05 * ± 52.47
	660	0.33 * ± 0.20	136.13 ± 76.44	5.86 * ± 2.90	75.24 ± 7.33	157.91 * ± 45.49
LN–LC (g LC ha^{-1})	0	2.16 ± 0.83	63.99 ± 27.73	4.92 ± 0.50	109.07 ± 22.63	234.06 ± 18.67
	7	1.74 ± 0.62	71.65 ± 48.31	3.22 ± 0.43	92.38 ± 13.04	227.93 ± 14.64
	10	1.71 ± 0.64	99.73 ± 50.14	8.03 * ± 2.11	106.05 ± 17.68	229.61 ± 15.81
	14	1.71 ± 0.48	149.98 * ± 45.90	8.91 * ± 1.52	114.85 ± 18.86	224.07 ± 12.68
	20	1.33 ± 0.58	107.49 ± 45.92	7.61 * ± 1.92	86.41 * ± 12.30	210.37 * ± 18.85
	28	1.21 * ± 0.36	121.39 * ± 41.45	6.00 ± 1.20	78.01 * ± 15.53	190.02 * ± 34.93

Regarding the enzymatic activity of CM-cellulase, it was found that the LC and unloaded LN did not significantly affect the enzyme activity, as shown in Table 3 and Supplementary Figure S2a,b, respectively ($p > 0.05$). However, after LC encapsulation in LN, an increase in this enzyme activity was observed being statistically significant at concentrations of 14 and 28 g LC ha^{-1} was observed, as indicated in Table 3 and Supplementary Figure S2c ($p < 0.05$; F (10, 24) = 5.925). The reason for an increase in the CM-cellulase activity after LN–LC could probably be due to the presence of the organic compound TegoCare 450, which can act as a glucose source since glucose is part of TegoCare constitution. Although this phenomenon was not observed for single LN, in some way, LC could have modified the LN structure, increasing the bioavailability of glucose from TegoCare. In other words, this would not be an increase in activity, but rather a quantification of glucose that did not result from the enzyme activity per se but rather from the LN themselves.

According to the enzymatic activity of urease, LN–LC significantly increased the enzyme activity compared to the control at 10, 14, and 20 g LC ha^{-1} ($p < 0.05$; Kruskal–Wallis statistics value = 68.29) (Table 3 and Supplementary Figure S3c). LN significantly increased the enzyme activity compared to the control at 480 and 660 g SL ha^{-1}. However, the inhibition of urease enzyme in relation to LC (10 g LC ha^{-1}) may have occurred by chance ($p < 0.05$; Kruskal–Wallis statistics value = 46.26) (Table 3 and Supplementary Figure S3a,b).

For arylsulfatase enzyme activity, it was observed that LN–LC caused a significant decrease in the activity compared to the control at the highest doses (20 and 28 g LC ha^{-1}) ($p < 0.05$; F (10.124) = 4.161) (Table 3 and Supplementary Figure S4c). The reduction in activity caused by the dose of 20 g LC ha^{-1} is in line with the significant reduction caused by unloaded LN compared with the control ($p < 0.05$; F (10, 124) = 4.904) (Table 3 and Supplementary Figure S4b). The activity of arylsulfatases increases with the decreasing of S in soil [49], and their activity can be affected by the presence of contaminants, pH changes, and organic matter content [50]. As shown in Table 2, after the addition of LN–LC, there was a significant increase of pH when compared with the control soil, and, thus, the pH change—although small—favored by the presence of LN–LC could have affected the arylsulfatase activity.

However, so far, there are no scientific reports evidencing the effect of LC and LN on the enzymatic activity of arylsulfatase.

Regarding acid phosphatase enzyme activity, LC at doses of 14 to 28 g LC ha^{-1} (Table 3 and Supplementary Figure S5a) induced a significant reduction in enzyme activity compared to the control group. LN significantly decreased the enzyme activity compared to the control at all concentrations tested (Table 3 and Supplementary Figure S5b) ($p < 0.05$; Kruskal–Wallis statistics value = 71.23), while LN–LC caused a significant decrease only at the highest doses, as illustrated by Supplementary Figure S5c) ($p < 0.05$; Kruskal–Wallis statistics value = 39.41). Phosphatases are produced in soil when the concentration of phosphorus decreases and alkaline soil can favor the availability of this element [51]. According to Nannipieri et al. (1982), acid phosphatases can be inhibited by high concentrations of protons due to the ionization of specific groups of enzymes [52]. Therefore, since the soil pH is low (~4.61 $\pm$ 0.010 for LC), in the acidic medium, the amino groups of LC may be protonated (pka > 9), and the insecticide becomes positively charged [53]. For this reason, the inhibition of enzyme activity by LC at the highest concentrations can be related to the presence of protonated LC. In addition, and likely more relevant for explaining the inhibition of the acid phosphatase activity by LN and LN–LC, another phenomenon can be involved. LN are produced also by using soy lecithin that is a phospholipid, which could have been metabolized as a phosphorus source, thus increasing P element in soil and leading to an inhibition of acid phosphatase activity after the addition of LN.

The data obtained for nitrogen mineralization indicated that there were no inhibitory effects for all formulations tested, as shown in Figure 7. In opposition, LN at the highest concentration ($p < 0.05$; Kruskal–Wallis statistics value = 56.00) and LN–LC at intermediate concentrations significantly increased ($p < 0.05$; F (10, 124) = 12.54) the mineralization of nitrogen (Figure 7b,c). To date, there are no scientific reports that show the effect of LC and LN on nitrogen mineralization. However, these findings may indicate that this significant stimulation may have resulted from the composition of synthesized LN, since it has soy lecithin (rich in nitrogen) in its composition. When mineralization occurs, there is an increase in ammonia that will be converted to nitrites and nitrate through the nitrification process.

Although an increase of nitrogen mineralization was observed, LN–LC caused an inhibition of the potential nitrification, especially at the highest concentrations, as shown in Figure 8c ($p < 0.05$; F (10, 116) = 16.11). It is well-known that ammonia oxidizing bacteria are highly sensitive to changing environmental conditions, and, thus, it is possible that the changes in soil pH favored the inhibition of the enzymes responsible for the nitrification process (Table 2) [54]. Therefore, a decrease in pH values, although smaller, may have resulted in the inhibition of the potential nitrification.

The LN–LC affected the activity of the microbial community, especially the metabolism of nitrogen, sulfur, and phosphorus; and apart from the activity of phosphatases, the effects seem to result mainly from the LN and not from the insecticide LC. The effects observed do not indicate a negative impact on the microbial community, except for that suggested by the inhibition of dehydrogenase at the highest concentration; rather, they are likely indicators of the availability of P and N by the presence of LN.

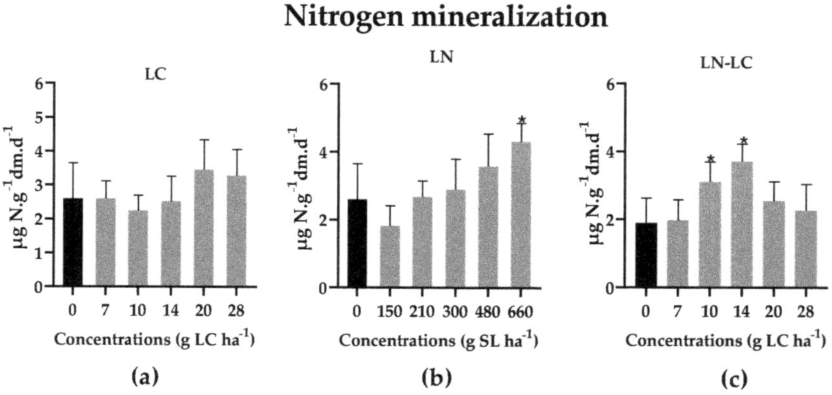

Figure 7. Nitrogen mineralization activity in soils exposed for 15 days to different concentrations of LC (**a**), LN (**b**), and LN–LC (**c**). The concentrations tested are based on the amount of LC. In the case of LN, the concentrations indicated are of solid lipid (SL) used in the synthesis of LN. The same concentrations of SL were tested for LN–LC. Results are represented by mean ± SD values. Asterisks mark the significant differences in relation to the control group (0 g LC ha^{-1}): (**c**) $p < 0.05$, Dunnett; (**a**,**b**) $p < 0.05$, Dunn's.

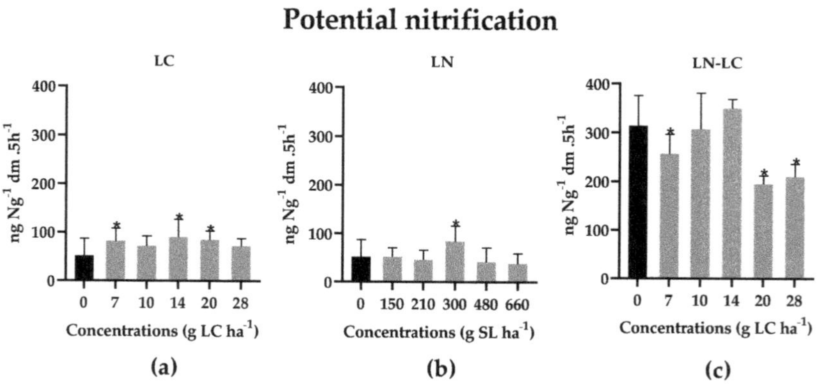

Figure 8. Nitrification potential activity in soils exposed for 15 days to different concentrations of LC (**a**), LN (**b**), and LN–LC (**c**). The concentrations tested are based on the amount of LC. In the case of LN, the concentrations indicated are of solid lipid (SL) used in the synthesis of LN. The same concentrations of SL were tested for LN–LC. Results are represented by mean ± SD values. Asterisks mark the significant differences in relation to the control group (0 g LC ha^{-1}) ($p < 0,05$, Dunnett).

4. Conclusions

Nanoparticles composed of physiological lipids (LN) have been proposed as prominent nanocarriers for LC delivery in agriculture practices, being successfully produced with a mean particle size of 165.4 ± 2.343 nm, narrow size distribution, and good physical stability for at least 4 months, at 25 °C. In addition, the high %EE values (98.44 ± 0.04%) indicate the excellent compatibility of LC and the lipid matrix of LN.

Regarding the environmental safety studies, LN showed no risk to the growth of *S. lycopersicum* and *Z. mays* species, as well as to the survival and reproduction of the soil invertebrates *F. candida* and *E. fetida*. However, the earthworms were sensitive to the presence of LN and LN–LC, avoiding its presence at least in the first 48 h of contact. This probably happened due to the lipid composition of both nanoparticles which has favored the interaction with cuticular sensorial cells. This avoiding effect is not necessarily negative, as it may prevent a more intensive contact of earthworms with the encapsulated insecticide; it should be noted that the highest concentration tested for the different formulations was higher than the application dose. Therefore, in the present study, these findings allow us to conclude that LN may be an alternative to encapsulate LC, as no significant effects are expected on soil biota. This was also reinforced by the long-term tests performed. These environmentally safe concentrations of LN–LC need now to be tested for their efficacy on target species of insects.

Supplementary Materials: https://www.mdpi.com/article/10.3390/nano12152576/s1, Figure S1: Average dehydrogenases activity in soils exposed for 15 days to different concentrations of LC (**a**), LN (**b**) and LN–LC (**c**). The concentrations tested are based on the amount of the LC. In the case of LN, the concentrations indicated are of solid lipid (SL) used in the synthesis of LN. The same concentrations of SL were tested for LN–LC. Results are represented by mean ± SD values. The asterisks mark the significant differences from the control group (0 g LC ha^{-1}) ($p < 0.05$, Dunnett); Figure S2: CM–cellulase activity in soils exposed for 15 days to different concentrations of LC (**a**), LN (**b**) and LN–LC (**c**). The concentrations tested are based on the amount of LC. In the case of LN, the concentrations indicated are of solid lipid (SL) used in the synthesis of LN. The same concentrations of SL were tested for LN–LC. Results are represented by mean ± SD values. Asterisks mark the significant differences in relation to the control group (0 g LC ha^{-1}) ($p < 0.05$, Dunnett); Figure S3: Urease activity in soils exposed for 15 days to different concentrations of LC (**a**), LN (**b**) and LN–LC (**c**). The concentrations tested are based on the amount of LC. In the case of LN, the concentrations indicated are of solid lipid (SL) used in the synthesis of LN. The same concentrations of SL were tested for LN–LC. Results are represented by mean ± SD values. Asterisks mark the significant differences in relation to the control group (0 g LC ha^{-1}) ($p < 0.05$, Dunn); Figure S4: Arylsulfatase activity in soils exposed for 15 days to different concentrations of LC (**a**), LN (**b**) and LN–LC (**c**). The concentrations tested are based on the amount of LC. In the case of LN, the concentrations indicated are of solid lipid (SL) used in the synthesis of LN. The same concentrations of SL were tested for LN–LC. Results are represented by mean ± SD values. Asterisks mark the significant differences in relation to the control group (0 g LC ha^{-1}) ($p < 0.05$, Dunnett); Figure S5: Soil acid phosphatase acid enzymes activity in soils exposed for 15 days to different concentrations of LC (**a**), LN (**b**) and LN–LC (**c**). The concentrations tested are based on the amount of LC. In the case of LN, the concentrations indicated are of solid lipid (SL) used in the synthesis of LN. The same concentrations of SL were tested for LN–LC. Results are represented by mean ± SD values. Asterisks mark the significant differences in relation to the control group (0 g LC ha^{-1}) ($p < 0.05$, Dunn).

Author Contributions: Conceptualization, T.A., R.P. and C.M.P.; methodology: C.G., M.B.d.S., C.P., T.I.d.M., M.R.d.S., C.M.P. and T.A.; investigation, C.G., M.B.d.S., C.P., T.I.d.M., M.R.d.S., C.M.P., R.P. and T.A.; data curation, C.G., M.B.d.S. and R.P.; writing—original draft preparation, C.G. and T.A.; writing—review and editing, T.A. and R.P.; supervision, R.P., C.M.P. and T.A.; funding acquisition, T.A., R.P. and C.M.P. All authors have read and agreed to the published version of the manuscript.

Funding: This work was supported by Fundação para a Ciência e Tecnologia (FCT) (P2020/COMPETE), through the research project SaFe NPest—Synthesis and Environmental Safety of Nanopesticides (POCI-01-0145-FEDER-029343). This work was also partially supported by the Foundation of Science and Technology, through the Strategic Projects of GreenUPorto (UIDB/05748/2020 and UIDP/05748/2020), CIQUP (UIDB/00081/2020) and LA/P/0056/2020.

Institutional Review Board Statement: Not applicable.

Informed Consent Statement: Not applicable.

Data Availability Statement: Not applicable.

Conflicts of Interest: The authors declare no conflict of interest.

References

1. Grillo, R.; Fraceto, L.F.; Amorim, M.J.B.; Scott-Fordsmand, J.J.; Schoonjans, R.; Chaudhry, Q. Ecotoxicological and Regulatory Aspects of Environmental Sustainability of Nanopesticides. *J. Hazard. Mater.* **2021**, *404*, 124148. [CrossRef]
2. Khandelwal, N.; Barbole, R.S.; Banerjee, S.S.; Chate, G.P.; Biradar, A.V.; Khandare, J.J.; Giri, A.P. Budding Trends in Integrated Pest Management Using Advanced Micro- and Nano-Materials: Challenges and Perspectives. *J. Environ. Manag.* **2016**, *184*, 157–169. [CrossRef] [PubMed]
3. Ghormade, V.; Deshpande, M.V.; Paknikar, K.M. Perspectives for Nano-Biotechnology Enabled Protection and Nutrition of Plants. *Biotechnol. Adv.* **2011**, *29*, 792–803. [CrossRef] [PubMed]
4. Nair, R.; Varghese, S.H.; Nair, B.G.; Maekawa, T.; Yoshida, Y.; Kumar, D.S. Nanoparticulate Material Delivery to Plants. *Plant Sci.* **2010**, *179*, 154–163. [CrossRef]
5. Moore, M.T.; Locke, M.A. Experimental Evidence for Using Vegetated Ditches for Mitigation of Complex Contaminant Mixtures in Agricultural Runoff. *Water. Air. Soil Pollut.* **2020**, *231*, 140. [CrossRef]
6. Kim, D.Y.; Kadam, A.; Shinde, S.; Saratale, R.G.; Patra, J.; Ghodake, G. Recent Developments in Nanotechnology Transforming the Agricultural Sector: A Transition Replete with Opportunities. *J. Sci. Food Agric.* **2018**, *98*, 849–864. [CrossRef] [PubMed]
7. Walker, G.W.; Kookana, R.S.; Smith, N.E.; Kah, M.; Doolette, C.L.; Reeves, P.T.; Lovell, W.; Anderson, D.J.; Turney, T.W.; Navarro, D.A. Ecological Risk Assessment of Nano-Enabled Pesticides: A Perspective on Problem Formulation. *J. Agric. Food Chem.* **2018**, *66*, 6480–6486. [CrossRef] [PubMed]
8. Food and Drug Administration Guidance (FDA). FDA-2017-D-0759: Drug Products, Including Biological Products, That Contain Nanomaterials—Guidance for Industry. 2022. Available online: https://www.fda.gov/regulatory-information/search-fda-guidance-documents/drug-products-including-biological-products-contain-nanomaterials-guidance-industry (accessed on 20 June 2022).
9. Sharma, A.; Baldi, A. Nanostructured Lipid Carriers: A Review Journal of Developing Drugs. *J. Dev. Drugs* **2018**, *7*, 1000191. [CrossRef]
10. Severino, P.; Andreani, T.; Jäger, A.; Chaud, M.V.; Santana, M.H.A.; Silva, A.M.; Souto, E.B. Solid Lipid Nanoparticles for Hydrophilic Biotech Drugs: Optimization and Cell Viability Studies (Caco-2 & HEPG-2 Cell Lines). *Eur. J. Med. Chem.* **2014**, *81*, 28–34. [CrossRef] [PubMed]
11. Pallerla, S.M.; Prabhakar, B. A Review on Solid Lipid Nanoparticles. *Int. J. Pharm. Sci. Rev. Res.* **2013**, *20*, 196–206. [CrossRef]
12. Graily-Moradi, F.; Hejazi, M.J.; Enayati, A.A.; Hamishehkar, H. Evaluation of Co-Nanoencapsulation Process on the Toxicity and Biochemical Metabolism of Imidacloprid and Lambda-Cyhalothrin in Myzus Persicae (Sulzer). *Biocatal. Agric. Biotechnol.* **2021**, *33*, 101974. [CrossRef]
13. Oudou, H.C.; Hansen, H.C.B. Sorption of Lambda-Cyhalothrin, Cypermethrin, Deltamethrin and Fenvalerate to Quartz, Corundum, Kaolinite and Montmorillonite. *Chemosphere* **2002**, *49*, 1285–1294. [CrossRef]
14. European Food Safety Authority (EFSA). *Outcome of the Consultation with Member States, the Applicant and EFSA on the Pesticide Risk Assessment for Lambda-Cyhalothrin in Light of Confirmatory Data*; EFSA Supporting Publication: Parma, Italy, 2020; Volume 17. [CrossRef]
15. De Souza, A.L.R.; Andreani, T.; De Oliveira, R.N.; Kiill, C.P.; Dos Santos, F.K.; Allegretti, S.M.; Chaud, M.V.; Souto, E.B.; Silva, A.M.; Gremião, M.P.D. In Vitro Evaluation of Permeation, Toxicity and Effect of Praziquantel-Loaded Solid Lipid Nanoparticles against Schistosoma Mansoni as a Strategy to Improve Efficacy of the Schistosomiasis Treatment. *Int. J. Pharm.* **2014**, *463*, 31–37. [CrossRef]
16. *ISO 10390:2020*; Soil, Treated Biowaste and Sludge—Determination of pH. International Organization for Standardization Guideline: Geneva, Switzerland, 2020.
17. FAOUN (Food and Agriculture Organization of the United Nations). *Physical and Chemical Methods of Soil and Water Analysis*; FAOUN: Rome, Italy, 1984; Volume 10, pp. 1–275.
18. Soil and Plant Analysis Council. *Soil Analysis: Handbook of Reference Methods*; Soil and Plant Analysis Council: Boca Raton, FL, USA, 2000. [CrossRef]
19. *ISO 11268-2:2012*; Soil Quality—Effects of Pollutants on Earthworms—Part 2: Determination of Effects on Reproduction of *Eisenia Fetida/Eisenia Andrei*. International Organization for Standardization Guideline: Geneva, Switzerland, 2012.
20. OECD. *Test No. 227: Terrestrial Plant Test: Vegetative Vigour Test, OECD Guidelines for the Testing of Chemicals*; Section 2; OECD Publishing: Paris, France, 2006. [CrossRef]
21. OECD. *Test No. 232: Collembolan Reproduction Test in Soil, OECD Guidelines for the Testing of Chemicals*; Section 2; OECD Publishing: Paris, France, 2016. [CrossRef]
22. *ISO 17512-1:2008*; Soil Quality—Avoidance Test for Determining the Quality of Soils and Effects of Chemicals on Behaviour—Part 1: Test With Earthworms (*Eisenia Fetida* and *Eisenia Andrei*). International Organization for Standardization Guideline: Geneva, Switzerland, 2008.
23. OECD. *Test No. 222: Earthworm Reproduction Test (Eisenia Fetida/Eisenia Andrei), OECD Guideline for the Testing of Chemicals*; Section 2; OECD Publishing: Paris, France, 2016. [CrossRef]
24. Robinson, C.; Schinner, F.; Ohlinger, R.; Kandeler, E.; Margesin, R. Methods in Soil Biology. *J. Ecol.* **1997**. [CrossRef]
25. Eivazi, F.; Tabatabai, M.A. Phosphatases in Soils. *Soil Biol. Biochem.* **1977**, *9*, 167–172. [CrossRef]

26. Tabatabai, M.A.; Bremner, J.M. Use of P-Nitrophenyl Phosphate for Assay of Soil Phosphatase Activity. *Soil Biol. Biochem.* **1969**, *1*, 301–307. [CrossRef]
27. Schinner, F.; von Mersi, W. Xylanase-, CM-Cellulase- and Invertase Activity in Soil: An Improved Method. *Soil Biol. Biochem.* **1990**, *22*, 511–515. [CrossRef]
28. Kandeler, E.; Gerber, H. Short-Term Assay of Soil Urease Activity Using Colorimetric Determination of Ammonium. *Biol. Fertil. Soils* **1988**, *6*, 68–72. [CrossRef]
29. Berg, P.; Rosswall, T. Ammonium Oxidizer Numbers, Potential and Actual Oxidation Rates in Two Swedish Arable Soils. *Biol. Fertil. Soils* **1985**, *1*, 131–140. [CrossRef]
30. Danaei, M.; Dehghankhold, M.; Ataei, S.; Hasanzadeh Davarani, F.; Javanmard, R.; Dokhani, A.; Khorasani, S.; Mozafari, M.R. Impact of Particle Size and Polydispersity Index on the Clinical Applications of Lipidic Nanocarrier Systems. *Pharmaceutics* **2018**, *10*, 57. [CrossRef]
31. Iqbal, M.A.; Md, S.; Sahni, J.K.; Baboota, S.; Dang, S.; Ali, J. Nanostructured Lipid Carriers System: Recent Advances in Drug Delivery. *J. Drug Target.* **2012**, *20*, 813–830. [CrossRef]
32. Shah, R.; Eldridge, D.; Palombo, E.; Harding, I. Optimisation and Stability Assessment of Solid Lipid Nanoparticles Using Particle Size and Zeta Potential. *J. Phys. Sci.* **2014**, *25*, 59–75.
33. Uner, M. Preparation, Characterization and Physico-Chemical Properties of Solid Lipid Nanoparticles (SLN) and Nanostructured Lipid Carriers (NLC): Their Benefits as Colloidal Drug Carrier Systems. *Pharmazie* **2006**, *61*, 375–386. [PubMed]
34. Riyaz, M.; Ahmad Shah, R.; Sivasankaran, K. Pesticide Residues: Impacts on Fauna and the Environment. In *Biodegradation Technology of Organic and Inorganic Pollutants*; IntechOpen: London, UK, 2021. [CrossRef]
35. Bouguerra, S.; Gavina, A.; da Graça Rasteiro, M.; Rocha-Santos, T.; Ksibi, M.; Pereira, R. Effects of Cobalt Oxide Nanomaterial on Plants and Soil Invertebrates at Different Levels of Biological Organization. *J. Soils Sediments* **2019**, *19*, 3018–3034. [CrossRef]
36. Rodríguez-Seijo, A.; Soares, C.; Ribeiro, S.; Amil, B.F.; Patinha, C.; Cachada, A.; Fidalgo, F.; Pereira, R. Nano-Fe_2O_3 as a Tool to Restore Plant Growth in Contaminated Soils—Assessment of Potentially Toxic Elements (Bio)Availability and Redox Homeostasis in *Hordeum Vulgare* L. *J. Hazard. Mater.* **2022**, *425*, 127999. [CrossRef]
37. Bragança, I.; Lemos, P.C.; Barros, P.; Delerue-Matos, C.; Domingues, V.F. Phytotoxicity of Pyrethroid Pesticides and Its Metabolite towards Cucumis Sativus. *Sci. Total Environ.* **2018**, *619–620*, 685–691. [CrossRef]
38. Liu, T.F.; Wang, T.; Sun, C.; Wang, Y.M. Single and Joint Toxicity of Cypermethrin and Copper on Chinese Cabbage (Pakchoi) Seeds. *J. Hazard. Mater.* **2009**, *163*, 344–348. [CrossRef] [PubMed]
39. Pereira, J.L.; Antunes, S.C.; Ferreira, A.C.; Gonçalves, F.; Pereira, R. Avoidance Behavior of Earthworms under Exposure to Pesticides: Is It Always Chemosensorial? *J. Environ. Sci. Health.-Part B Pestic. Food Contam. Agric. Wastes* **2010**, *45*, 229–232. [CrossRef]
40. Gavina, A.; Bouguerra, S.; Lopes, I.; Marques, C.R.; Rasteiro, M.G.; Antunes, F.; Rocha-Santos, T.; Pereira, R. Impact of Organic Nano-Vesicles in Soil: The Case of Sodium Dodecyl Sulphate/Didodecyl Dimethylammonium Bromide. *Sci. Total Environ.* **2016**, *547*, 413–421. [CrossRef] [PubMed]
41. Wang, Y.; Cang, T.; Zhao, X.; Yu, R.; Chen, L.; Wu, C.; Wang, Q. Comparative Acute Toxicity of Twenty-Four Insecticides to Earthworm, Eisenia Fetida. *Ecotoxicol. Environ. Saf.* **2012**, *79*, 122–128. [CrossRef]
42. Maddela, N.R.; Venkateswarlu, K. Soil Enzymes: Indicators of Soil Pollution. In *Insecticides—Soil Microbiota Interactions*; Springer: Cham, Switzerland, 2018. [CrossRef]
43. Lin, J.; Ma, K.; Chen, H.; Chen, Z.; Xing, B. Influence of Different Types of Nanomaterials on Soil Enzyme Activity: A Global Meta-Analysis. *Nano Today* **2022**, *42*, 101345. [CrossRef]
44. Bouguerra, S.; Gavina, A.; Natal-da-Luz, T.; Sousa, J.P.; Ksibi, M.; Pereira, R. The Use of Soil Enzymes Activity, Microbial Biomass, and Basal Respiration to Assess the Effects of Cobalt Oxide Nanomaterial in Soil Microbiota. *Appl. Soil Ecol.* **2022**, *169*, 104246. [CrossRef]
45. Kondal, R.; Kalia, A.; Krejcar, O.; Kuca, K.; Sharma, S.P.; Luthra, K.; Dheri, G.S.; Vikal, Y.; Taggar, M.S.; Abd-Elsalam, K.A.; et al. Chitosan-Urea Nanocomposite for Improved Fertilizer Applications: The Effect on the Soil Enzymatic Activities and Microflora Dynamics in N Cycle of Potatoes (*Solanum tuberosum* L.). *Polymers* **2021**, *13*, 2887. [CrossRef] [PubMed]
46. Riah, W.; Laval, K.; Laroche-Ajzenberg, E.; Mougin, C.; Latour, X.; Trinsoutrot-Gattin, I. Effects of Pesticides on Soil Enzymes: A Review. *Environ. Chem. Lett.* **2014**, *12*, 257–273. [CrossRef]
47. Kumar, S.; Chaudhuri, S.; Maiti, S.K. Soil Dehydrogenase Enzyme Activity in Natural and Mine Soil—A Review. *Middle East J. Sci. Res.* **2013**, *13*, 898–906. [CrossRef]
48. Hafida, Z.; Caron, J.; Angers, D.A. Pore Occlusion by Sugars and Lipids as a Possible Mechanism of Aggregate Stability in Amended Soils. *Soil Sci. Soc. Am. J.* **2007**, *71*, 1831–1839. [CrossRef]
49. Fageria, N.K.; Baligar, V.C.B.T.-A. *A Enhancing Nitrogen Use Efficiency in Crop Plants*; Academic Press: Cambridge, MA, USA, 2005; Volume 88, pp. 97–185. [CrossRef]
50. Makoi, J.H.J.R.; Ndakidemi, P.A. Selected Soil Enzymes: Examples of Their Potential Roles in the Ecosystem. *Afr. J. Biotechnol.* **2008**, *7*, 181–191. [CrossRef]
51. Bot, A.; Benites, J. *The Importance of Soil Organic Matter Key to Drought-Resistant Soil and Sustained Food and Production*; FAO: Rome, Italy, 2005.

52. Nannipieri, P.; Ceccanti, B.; Conti, C.; Bianchi, D. Hydrolases Extracted from Soil: Their Properties and Activities. *Soil Biol. Biochem.* **1982**, *14*, 257–263. [CrossRef]
53. Al-Qodah, Z.; Shawaqfeh, A.T.; Lafi, W.K. Adsorption of Pesticides from Aqueous Solutions Using Oil Shale Ash. *Desalination* **2007**, *208*, 294–305. [CrossRef]
54. Hayatsu, M.; Katsuyama, C.; Tago, K. Overview of Recent Researches on Nitrifying Microorganisms in Soil. *Soil Sci. Plant Nutr.* **2021**, *67*, 619–632. [CrossRef]

nanomaterials

MDPI

Article

Low Concentrations of Silver Nanoparticles Inhibit Spore Germination and Disturb Gender Differentiation of *Ceratopteris thalictroides* (L.) Brongn

Zhenwei Lu [1,2,3], Liyan Yin [1,4,*], Wei Li [3] and Hong-Sheng Jiang [3,*]

1 Hainan Key Laboratory for Sustainable Utilization of Tropical Bioresources, School of Life Sciences, Hainan University, Haikou 570228, China; luzhenwei599@163.com
2 Jiangsu Coastal Area Institute of Agricultural Sciences, Yancheng 224002, China
3 Key Laboratory of Aquatic Botany and Watershed Ecology, Wuhan Botanical Garden, Chinese Academy of Sciences, Wuhan 430074, China; liwei@wbgcas.cn
4 One Health Institute, Hainan University, Haikou 570228, China
* Correspondence: lyyin@163.com (L.Y.); jhs@wbgcas.cn (H.-S.J.); Tel.: +86-898-6616-0721 (L.Y.); +86-27-8770-0855 (H.-S.J.)

Abstract: Because of their excellent antibacterial properties, silver nanoparticles (AgNPs) are widely used in all walks of life, which has caused them to be discharged into aquatic environments with possible negative effects on aquatic plants. In the present study, we used an aquatic fern, *Ceratopteris thalictroides*, as a model to investigate the effects of AgNPs on its spore germination, gametophytes, sex differentiation, and growth. The results demonstrated that AgNPs significantly inhibited spore germination of *C. thalictroides* at a AgNP concentration higher than 0.02 mg/L. Additionally, we found sex-dependent effects of AgNPs on the development and growth of the gametophyte of *C. thalictroides*. The proportion of hermaphrodites in the gametophytes and the area of gametophytes significantly decreased under AgNP treatment, while no significant effect was observed in the male gametophytes. Using the AgNP filtrate (without nanoparticles) and AgNPs plus cysteine (Ag^+ chelator), we found that the release of Ag^+ from nanoparticles was not the cause of the toxicity of AgNPs on *C. thalictroides*. The EC_{50} of AgNPs on spore germination was 0.0492 mg/L, thus indicating an ecological risk of AgNPs on this species even at concentrations lower than the Ag element concentration of the WHO guidelines for drinking-water quality.

Keywords: silver nanoparticles; aquatic plants; *Ceratopteris thalictroides*; sex-dependent response; spore germination; aquatic fern

Citation: Lu, Z.; Yin, L.; Li, W.; Jiang, H.-S. Low Concentrations of Silver Nanoparticles Inhibit Spore Germination and Disturb Gender Differentiation of *Ceratopteris thalictroides* (L.) Brongn. *Nanomaterials* **2022**, *12*, 1730. https://doi.org/10.3390/nano12101730

Academic Editor: Alexey Pestryakov

Received: 30 March 2022
Accepted: 17 May 2022
Published: 18 May 2022

1. Introduction

Due to their excellent antibacterial properties, silver nanoparticles (AgNPs) are widely used in various fields, including antibacterial coatings, medical machinery, cosmetics, clothing, food packaging, plastic products, detergent, paint, and some porous structures [1–9]. In the process of production, application, and recovery, AgNPs can be discharged into environments in many ways. According to a survey, about 60 tons of AgNPs are discharged into the aquatic environment each year [10]. Previous studies reported that the concentration of AgNPs in surface water in Europe was 0.5–2 ng/L, and the content of AgNPs in the final water of sewage-treatment plants was 32–111 ng/L. The content of AgNPs in activated sludge was 1.3–4.4 mg/kg [11,12]. The World Health Organization (WHO) guidelines for drinking-water quality state that the concentration of silver used to control bacteria in drinking water is 0.1 mg/L, which does not pose a risk to human health [13]. As the use of AgNPs increases, the content of AgNPs in water and sediment will increase, with potential negative effects to the ecosystems.

In recent years, an increasing number of studies have confirmed the toxicity of AgNPs to various organisms, including bacteria, algae, invertebrates, and higher plants, etc. [14–17].

The biological toxicity mechanism of AgNPs has been debated for decades, because it was not clear whether the toxicity comes from nanoparticles themselves or the release of Ag^+ by AgNPs (Ag^+_{rel}). Previous studies considered that the Ag^+_{rel} might be the main cause of the toxicity of AgNPs through the influence of AgNPs and $AgNO_3$ on *Daphnia magna* and *Escherichia coli* [18,19]. Due to the small particle size and large specific surface area, the silver atoms are exposed to the outside of AgNPs, which leads to the release of more Ag^+ into the environment. However, studies on *Lolium multiflorum* and *Spirodela polyrhiza* both indicated that the Ag^+_{rel} may only be one reason for the toxicity of AgNPs [20,21]. Some studies showed that AgNPs absorbed by plants released Ag^+ inside the cells, causing effects [22–24]. Recent studies have shown that the root system of wheat absorbed both the Ag^+_{rel} and the AgNPs particles, resulting in toxic effects on wheat [25]. In vitro, Ag^+_{rel} can interact with thiol groups in proteins, resulting in toxicity by enzyme inactivation or protein denaturation [26]. Although the extracellular oxidative release of Ag^+ is the main cause of AgNPs toxicity in some cells and organisms, it still does not fully explain the toxicity mechanism of AgNPs [27–30]. Therefore, the study of the toxic mechanism of AgNPs requires further exploration. Additionally, at present, most of the studies on the phytotoxicology of AgNPs have focused on plant growth, but the research on their effects on plant reproduction remains limited. To date, only a few studies have reported that AgNPs reduce the germination rate and quality of *Arabidopsis thaliana* seeds [31]. AgNPs delay the bolting and flowering time of *A. thaliana* and affect its fruit and pod development [32].

Ceratopteris thalictroides (L.) Brongn is a pteridophyte that grows rooted or floating in shallow water, and is an endangered freshwater plant in China. *C. thalictroides* is one of the only vascular spore-bearing homosporous model species, which produce one type of spore, germinating into a gametophyte capable of producing both eggs and sperm. In the early stage of fern development, the environmental conditions to which spores are exposed can have a direct and vital impact on survival and reproduction. The main goal of the present study is to provide information on the phytotoxicity of AgNPs on pteridophytes. In particular, this study assesses the influence of environmentally relevant concentrations of AgNP on spore germination and the growth and development of *C. thalictroides*' gametophytes. We hypothesized that: (1) Environmentally relevant concentrations of AgNPs could inhibit spore germination and disturb gender differentiation of *C. thalictroides*; (2) the particles themselves, but not Ag^+_{rel}, are mainly responsible for the toxicity of AgNPs.

2. Materials and Methods

2.1. Plant Material

The spores of *C. thalictroides* were collected from plants in the Wuhan Botanical Garden, Chinese Academy of Sciences in November 2018 and stored at 4 °C, protected from light. The shape and size of spores were observed by a scanning electron microscope (SEM, DMi8, Leica, Germany). The specific process was as follows: spores with plump grains were randomly selected, repeated 5 times, and photos were taken. The software Image J (version 1.50) was used to measure spore size. The TEM image showed that the spore of *C. thalictroides* was nearly round or fan-shaped, and the spore surface has periodical ornamentation, which was formed by the uplift of the outer wall (Figure 1) and the spore diameter was 105.68 ± 3.83 µm (Table 1). 10 mg of spores were added into 10 mL distilled water, and the number of spores per mg was counted under microscopy. The density of spore was obtained as follows: approximately 1 mg spores were added into 1 mL distilled water, then a saturated sodium chloride solution was added drop by drop until the spores were completely suspended, 1 mL solution was weighted using a balance (SQP, with an accuracy of ± 0.0001 g), and the density was calculated with three replicates. The spore density was 1.05 ± 0.04 g/cm^3, the weight of individual spore was $0.41 \pm 0.03 \times 10^{-6}$ mg, and there were $2.45 \pm 0.15 \times 10^6$ spores per mg. (Table 1).

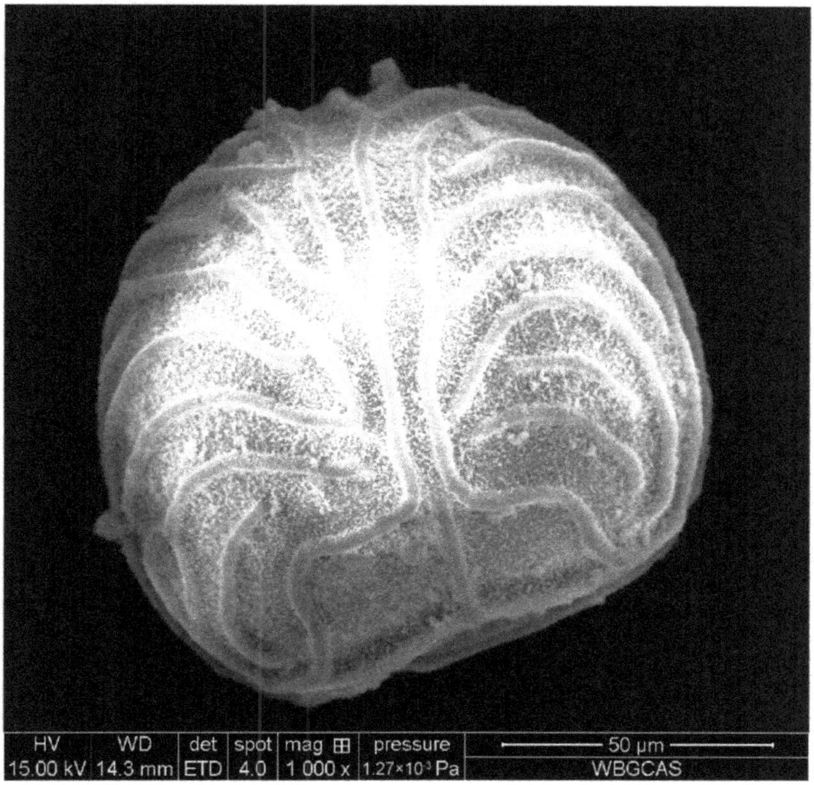

Figure 1. SEM image of spores of *Ceratopteris thalictroides*.

Table 1. Diameter, density, and weight of *Ceratopteris thalictroides*. Data represent the mean ± SD (n = 5).

Diameter (μm)	Density (g/cm^3)	Weight (mg)	Number of Spores per Milligram
105.68 ± 3.83	1.05 ± 0.04	$0.41 \pm 0.03 \times 10^{-6}$	$2.45 \pm 0.15 \times 10^{6}$

2.2. Characterization of AgNPs

The polyvinyl pyrrolidone (PVP) coating AgNPs stock (5030 mg/L) purchased from NanoComposix (San Diego, CA, USA) had an average diameter of 6.3 ± 1.4 nm and was stored at 4 °C, protected from light. To explore the effects of culture solution on AgNPs, the particle size and distribution of AgNPs in pure water and 10% Hoagland's solution at 2.5 mg/L photomicrographs taken using a transmission electron microscope (TEM, HT-7700) were analyzed using Image J software. Additionally, the hydrodynamic diameter, polydispersity index (PDI), and Zeta potential of AgNPs at 5 mg/L were determined by using a surface potential analyzer (Dynamic Light Scattering, Nano ZS ZEN3600, Malvern, UK) in pure water and in 10% Hoagland's solution. In the study, the PVP concentration of the coating substance of AgNPs was 2.85 mg/mL, and the Ag^{+}_{rel} was determined as described in the previous study [26]. The measured Ag^{+}_{rel} was 4.08 mg/L in 100 mg/L AgNPs suspension, meaning that AgNPs contained ~4.1% Ag^{+}_{rel}.

2.3. Effects of AgNPs on Spore Germination and Gametophyte Differentiation in C. thalictroides

AgNPs were diluted into 0, 0.02, 0.04, 0.06, 0.08, and 0.1 mg/L with 10% Hoagland's solution in a step-by-step dilution method. $AgNO_3$ was prepared by using 10% Hoagland's solution with concentrations of 0, 0.02, 0.03, 0.04, 0.05, and 0.06 mg/L in a step-by-step dilution method. 5 mL AgNPs and $AgNO_3$ solutions of each concentration were added to the 6-well cell-culture plate (Corning 3516, Corning NY, USA), and then approximately 100 spores were dropped into each well. Each treatment had 5 replicates. The cell-culture plates were incubated in a chamber under 16 h of light, 8 h of darkness, 25 ± 0.1 °C, and 20 μmol photons m^{-2} s^{-1} photosynthetically active radiation. After 8 days, the spore germination rate of *C. thalictroides* was observed and calculated under an electric stereoscopic fluorescence microscopy (SMZ25, Nikon Corporation, Tokyo, Japan). The production of false roots was used to assess spore germination. After that, the spore germination rate was observed every 2 days until the germination rate had not changed markedly. After 19 days, the numbers of hermaphrodites and male gametophytes were counted, and after 21 days, the area of the two kinds of gametophytes was counted. Concentration for 50% of maximal effect (EC_{50}) of spore gemination rate at day 17 was calculated by a two-parameter logistic model with R software (R3.5.3 for Windows, R Foundation for Statistical Computing, Vienna, Austria).

2.4. Toxicity Sources of AgNPs on Spore Germination and Gametophyte Differentiation in C. thalictroides

To investigate whether the effect of AgNPs on spore germination was caused by particles or Ag^{+}_{rel}, the following 5 treatments were conducted: (1) 5 mg/L AgNPs were added into an ultrafiltration tube (0.5 mL, 30 KD, Millipore, Bedford, MA, USA) and centrifuged (3K15, Sigma, Osterode, Niedersachsen, Germany) at 8000 g at 4 °C for 10 min to obtain the AgNPs filtrate, which was diluted 62.5 times, which was equivalent to the concentration of 0.08 mg/L AgNPs filtrate; (2) 0.004 mg/L $AgNO_3$ (AgNPs releases about 4.1% Ag^+, which is about the same concentration as 0.08 mg/L AgNPs in the test); (3) 0.08 mg/LAgNPs + 0.09 mg/L cysteine (AgNPs + cys all, this concentration of cysteine can chelate all Ag^+ even if all AgNPs is converted to Ag^+); (4) 0.08 mg/L AgNPs + 0.0045 mg/L cysteine (AgNPs + cys part, this concentration of cysteine only can chelate the Ag^{+}_{rel}); (5) 0.09 mg/L cysteine (as the control, to determine whether cysteine affects spore germination of *C. thalictroides*). Approximately 5 mL of the solutions were added to the 6-well cell-culture plates, and then a drop of spore suspension, with about 100 spores of *C. thalictroides* per suspension drop, was added into each well. Each treatment was replicated five times. The plates were then incubated in a light chamber under 16 h of light, 8 h of darkness, 25 ± 0.1 °C, and 20 μmol photons $m^{-2} \cdot s^{-1}$ photosynthetically active radiation. After 8 days of cultivation, the spore germination rate of *C. thalictroides* was observed under a stereomicroscope. The germination rate of *C. thalictroides* spores was observed every 2 days until the germination rate had not changed markedly. After 19 days, the number of hermaphrodites and male gametophytes was counted, and after 21 days, the area of the two gametophytes was counted.

2.5. Statistics Analysis

All data are presented as mean with one standard deviation. SPSS Statistics 20 was used for one-way ANOVA. If $p < 0.05$, the Tukey method was used to compare differences between groups. The particle size and potential of silver nanoparticles in pure water and culture medium were tested by independent sample *t*-test.

3. Results

3.1. Characterization of AgNPs

The shape of AgNPs was spherical under TEM both in distilled water and 10% Hoagland's solution (Figure 2), and the measured core diameter was 6.2 ± 2.0 nm in distilled water and 7.8 ± 2.7 nm in 10% Hoagland's solution, respectively (Table 2). From the TEM, most AgNPs were 4–6 nm in diameter both in distilled water and 10% Hoagland's solution (Figure 3A,B). The DLS results showed that the hydrodynamic diameter of AgNPs increased from 20.3 nm in distilled water to 27.1 nm in 10% Hoagland's solution (Table 2). Similarly, the DPI of AgNPs increased from 0.637 in distilled water to 0.791 in Hoagland's solution (Table 2), while the Zeta potential of AgNPs in distilled water was -10.7 ± 0.4 mV which decreased significantly to -2.1 ± 0.4 mV in 10% Hoagland's solution (Table 2).

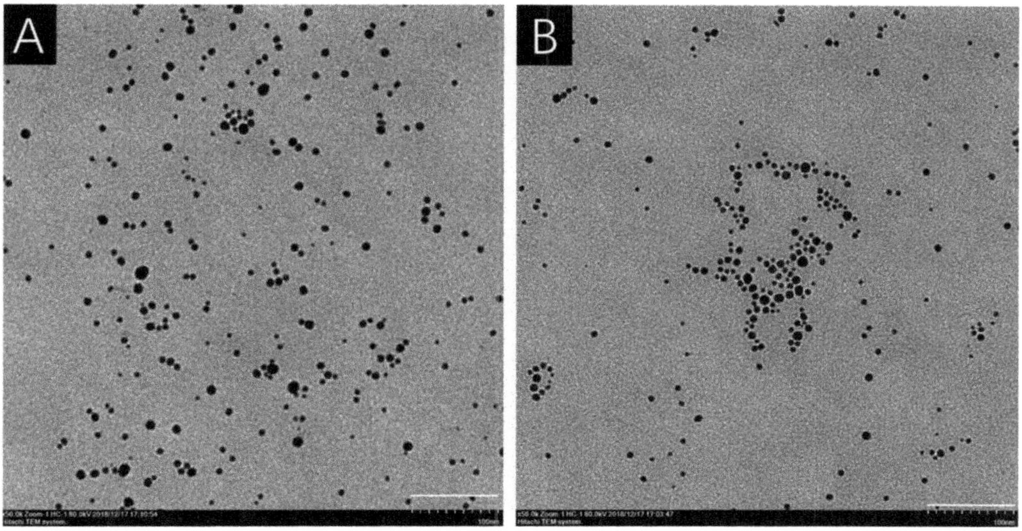

Figure 2. TEM image of AgNPs in distilled water (**A**) and 10% Hoagland's solution (**B**). The white bars indicate 100 nm.

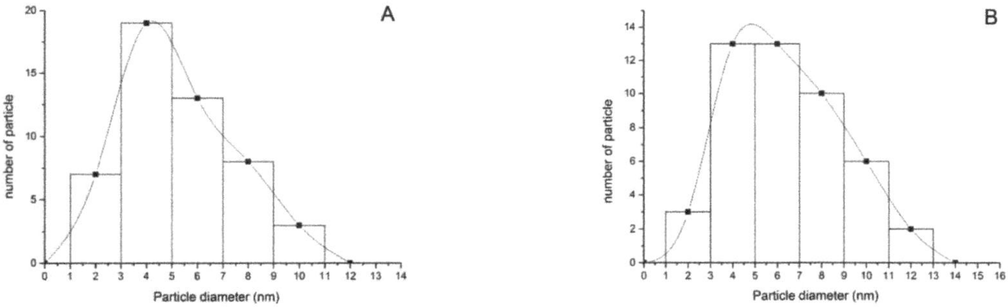

Figure 3. Size distribution of AgNPs in distilled water (**A**) and 10% Hoagland's solution (**B**).

Table 2. Core diameter, hydrodynamic diameter, polydispersity index (PDI) and Zeta potential of AgNPs in distilled water (A) and 10% Hoagland's solution (B). Data represent the mean ± SD (n = 5). Data with different letters are significantly different ($p < 0.05$).

Solution	Core Diameter * (nm)	Hydrodynamic Diameter * (nm)	PDI	Zeta Potential (mV)
H_2O	6.2 ± 2.0	20.3 ± 3.1	0.637	-10.7 ± 0.4 a
10% Hoagland's solution	7.8 ± 2.7	27.1 ± 4.1	0.791	-2.1 ± 0.4 b

* Core diameter was obtained by using a TEM; hydrodynamic diameter was obtained by using dynamic light scatter.

3.2. *Effects of AgNPs and Ag^+ on the Spore Germination Rate of C. thalictroides*

In the control, more than 80% of spores had germinated by day 8. Although some spores continued to germinate after day 8, numbers were not significantly different between days 8 and day 17. The spore germination rate of *C. thalictroides* was significantly inhibited under AgNPs treatments and decreased significantly with increasing concentration of AgNPs (Figure 4A). Though 0.02 mg/L AgNPs did not significantly decrease the final germination rate, it delayed spore germination. When the concentration of AgNPs reached 0.06 mg/L, the spore germination rate was less than 30%. When the concentration of AgNPs reached 0.1 mg/L, no spore germination was observed. Ag^+ showed a similar effect on spore germination as AgNPs (Figure 4B). After 17 days of exposure, the EC_{50} of AgNPs on spore germination was 0.0492 ± 0.0012 mg/L and the EC_{50} of Ag^+ on spore germination was 0.0349 ± 0.0007 mg/L (Figure 5).

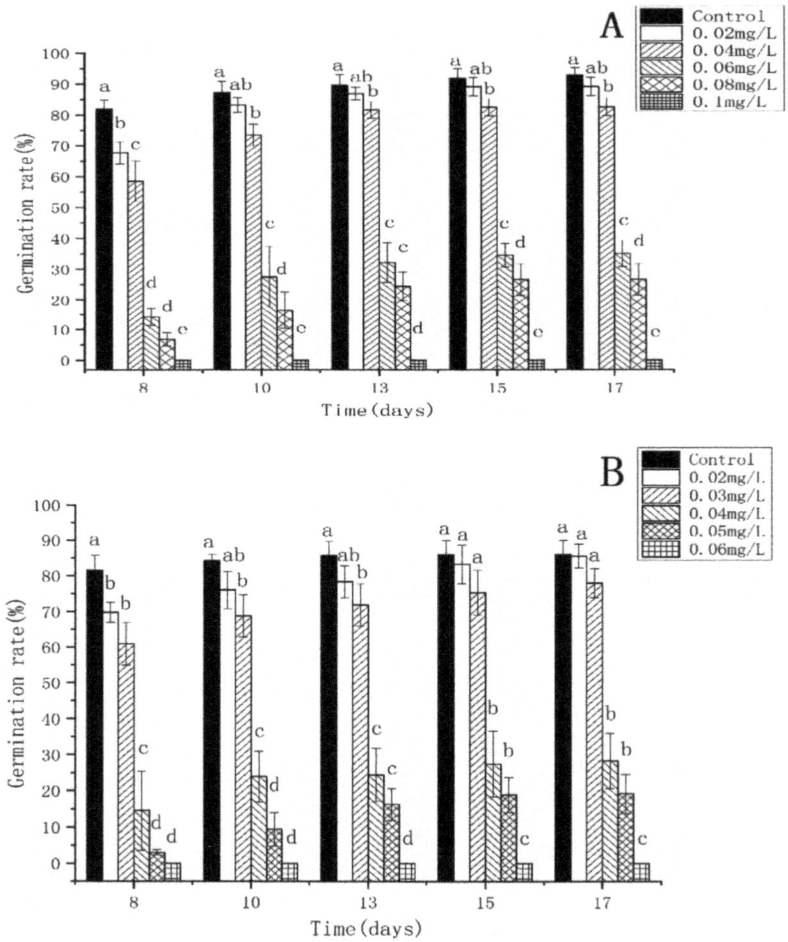

Figure 4. Effect of AgNPs (**A**) and Ag^+ (**B**) on the spore germination of *Ceratopteris thalictroides*. Data represent the mean ± SD (n = 5). Data with different letters are significantly different ($p < 0.05$).

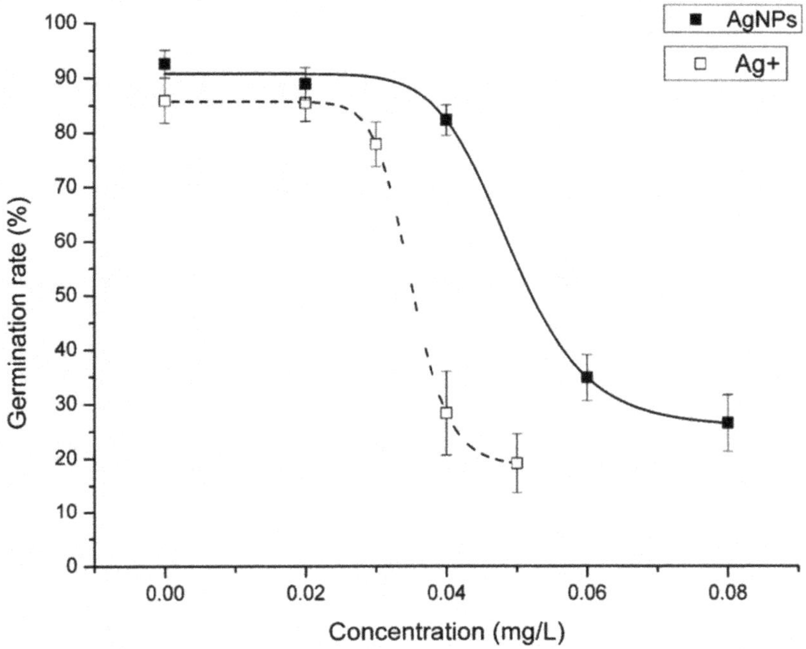

Figure 5. Concentration courses of effect of AgNPs and Ag^+ on the spore germination rate of *Ceratopteris thalictroides* after 17 d exposure. Data represent the mean ± SD (n = 5).

3.3. *Effects of AgNPs and* Ag^+ *on Gametophytes of C. thalictroides*

After 19 days exposure to AgNPs, the proportion of hermaphrodite gametophytes decreased significantly with the increase in AgNP concentrations (Figure 6). When the AgNP concentration was at 0.02 and 0.04 mg/L, the proportion of hermaphrodites was 83% and 67%, which did not significantly differ from the control (75%). However, the hermaphrodites in the 0.06 and 0.08 mg/L AgNP treatments were 24% and 5% of the gametophytes, a decrease of 51% and 70% compared with the control. In comparison to the AgNP treatments, the hermaphrodites in the 0.05 mg/L Ag^+ treatment group were 50% of the gametophytes, which decreased by 20% compared with the control group (Figure 6).

After 21 days of exposure to AgNPs, the area of hermaphrodites decreased significantly with the increase in AgNPs concentration, while the area of male gametophytes did not change significantly (Figure 7A). When the concentration of AgNPs was 0.04 mg/L, the growth area of hermaphrodites decreased significantly compared with the control group, which was about 62.5% of it. When the concentration of AgNPs was 0.06 mg/L and 0.08 mg/L, the growth area of hermaphrodites was about 25% of that of the control. Similar to the AgNPs treatments, after 21 days of exposure to Ag^+, the area of hermaphrodites decreased significantly with the increase of Ag^+ concentration, while the area of male gametophytes did not change significantly (Figure 7B). When Ag^+ concentration was 0.02 mg/L, no significant change was found in the area of hermaphrodites compared with the control group. When Ag^+ concentration was 0.04 mg/L, the area of hermaphrodites decreased significantly compared with the control group, which was about 50% of it. When the concentration of Ag^+ was 0.05 mg/L, the area of hermaphrodites was significantly reduced, which was about 16% of that of the control group.

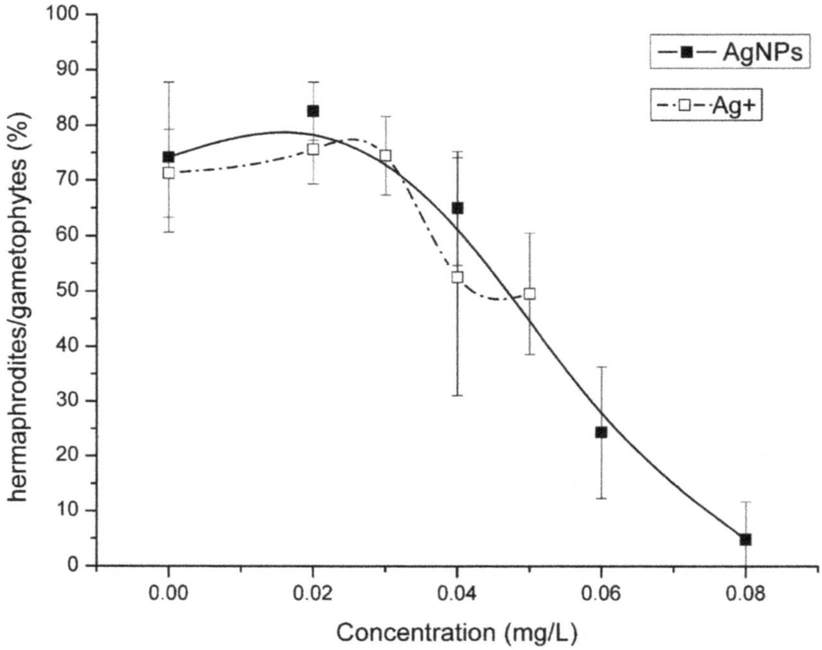

Figure 6. Effect of AgNPs and Ag^+ on the proportion of hermaphrodites after 19 d exposure. Data represent the mean ± SD (n = 5).

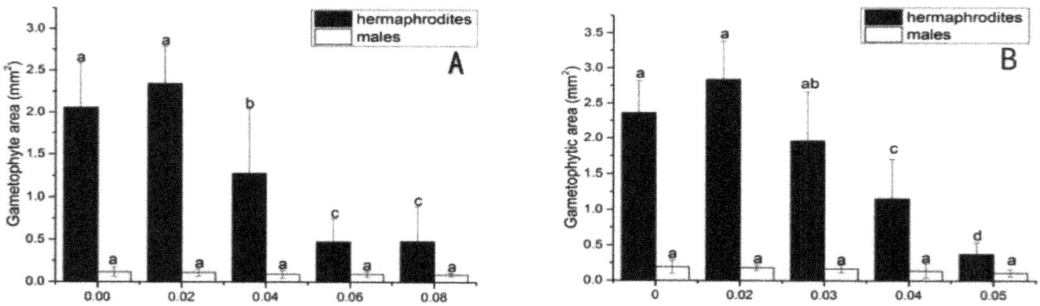

Figure 7. Effect of AgNPs (**A**) and Ag^+ (**B**) on the area of male gametophytes and hermaphrodites of *Ceratopteris thalictroides* after 21 d exposure. Data represent the mean ± SD (n = 5). Data with different letters are significantly different among treatment within the same sexual gametophyte ($p < 0.05$).

4. Discussion

Previous studies have indicated that the toxicity of AgNPs was related to their particle size, and small particles were more toxic to organisms [9,20,33]. In our experiments, the core size of AgNPs did not change significantly, while the hydrodynamic diameter increased in 10% of Hoagland's solution, indicating an AgNP aggregation in the solution. This aggregation was confirmed by the increase in PDI in our present study. Previous studies reported that cations in solution could be responsible for the aggregation of AgNPs by decreasing the zeta potential of nanoparticles, especially the divalent cation [26,34]. There were approximately 20 mg/L Ca^{2+}, 5 mg/L Mg^{2+}, 30 mg/L K^+ and other trace metal ions in 10% Hoagland's solution that could explain the aggregation of AgNPs. Though AgNPs aggregated in 10% of Hoagland's solution, the size distribution showed that there were still some AgNPs with particle sizes of less than 5 nm. Dietz et al. reported that AgNPs with a particle size of 5 nm or smaller may penetrate plant cell walls and enter plant cells [35]. In this study, AgNPs with a particle size less than 5 nm may enter spores of *C. thalictroides* and interact with important bioactive substances such as proteins and/or nucleic acids in the cell, thus affecting spore germination and sex differentiation of gametophytes.

In the present study, the spore morphology and diameter of *C. thalictroides* were consistent with previous studies, with a diameter of 90–150 μm [36–38], indicating that the spores used in this study were mature spores with normal development. Our results showed that AgNPs or Ag^+ significantly inhibited spore germination of *C. thalictroides*, and the EC_{50} of AgNPs and Ag^+ on *C. thalictroides* spore germination was 0.0492 ± 0.0012 mg/L and 0.0349 ± 0.0007 mg/L, respectively. According to the World Health Organization (WHO) guidelines, the concentration of silver used to control bacteria in drinking water is 0.1 mg/L [13]. In our study, 0.1 mg/L AgNPs or Ag^+ can significantly inhibit the spore germination of *C. thalictroides* to achieve bacteriostatic purposes. Spore germination is crucial in the life history of *C. thalictroides*. Even if the Ag element in the water was within the recommended range of drinking-water guidelines, it could significantly inhibit the spore germination of *C. thalictroides*, which may have a significant impact on their reproduction. Many studies have shown that AgNPs had toxic effects on animals, higher plants, bacteria, and fungi [39–49]; however, the concentration used was rather high, and the EC_{50} of AgNPs was usually more than 1 mg/L. *Ceratopteris thalictroides* is one of the most sensitive species to the toxicity of AgNPs or Ag^+, and thus could be used as a Ag-toxicity indicator.

In this study, AgNPs significantly affected the sex differentiation of gametophytes of *C. thalictroides*. AgNPs significantly inhibited the differentiation of gametophytes into hermaphrodites, and at a concentration of 0.08 mg/L the proportion of hermaphrodites decreased by 70% compared with the control. In contrast, AgNPs had no significant effect on the differentiation into male gametophytes (Figure S1A). These results indicate that the sensitivity of gametophytes from *C. thalictroides* to AgNPs was sex-dependent. A previous study reported that delayed spore germination of *C. thalictroides* could affect the sex differentiation of *C. thalictroides*' gametophytes [50], which could explanation our sex-dependent result. However, the effect of Ag^+ on sex differentiation of gametophytes was different from that of AgNPs. Although Ag^+ significantly decreased the proportion of hermaphrodites as AgNPs, both numbers of hermaphrodites and male gametophytes decreased; hermaphrodites decreased faster than males and resulted in a decrease in the proportion of hermaphrodites (Figure S1B). Previous studies showed that 0.1 mg /L microplastics resulted in a 30% decrease in the proportion of hermaphrodites in *C. thalictroides* compared to the control group and that 4 mg/L quinclorac resulted in a 30% decrease in the proportion of hermaphrodites compared with the control group [51,52]. Compared with Ag^+, the traditional herbicide quinclorac, and new pollutant microplastics, AgNPs had a greater effect on the sex differentiation of gametophytes of *C. thalictroides*. The sex-dependent response also showed in the growth of gametophytes. In this study, the area of gametophytes of different genders responded differently to the toxicity of AgNPs. The area of hermaphroditic gametophytes induced by AgNPs at 0.08 mg/L was 25% of that in the control group. The results of this experiment showed that AgNPs significantly reduced the area of hermaphrodites, which may lead to a decrease in the biomass of *C. thalictroides*, affecting its normal growth, development, and breeding populations [53]. Ag^+ significantly affected the area of gametophytes of *C. thalictroides*; the area of hermaphrodites exposed to 0.05 mg/L Ag^+ was 16% that of control hermaphrodites. However, the area of male gametophytes in the 0.08 mg/L AgNP treatment and the 0.05 mg/L Ag^+ treatment was not significantly different from that in the control. Different gender sensitivity to the toxicity of AgNPs has also been observed in animals. Studies showed that AgNPs accumulated in the ovaries, resulting in abnormal follicular development of female fish, resulting in the loss of their reproductive ability, and also resulting in embryonic dysplasia of female fish, reducing the survival rate of embryos [53–59]. When AgNPs interacted with serum proteins of *Micropterus dolomieu*, the formation of protein corona of AgNPs was sex-dependent, mainly because female serum contained some egg-specific proteins such as vitellogenin and zona pellucida, etc. This may lead to differences in the response of female and male smallmouth bass (*Micropterus dolomieu*) to AgNPs toxicity [60]. The sex-dependent response of *C. thalictroides*' gametophytes differentiation to AgNPs may be caused by the differential expression of sex-specific proteins during the process of gametophyte differentiation.

It is a hot topic in the toxicity of AgNPs whether it depends on Ag^{+}_{rel} or the particle itself. In the present study, the EC50 of AgNPs and Ag^+ on spore germination by *C. thalictroides* was 0.0492 ± 0.0012 mg/L and 0.0349 ± 0.0007 mg/L, respectively, which indicated that Ag^+ is more toxic than AgNPs, but the toxicity was in the same order of magnitude. Thus, the low concentration of Ag^{+}_{rel} could not explain the high toxicity of AgNPs. In the present study, the effects of the 0.08 mg/L AgNP filtrate treatment on spore germination (Figure S2), sex differentiation of gametophytes (Figure S3), and area of gametophytes (Figure S4) were not significantly different from those of the control group, and they significantly differed from the 0.08 mg/L AgNP treatment. About 4.1% Ag^+ was released by 0.08 mg/L AgNPs, so about 0.004 mg/L Ag^+ was released in the 0.08 mg/L treatment. The effects of 0.004 mg/L Ag^+ treatment on spore germination, sex differentiation of gametophytes, and area of gametophytes were not significantly different from those of the control. In addition, 0.004 mg/L Ag^+ was much lower than the EC50 of Ag^+ on spores of *C. thalictroides*. These results indicate that the Ag^+ released by AgNPs did not have a significant effect on the germination and growth of spores of *C. thalictroides*, and the toxicity of AgNPs may not be caused by the Ag^{+}_{rel}. The effects of 0.08 mg/L AgNPs + 0.09 mg/L cysteine, 0.08 mg/L AgNPs + 0.045 mg/L cysteine and 0.09 mg/L cysteine on spore germination, sex differentiation of gametophytes, and area of gametophytes were not significantly different from those of the control group. The results showed that when cysteine was mixed with Ag^+, AgNPs had no significant effect on spore germination, sex differentiation of gametophytes, and growth area of gametophytes, suggesting that the toxicity of AgNPs to the spores of *C. thalictroides* might be changed by cysteine. AgNPs interacting with organic matter in solution should be considered in its toxicity. When the cysteine was combined with the Ag^{+}_{rel} or AgNPs itself, it formed a passivation layer on the surface of AgNPs. This prevents the further release of Ag^+ and also prevents interaction between AgNP particles and spores of *C. thalictroides* (Figures S2–S4). Some studies have shown that Cl^- can react with dissolved Ag^+ to generate a silver chloride passivation layer on AgNPs, reducing the toxicity of AgNPs [61–63]. Additionally, in the present study, the cysteine as a chelator of Ag^+ could also be one of the antioxidants [64] that had a possibility to alleviate the toxicity of AgNPs on *C. thalictroides*. In conclusion, the Ag^{+}_{rel} cannot explain the highly toxic effect of AgNPs on the spores of *C. thalictroides*. The interaction among AgNPs, material in solution, and the spores of *C. thalictroides* could be important in the toxic mechanisms of AgNPs. Thirty percent of the AgNPs used in this study had a particle size of less than 5 nm, which could be likely to enter the spores of *C. thalictroides* and produce effects. Previous studies have shown that AgNPs entered animal and plant cells where they released Ag^+, thus producing toxic effects [23–26]. This is a further direction for our work on the toxic effects of AgNPs on spores of *C. thalictroides*.

5. Conclusions

In this study, it was found that AgNPs significantly inhibited the spore germination of *C. thalictroides*. Moreover, AgNPs also significantly affected further gametophyte development, and these effects were sex-dependent. AgNPs significantly inhibited the development of hermaphrodites and significantly inhibited their growth. According to the World Health Organization (WHO) guidelines, the concentration of silver used to control bacteria in drinking water is 0.1 mg/L. In our study, 0.1 mg/L AgNPs or Ag^+ can completely inhibit the spore germination of *C. thalictroides*.

By comparing the toxicity of AgNPs and Ag^+, it was found that the toxicity of AgNPs to *C. thalictroides* was slightly lower than that of Ag^+, but the toxicity of the two forms of silver was in the same order of magnitude. It was found that the Ag^{+}_{rel} was not the main source of the toxicity of the AgNPs.

The sensitivity of spore of *C. thalictroides*' to AgNPs was greater than in other studied higher plants, animals, and even some bacteria, fungi, algae. It was particularly sensitive to the toxicity of AgNPs, implying that the real concentration of AgNPs in some aquatic environments may affect spore germination of *C. thalictroides* which may affect its reproduction. Since *C. thalictroides* is one of the most sensitive species to the toxic effects of AgNPs or Ag^+, it could be used as an Ag toxicity indicator.

Supplementary Materials: The following supporting information can be downloaded at: https://www.mdpi.com/article/10.3390/nano12101730/s1, Figure S1: Effect of AgNPs (A) and Ag^+ (B) on the number of gametophytes of *Ceratopteris thalictroides* after 19 d exposure. Data represent the mean ± SD (n = 5). Data with different letters are significantly different ($p < 0.05$); Figure S2: Effect of various comparison groups on the spore germination of *Ceratopteris thalictroides* after 17 d exposure. (Control; AgNPs filtrate: 0.08 mg/L AgNP filtrate; Ag^+: 0.004 mg/L Ag^+; AgNPs + cys(all): 0.08 mg/L AgNPs + 0.09 mg/L cys; AgNPs + cys(part): 0.08 mg/L AgNPs + 0.0045 mg/L cys; Cys: 0.09 mg/L cys) Data represent the mean ± SD (n = 5). Data with same letters are not significantly different (p = 0.05); Figure S3: The proportion of hermaphrodites of various comparison groups after 19 d exposure. (Control; AgNPs filtrate: 0.08 mg/L AgNP filtrate; Ag^+: 0.004 mg/L Ag^+; AgNPs + cys(all): 0.08 mg/L AgNPs + 0.09 mg/L cys; Cys: 0.09 mg/L cys; AgNPs + cys(part): 0.08 mg/L AgNPs + 0.0045 mg/L cys) Data represent the mean ± SD (n = 5). Data with same letters are not significantly different (p = 0.05); Figure S4: Effect of various comparison groups on area of male gametophytes and hermaphrodites of *Ceratopteris thalictroides* after 21 d exposure. (Control; AgNP filtrate: 0.08 mg/L AgNP filtrate; Ag^+: 0.004 mg/L Ag^+; AgNPs + cys(all): 0.08 mg/L AgNPs + 0.09 mg/L cys; Cys: 0.09 mg/L cys; AgNPs + cys(part): 0.08 mg/L AgNPs + 0.0045 mg/L cys) Data represent the mean ± SD (n = 5). Data with same letters are not significantly different (p = 0.05).

Author Contributions: Conceptualization, H.-S.J., W.L. and L.Y.; methodology, H.-S.J. and Z.L.; formal analysis, Z.L.; investigation, Z.L.; writing—original draft preparation, Z.L.; writing—review and editing, H.-S.J. and L.Y.; supervision, H.-S.J. and L.Y.; funding acquisition, H.-S.J. and L.Y. All authors have read and agreed to the published version of the manuscript.

Funding: This research and APC were funded by the National Natural Science Foundation of China (41907223, 31971520) and the Youth Innovation Promotion Association CAS (2021340).

Institutional Review Board Statement: Not applicable.

Informed Consent Statement: Not applicable.

Data Availability Statement: Not applicable.

Acknowledgments: The authors would like to thank the Public Laboratory Platform, WBGCAS (http://www.wbg.cas.cn/gglabplat/index.html, accessed on 1 March 2022) for providing instruments. We would like to thank S.C. Maberly for improving the English of the manuscript and we also thank Jingzhe Zhou for the proofreading of the manuscript.

Conflicts of Interest: The authors declare that they have no known competing financial interest or personal relationship that could have appeared to influence the work reported in this paper.

References

1. Capek, I. Preparation of metal nanoparticles in water-in-oil (w/o) microemulsions. *Adv. Colloid Interface Sci.* **2004**, *110*, 49–74. [CrossRef] [PubMed]
2. Frattini, A.; Pellegri, N.; Nicastro, D.; Sanctis, O.D. Effect of amine groups in the synthesis of Ag nanoparticles using amino silanes. *Mater. Chem. Phys.* **2005**, *94*, 148–152. [CrossRef]
3. Mueller, N.C.; Nowack, B. Exposure modeling of engineered nanoparticles in the environment. *Environ. Sci. Technol.* **2008**, *42*, 4447–4453. [CrossRef] [PubMed]
4. Navarro, E.; Piccapietra, F.; Wagner, B.; Marconi, F.; Kaegi, R.; Odzak, N.; Sigg, L.; Behra, R. Toxicity of silver nanoparticles to *Chlamydomonas reinhardtii*. *Environ. Sci. Technol.* **2008**, *42*, 8959–8964. [CrossRef]

5. Piccinno, F.; Gottschalk, F.; Seeger, S.; Nowack, B. Industrial production quantities and uses of ten engineered nanomaterials in Europe and the world. *J. Nanopart. Res.* **2012**, *14*, 1109. [CrossRef]
6. Zhang, Y.; Yu-Rui, L.; Robert, A.; Michael, R. Inventory of engineered nanoparticle-containing consumer products available in the Singapore retail market and likelihood of release into the aquatic environment. *Int. J. Environ. Res. Public Health* **2015**, *12*, 8717. [CrossRef]
7. Huang, S.M.; Liu, S.M.; Ko, C.L.; Chen, W.C. Advances of Hydroxyapatite Hybrid Organic Composite Used as Drug or Protein Carriers for Biomedical Applications: A Review. *Polymers* **2022**, *14*, 976. [CrossRef]
8. Baldino, L.; Aragón, J.; Mendoza, G.; Irusta, S.; Cardea, S.; Reverchon, E. Production, characterization and testing of antibacterial PVA membranes loaded with HA-Ag_3PO_4 nanoparticles, produced by SC-CO_2 phase inversion. *J. Chem. Technol. Biotechnol.* **2019**, *94*, 98–108. [CrossRef]
9. Prihandana, G.S.; Sriani, T.; Muthi'ah, A.D.; Machmudah, A.; Mahardika, M.; Miki, N. Study Effect of nAg Particle Size on the Properties and Antibacterial Characteristics of Polysulfone Membranes. *Nanomaterials* **2022**, *12*, 388. [CrossRef]
10. Handy, R.D.; Cornelis, G.; Fernandes, T.; Tsyusko, O.; Decho, A.; Sabo-Attwood, T.; Metcalfe, C.; Steevens, J.A.; Klaine, S.J.; Koelmans, A.A. Ecotoxicity test methods for engineered nanomaterials: Practical experiences and recommendations from the bench. *Environ. Toxicol. Chem.* **2012**, *31*, 15–31. [CrossRef]
11. Geranio, L.; Heuberger, M.; Nowack, B. The behavior of silver nanoparticles during washing. *Environ. Sci. Technol.* **2009**, *43*, 8113–8118. [CrossRef] [PubMed]
12. Gottschalk, F.; Sonderer, T.; Scholz, R.W.; Nowack, B. Modeled environmental concentrations of engineered nanomaterials (TiO_2, ZnO, Ag, CNT, fullerenes) for different regions. *Environ. Sci. Technol.* **2009**, *43*, 9216–9222. [CrossRef] [PubMed]
13. Edition, F. Guidelines for drinking-water quality. *WHO Chron.* **2011**, *38*, 104–108.
14. Hund-Rinke, K.; Simon, M. Ecotoxic effect of photocatalytic active nanoparticles (TiO_2) on algae and daphnids (8 pp). *Environ. Sci. Pollut. Res.* **2006**, *13*, 225–232. [CrossRef]
15. Lee, W.F.; Huang, Y.C. Swelling and antibacterial properties for the superabsorbent hydrogels containing silver nanoparticles. *J. Appl. Polym. Sci.* **2007**, *106*, 1992–1999. [CrossRef]
16. Shahverdi, A.R.; Fakhimi, A.; Shahverdi, H.R.; Minaian, S. Synthesis and effect of silver nanoparticles on the antibacterial activity of different antibiotics against *Staphylococcus aureus* and *Escherichia coli*. *Nanomed. Nanotechnol. Biol. Med.* **2007**, *3*, 168–171. [CrossRef]
17. Shoults-Wilson, W.A.; Reinsch, B.C.; Tsyusko, O.V.; Bertsch, P.M.; Lowry, G.V.; Unrine, J.M. Effect of silver nanoparticle surface coating on bioaccumulation and reproductive toxicity in earthworms (*Eisenia fetida*). *Nanotoxicology* **2011**, *5*, 432–444. [CrossRef]
18. Xiu, Z.; Zhang, Q.; Puppala, H.L.; Colvin, V.L.; Alvarez, P. Negligible particle-specific antibacterial activity of silver nanoparticles. *Nano Lett.* **2012**, *12*, 4271–4275. [CrossRef]
19. Zhao, C.M.; Wang, W.X. Importance of surface coatings and soluble silver in silver nanoparticles toxicity to *Daphnia magna*. *Nanotoxicology* **2012**, *6*, 361–370. [CrossRef]
20. Yin, L.Y.; Cheng, Y.; Colman, B.P.; Auffan, M.; Wiesner, M.; Bernhardt, E.S. More than the ions: The effects of silver nanoparticles on *Lolium multiflorum*. *Environ. Sci. Technol.* **2011**, *45*, 2360–2367. [CrossRef]
21. Jiang, H.S.; Ming, L.; Chang, F.Y.; Wei, L.; Yin, L.Y. Physiological analysis of silver nanoparticles and $AgNO_3$ toxicity to *Spirodela polyrhiza*. *Environ. Toxicol. Chem.* **2012**, *31*, 1880–1886. [CrossRef] [PubMed]
22. Jiang, H.S.; Qiu, X.N.; Li, G.B.; Yin, L.Y. Silver nanoparticles induced the accumulation of reactive oxygen species and alteration of antioxidant systems in the aquatic plant *Spirodela polyrhiza*. *Environ. Toxicol. Chem.* **2014**, *33*, 1398–1405. [CrossRef] [PubMed]
23. Jiang, H.S.; Yin, L.Y.; Ren, N.N.; Zhao, S.T.; Li, Z.; Zhi, Y.; Shao, H.; Li, W.; Gontero, B. Silver nanoparticles induced reactive oxygen species via photosynthetic energy transport imbalance in an aquatic plant. *Nanotoxicology* **2017**, *11*, 157–167. [CrossRef]
24. Jiang, X.; Miclăuş, T.; Wang, L.; Foldbjerg, R.; Sutherland, D.S.; Autrup, H.; Chen, C.; Beer, C. Fast intracellular dissolution and persistent cellular uptake of silver nanoparticles in CHO-K1 cells: Implication for cytotoxicity. *Nanotoxicology* **2015**, *9*, 181–189. [CrossRef] [PubMed]
25. Dang, F.; Wang, Q.; Cai, W.; Zhou, D.M.; Xing, B. Uptake kinetics of silver nanoparticles by plant: Relative importance of particles and dissolved ions. *Nanotoxicology* **2020**, *14*, 654–666. [CrossRef] [PubMed]
26. Jiang, H.S.; Zhang, Y.Z.; Lu, Z.W.; Lebrun, R.; Li, W. Interaction between Silver Nanoparticles and Two Dehydrogenases: Role of Thiol Groups. *Small* **2019**, *15*, 1900860. [CrossRef]
27. Van Aerle, R.; Lange, A.; Moorhouse, A.; Paszkiewicz, K.; Ball, K.; Johnston, B.D.; De-Bastos, E.; Booth, T.; Tyler, C.R.; Santons, E.M. Molecular mechanisms of toxicity of silver nanoparticles in zebrafish embryos. *Environ. Sci. Technol.* **2013**, *47*, 8005–8014. [CrossRef]
28. Bouwmeester, H.; Poortman, J.; Peters, R.J.; Wijma, E.; Kramer, E.; Makama, S.; Puspitaninganindita, K.; Marvin, H.J.P.; Peijnenburg, A.A.C.M.; Hendriksen, P.J.M. Characterization of translocation of silver nanoparticles and effects on whole-genome gene expression using an in vitro intestinal epithelium coculture model. *ACS Nano* **2011**, *5*, 4091–4103. [CrossRef]
29. Sakamoto, M.; Ha, J.Y.; Yoneshima, S.; Kataoka, C.; Tatsuta, H.; Kashiwada, S. Free silver ion as the main cause of acute and chronic toxicity of silver nanoparticles to cladocerans. *Arch. Environ. Contam. Toxicol.* **2015**, *68*, 500–509. [CrossRef]
30. Shen, M.H.; Zhou, X.X.; Yang, X.Y.; Chao, J.B.; Liu, R.; Liu, J.F. Exposure medium: Key in identifying free Ag^+ as the exclusive species of silver nanoparticles with acute toxicity to *Daphnia magna*. *Sci. Rep.* **2015**, *5*, 9674. [CrossRef]

31. Wang, Q.D. *Phytotoxicity of Silver Nanoparticles to* Arabidopsis thaliana *in Hydroponic and Soil Systems*; Southern Illinois University at Carbondale: Carbondale, IL, USA, 2011.
32. Peng, X.F. The Effects of Silver Nanoparticles on the Vegetative Growth and Flowing of *Arabidopsis thaliana*. Master's Thesis, Zhejiang University of Technology, Hangzhou, China, 2013.
33. Wu, D.; Fan, W.; Kishen, A.; Gutmann, J.L.; Bing, F. Evaluation of the antibacterial efficacy of silver nanoparticles against *Enterococcus faecalis*, biofilm. *J. Endod.* **2014**, *40*, 285–290. [CrossRef] [PubMed]
34. Akaighe, N.; Depner, S.W.; Banerjee, S.; Gutmann, J.L.; Bing, F. The effects of monovalent and divalent cations on the stability of silver nanoparticles formed from direct reduction of silver ions by Suwannee River humic acid/natural organic matter. *Sci. Total Environ.* **2012**, *441*, 277–289. [CrossRef] [PubMed]
35. Dietz, K.J.; Herth, S. Plant nanotoxicology. *Trends Plant Sci.* **2011**, *16*, 582–589. [CrossRef] [PubMed]
36. Chen, W.H.; Xiao, W.B. On the development of gametophyte and sporophyte of *Ceratopteris*. *J. Hunan Agric. Univ.* **2008**, *34*, 306–310.
37. Dai, X.L.; Jin, Q.; Wang, Q.X. Studies on the development of gametophyte of *Ceratopteris thalictroides*. *Bull. Bot. Res.* **2005**, *25*, 274–276.
38. Zhan, Z.G.; Li, Y.J. Research progress on spore germination and gametophyte development of Water Fern. *North Hortic.* **2011**, *9*, 181–185.
39. Chen, X. Study on the Interaction between Nano-Silver Particles and *Chlamydomonas reinhardtii* in Water Environment. Master's Thesis, Shandong Agricultural University, Taian, China, 2019.
40. Becaro, A.A.; Jonsson, C.M.; Puti, F.C.; Siqueira, M.C. Toxicity of PVA-stabilized silver nanoparticles to algae and microcrustaceans. *Environ. Nanotechnol. Monit. Manag.* **2015**, *3*, 22–29. [CrossRef]
41. Liu, F.H.; Xu, W.Y.; Jiang, M.S.; Chen, C. The toxicity of nano-silver to 15 varieties plant pathogenic fungi. *Chin. Agric. Sci. Bull.* **2007**, *23*, 220–224.
42. Yu, P.; Liu, H.; Mou, Z.; Lu, Z.; Li, H.; Zhuang, D.; He, H. Acute Toxic Effects of Silver Nanoparticles and Silver Ion on Two Microalgae. *Asian J. Ecotoxicol.* **2017**, *12*, 188–198.
43. Pereira, S.; Jesus, F.; Aguiar, S.; Oliveira, R.D.; Nogueira, A. Phytotoxicity of silver nanoparticles to *Lemna minor*: Surface coating and exposure period-related effects. *Sci. Total Environ.* **2018**, *618*, 1389–1399. [CrossRef]
44. Pham, T.L. Toxicity of Silver Nanoparticles to Tropical Microalgae *Scenedesmus acuminatus*, *Chaetoceros gracilis* and Crustacean *Daphnia lumholtzi*. *Turk. J. Fish. Aquat. Sci.* **2019**, *19*, 1009–1016. [CrossRef]
45. Wang, Y. Neurotoxic Effects of Silver Nanoparticles on Early Life-Stage Development of Zebrafish (*Danio rerio*). Master's Thesis, Shanxi University, Taiyuan, China, 2019.
46. Lalau, C.M.; Simioni, C.; Vicentini, D.S.; Ouriques, L.C.; Matias, W. Toxicological effects of AgNPs on duckweed (*Landoltia punctata*). *Sci. Total Environ.* **2020**, *710*, 136318. [CrossRef] [PubMed]
47. Lu, M.T. Toxic Effects of Liposome-Encapsulated Silver (AgNPs, Ag^{+}) on *Daphnia magna*. Master's Thesis, China Mining University, Xuzhou, China, 2019.
48. Lv, Y. A Study on the Cytotoxicity of Two Types of Nanoparticles on Ciliated Protozoa *Pseudourostyla cristata*. Master's Thesis, East China Normal University, Shanghai, China, 2019.
49. Tan, C.L. The Biological Effect of Nanomaterials on Bacteria. Master's Thesis, Tianjin University of Technology, Tianjin, China, 2012.
50. Chasan, R. Ceratopteris: A model plant for the 90s. *Plant Cell* **1992**, *4*, 113–115. [CrossRef] [PubMed]
51. Tao, L. Effects of Herbicides on Sexual Reproduction of *Ceratopteris pteridoides* and *Vallisneria natans*. Master's Thesis, Graduate University of Chinese Academy of Sciences (Wuhan Botanical Garden), Wuhan, China, 2007.
52. Yuan, W.; Zhou, Y.; Liu, X.; Wang, J. New Perspective on the Nanoplastics Disrupting the Reproduction of an Endangered Fern in Artificial Freshwater. *Environ. Sci. Technol.* **2019**, *53*, 12715–12724. [CrossRef]
53. Austin, C.A.; Umbreit, T.H.; Brown, K.M.; Barber, D.S.; Dair, B.J.; Francke-Carroll, S.; Feswick, A.; Saint-Louis, M.A.; Hikawa, H.; Siebein, K.N. Distribution of silver nanoparticles in pregnant mice and developing embryos. *Nanotoxicology* **2012**, *6*, 912–922. [CrossRef]
54. Chatterjee, N.; Bhattacharjee, B. Revelation of ZnS Nanoparticles Induces Follicular Atresia and Apoptosis in the Ovarian Preovulatory Follicles in the Catfish *Mystus tengara* (Hamilton, 1822). *Scientifica* **2016**, *2016*, 3927340. [CrossRef]
55. Yeon-Jin, L.; Jong-Hye, C.; Pil-Je, K.; Kyung-Hee, C.; Su-Hyon, K.; Woo-Chan, S.; Kwang-Sik, P. A transfer of silver nanoparticles from pregnant rat to offspring. *Toxicol. Res.* **2012**, *28*, 139–141.
56. Liu, X.Q.; Zhang, H.F.; Zhang, W.D.; Zhang, P.F.; Hao, Y.N.; Song, R.; Li, L.; Feng, Y.N.; Hao, Z.H.; Shen, W. Regulation of neuroendocrine cells and neuron factors in the ovary by zinc oxide nanoparticles. *Toxicol. Lett.* **2016**, *256*, 19–32. [CrossRef]
57. Morishita, Y.; Yoshioka, Y.; Takimura, Y.; Shimizu, Y.; Namba, Y.; Nojiri, N.; Ishizaka, T.; Takao, K.; Yamashita, F.; Takuma, K. Distribution of silver nanoparticles to breast milk and their biological effects on breast-fed offspring mice. *ACS Nano* **2016**, *10*, 8180–8191. [CrossRef]
58. Tabatabaei, S.R.F.; Moshrefi, M.; Askaripour, M. Prenatal exposure to silver nanoparticles causes depression-like responses in mice. *Indian J. Pharm. Sci.* **2015**, *77*, 681.

59. Zhang, W.D.; Zhao, Y.; Zhang, H.F.; Wang, S.K.; Hao, Z.H.; Liu, J.; Yuan, Y.Q.; Zhang, P.F.; Yang, H.D.; Shen, W. Alteration of gene expression by zinc oxide nanoparticles or zinc sulfate in vivo and comparison with in vitro data: A harmonious case. *Theriogenology* **2016**, *86*, 850–861.e1. [CrossRef] [PubMed]
60. Gao, J.J.; Lin, L.; Wei, A.; Sepulveda, M.S. Protein Corona Analysis of Silver Nanoparticles Exposed to Fish Plasma. *Environ. Sci. Technol. Lett.* **2017**, *4*, 174–179. [CrossRef] [PubMed]
61. Ho, C.M.; Yau, S.K.W.; Lok, C.N.; So, M.H.; Che, C.M. Oxidative dissolution of silver nanoparticles by biologically relevant oxidants: A kinetic and mechanistic study. *Chem. Asian J.* **2010**, *5*, 285–293. [CrossRef] [PubMed]
62. Li, X.; Lenhart, J.J.; Walker, H.W. Dissolution-accompanied aggregation kinetics of silver nanoparticles. *Langmuir* **2010**, *26*, 16690–16698. [CrossRef]
63. Ma, R.; Levard, C.; Marinakos, S.M.; Cheng, Y.; Lowry, G.V. Size-controlled dissolution of organic-coated silver nanoparticles. *Environ. Sci. Technol.* **2012**, *46*, 752–759. [CrossRef]
64. Ferreira, L.A.; Dos Reis, S.B.; da Silva, E.D.N.; Cadore, S.; da Silva Bernardes, J.; Duran, N.; de Jesus, M.B. Thiol-antioxidants interfere with assessing silver nanoparticle cytotoxicity. *Nanomedicine* **2020**, *24*, 102130. [CrossRef]

nanomaterials

Article

Effect of Tube Diameters and Functional Groups on Adsorption and Suspension Behaviors of Carbon Nanotubes in Presence of Humic Acid

Mengyuan Fang [1,2,†], Tianhui Zhao [1,†], Xiaoli Zhao [1], Zhi Tang [1,*], Shasha Liu [3], Junyu Wang [1], Lin Niu [1] and Fengchang Wu [1]

1 State Key Laboratory of Environmental Criteria and Risk Assessment, Chinese Research Academy of Environmental Sciences, Beijing 100012, China; fangmengyuanfmy@163.com (M.F.); zth2512@163.com (T.Z.); zhaoxiaoli_zxl@126.com (X.Z.); jy.wong@foxmail.com (J.W.); 18233271321@163.com (L.N.); wufengchang@mail.gyig.ac.cn (F.W.)
2 College of Geoexploration Science and Technology, Jilin University, Changchun 130026, China
3 School of Energy and Environmental Engineering, University of Science and Technology Beijing, Beijing 100083, China; liushashajida@163.com
* Correspondence: tzwork@hotmail.com; Tel.: +86-10-8493-1804
† These authors contributed equally to this work.

Abstract: The adsorption and suspension behaviors of carbon nanotubes (CNTs) in the water environment determine the geochemical cycle and ecological risk of CNTs and the compounds attached to them. In this study, CNTs were selected as the research object, and the effect of tube diameters and functional groups (multiwall CNTs (MWNTs) and hydroxylated MWNTs (HMWNTs)) on the adsorption and suspension behaviors of the CNTs in the presence of humic acid (HA) was systematically analyzed. The results indicate that HA adsorption decreased with the increase in the solution pH, and the adsorption amount and rate were negatively correlated with the tube diameter of the CNTs. The surface hydroxylation of the CNTs prevented the adsorption of HA, and the maximum adsorption amounts on the MWNTs and HMWNTs were 195.95 and 74.74 mg g^{-1}, respectively. HA had an important effect on the suspension of the CNTs, especially for the surface hydroxylation, and the suspension of the CNTs increased with the increase in the tube diameter. The characteristics of the CNTs prior to and after adsorbing HA were characterized by transmission electron microscopy, fluorescence spectroscopy, Fourier-transform infrared spectroscopy and Raman spectroscopy. The results indicate that surface hydroxylation of the CNTs increased the adsorption of aromatic compounds, and that the CNTs with a smaller diameter and a larger specific surface area had a disordered carbon accumulation microstructure and many defects, where the adsorption of part of the HA would cover the defects on the CNTs' surface. Density functional theory (DFT) calculations demonstrated that HA was more easily adsorbed on the CNTs without surface hydroxylation. This investigation is helpful in providing a theoretical basis for the scientific management of the production and application of CNTs, and the scientific assessment of their geochemical cycle and ecological risk.

Keywords: carbon nanotubes; tube diameter; functional groups; adsorption; suspension/sedimentation; humic acid

check for updates

Citation: Fang, M.; Zhao, T.; Zhao, X.; Tang, Z.; Liu, S.; Wang, J.; Niu, L.; Wu, F. Effect of Tube Diameters and Functional Groups on Adsorption and Suspension Behaviors of Carbon Nanotubes in Presence of Humic Acid. *Nanomaterials* **2022**, *12*, 1592. https://doi.org/10.3390/nano12091592

Academic Editor: Zoltán Kónya

Received: 31 March 2022
Accepted: 4 May 2022
Published: 7 May 2022

1. Introduction

The invention and development of engineered nanomaterials have been aimed to fulfill their wide applications in product manufacturing and in providing a high quality of daily life. Significant attention has been paid to engineered nanomaterials due to their diverse structures and functions and complex environmental behavior, which may impose a significant impact on the ecological environment. Approximately one million tons of engineered nanomaterials is released into the ecosystem each year, while carbon nanotubes (CNTs) are one of the ten most used engineered materials [1]. CNTs are widely used in the fields of electronics, biomaterials, medicine, cosmetics, catalysis and environmental treatment due to their unique structures, excellent electrical conductivity and superior thermal and chemical stability [2–5], and they can be used as substitutes for the scarce metals for most technologies [6]. Studies have shown that CNTs are toxic to animals, microorganisms and embryonic stem cells [7,8]. CNTs may enter and accumulate in the human body through the respiratory tract, food chain and skin contact pathways and eventually pose threats to human health [7–9]. Therefore, it is necessary to reduce the release of CNTs into the environment, but the high aspect ratio, aromatic structure and size of CNTs make the degradation of CNTs very difficult [10]. Hence, more attention has to be paid to CNTs for their potential risk to human health and the ecological environment.

Once released into the aquatic environment, the environmental behaviors of CNTs such as adsorption, aggregation and sedimentation could directly affect their migration, transformation, fate and bioavailability [11–13], and they can be significantly influenced by the physical and chemical factors of the aquatic environment, humic acid (HA) and other compounds [11–13]. Previous studies demonstrated that the lower pH and higher ionic strength of the water environment were favorable for the adsorption of HA on CNTs' surfaces, but they were unfavorable for the CNTs' stabilization [14–17]. The adsorption of HA on the surface of CNTs and CNTs' suspension/sedimentation behaviors have been intensively studied [14–16]. HA plays a key role in the suspension and sedimentation of particles and colloids in the aquatic environment [18,19]. It can promote the dispersion, suspension and sedimentation of hydrophobic CNTs in the water via electrostatic repulsion, steric hindrance or solvation [11,16,18]. HA is a mixture of polyelectrolytes with different molecular weights. It contains a variety of functional groups, including carboxyl, hydroxyl, carbonyl, quinone and methoxy groups, resulting in both strong hydrophobicity and hydrophilicity [20]. HA can be adsorbed on the surface of CNTs through hydrogen bonding and hydrophobic, $\pi-\pi$ and electrostatic interactions [11,16,20,21]. The adsorption rate is affected by the oxidation degree, specific surface area and hydrophobic force of CNTs. Meanwhile, the intraparticle diffusion is also related to the adsorption but not as a sole rate-controlling step [22].

The adsorption and suspension/settlement behaviors of CNTs in the water environment are not only influenced by the physical and chemical elements of the water, but also their own physical and chemical properties. However, limited research has been reported about the effects of the tube diameters and functional groups of CNTs on the adsorption of HA and aggregation/sedimentation of CNTs. Previous studies have demonstrated that the saturated adsorption capacity of CNTs with different structures was different [11,13,23], and hydrophilic functional groups such as -COOH or -OH on the surface of CNTs could enhance the hydrophilicity and weaken the adsorption of hydrophobic organic molecules on CNTs, thus greatly increasing the dispersion of CNTs in the water. [15,16,24]. Meanwhile, the tube diameter of CNTs may directly influence the adsorption sites, the number of functional groups and the adsorption capacity of HA. Therefore, the type and number of functional groups on the CNTs' surface as well as the CNTs' tube diameter will affect both the adsorption of HA and the suspension/sedimentation behavior of CNTs, which ultimately influences their migration, transformation and fate in the aquatic environment [17,25,26], affecting their environmental geochemical fate.

In this study, multiwall carbon nanotubes (CNTs) were selected as the research object, the tube diameters and functional groups were used as internal factors and the solution pH, humic acid and other physical and chemical parameters were used as external factors. The adsorption, suspension and sedimentation behaviors of CNTs were systematically studied through adsorption isotherm and kinetic experiments. Density functional theory (DFT), fluorescence excitation–emission matrix (EEM) spectroscopy, Fourier-transform infrared (FTIR) spectroscopy and Raman spectroscopy were used to study the mechanisms of the adsorption, aggregation and sedimentation of CNTs. This investigation provides the theoretical basis for recognizing the environmental behaviors and potential ecological risks of CNTs with different diameters and functional groups in the natural aquatic environment.

2. Materials and Methodology

2.1. Materials and Reagents

All reagents were analytical grade. NaOH and HCl were purchased from Sinopharm Chemical Reagent Co., Ltd., (Shanghai, China). Multiwall CNTs (MWNTs) and hydroxylated MWNTs (HMWNTs) were purchased from Chengdu Organic Chemistry Co., Ltd., Chinese Academy of Sciences. MWNTs and HMWNTs with outer tube diameters of 4–6 nm (named MWNT-1 and HMWNT-1), 5–15 nm (named MWNT-2 and HMWNT-2) and 20–30 nm (named MWNT-3 and HMWNT-3) were used to study the impact of diameters on their environmental behaviors. Humic acid (HA) was purchased from the International Humic Substances Society (IHSS). The chemical compositions, functional groups and results of the NMR analysis of HA (Elliott Soil, 1S102H) are presented in Table 1 [27]. Deionized water (DI water) was supplied by an ultra-pure water system (Milli-Q Advantage System, Millipore, Boston, MA, USA). The water resistance value was $\geq$18.3 MΩ·cm, and the conductivity was $\leq$10 us·cm^{-1}.

Table 1. The elemental composition, functional groups and NMR analysis of HA.

			Carbon Distribution (mg L^{-1})					
Sample	Carbonyl 220–190	Carboxyl 190–165	Aromatic 165–110	Acetal 110–90	Hetero Aliphatic 90–60		Aliphatic 60–0	Aromatic/Aliphatic
HA	6	18	50	4	6		16	3.125
			Element Constitution %(w·w^{-1})					
	H_2O	Ash	C	H	O	N	S	P
HA	8.2	0.88	58.13	3.68	34.08	4.14	0.44	0.24
			Acid Functional Groups(m mol·g^{-1})					
	Carboxyl	Phenolic	Q_1	$LogK_1$	N_1	Q_2	$LogK_2$	N_2
HA	8.28	1.87	8.90	4.36	3.16	0.85	9.80	1.00

Notes: The data were sourced from the International Humic Substances Society (IHSS). Q_1 and Q_2 are the maximum charge densities of the two classes of binding sites; LogK1 and LogK2 are the mean logK values for proton binding by the two classes of sites; N is the number of fitted titration data points.

2.2. Preparation of HA Stock Solution and Determination of Standard Curve for HA Concentration

HA was dissolved in 0.1 M NaOH solution. Then, the pH of the solution was adjusted to 7.9 ± 0.2 with 0.1 M HCl. The solution was kept in a shaker at room temperature (22 ± 1 °C) for 24 h to dissolve completely and filtered through a 0.45 μm fiber membrane (MF Cat No: HAWP04700) prior to use. HA solutions with concentrations between 5.0 and 80.0 mg L^{-1} were prepared by diluting the HA stock solution. The concentration of the HA solution was measured using a UV–visible spectrophotometer (UV-vis 8453, Agilent, Palo Alto, CA, USA)with a 1 cm-light-path quartz cuvette. The light absorbance at a fixed wavelength of 254 nm was used for establishing the HA calibration curve.

2.3. Characterization of CNTs

Transmission electron microscopy (TEM; H-7500 Hitachi, Tokyo, Japan) was used to observe the morphology, tube diameter and length of CNTs and CNTs-HA. TEM images were recorded on an H7500 transmission electron micrograph (Hitachi, Tokyo, Japan) operated at 120 kV. All CNT samples used for TEM testing were prepared by cool drying of the solution containing suspended CNTs. Zeta potentials of CNTs were measured at various pH values using a Nano-ZS90 Zetasizer (Malvern Instruments, Worcestershire, UK).

2.4. Effect of Solution pH

An amount of 4.0 mg of CNTs was mixed with 20 mL of DI water in a 100 mL polycarbonate bottle aided by sonication for 30 min (240 W, 100 kHz). The temperature was maintained at 22 ± 1 °C using cooling water. Then, 20.0 mL of HA solution (40.0 mg L^{-1}) was added, while the solution pH was adjusted in the range of 2.0 to 10.0 using 0.1 M HCl or NaOH. The mixture was stirred on a rotary shaker for 48 h. Then, the supernatant was filtered through a 0.2 μm fiber membrane (PALL REF 4612) for the absorbance measurement using a UV–Vis spectrophotometer (UV-vis 8453, Agilent, Palo Alto, CA, USA) (manufacturer, city, (State or Province), country) at 254 nm. The experiment was repeated three times.

2.5. Adsorption Isotherm

The sonicated CNT suspension solution was prepared as described in Section 2.4. Then, 20.0 mL of HA solution (200.0 mg L^{-1}) was added to achieve final HA concentrations of 5.0, 10.0, 15.0, 20.0, 30.0, 40.0 and 50.0 mg L^{-1}. The solution pH was adjusted to 6.0 using 0.1 M HCl. The mixture was stirred in a rotary shaker for 48 h. Then, the supernatant was filtered through a 0.2 μm fiber membrane (PALL REF 4612) for the absorbance measurement using a UV–Vis spectrophotometer at 254 nm. The adsorption isotherms were fitted using Langmuir and Freundlich adsorption isotherm models. The experiment was repeated three times.

2.6. Adsorption Kinetics

An identical mixture solution was prepared as described in Section 2.4, and the initial concentration of HA was 20.0 mg L^{-1}. The solution pH was adjusted to 6.0 using 0.1 M HCl. The solution was then shaken for 5 min, 30 min, 60 min, 2 h, 4 h, 6 h, 12 h, 24 h, 48 h, 72 h, 96 h and 120 h. Then, the supernatant was filtered through a 0.2 μm fiber membrane (PALL REF 4612) for the absorbance measurement using a UV–Vis spectrophotometer at 254 nm. The experiment was repeated three times.

2.7. Sedimentation Test of CNTs

The sonicated CNT suspension solution was prepared as described in Section 2.4. Then, 20 mL of HA solution (200.0 mg L^{-1}) was added to achieve final HA concentrations of 5.0, 10.0, 20.0, 30.0, 40.0, 50.0 and 60.0 mg L^{-1}. The solution pH was adjusted to 6.0 using 0.1 M HCl. After shaking for 48 h, the sedimentation of the suspensions was measured with a UV–visible spectrophotometer at 800 nm [14,25]. The ratio of the absorbance at different times (C_e) to the initial absorbance (C_0) as a function of time was used to determine the sedimentation dynamics. A lower C_e/C_0 ratio indicates a higher sedimentation performance of the nanotubes and, in turn, much easier aggregation and settlement of the CNTs [28].

2.8. Fluorescence Spectral Analysis

The initial concentration of HA was 20.0 mg L^{-1}, and the fluorescence spectra of the remaining HA after being adsorbed by CNTs at 24 h, 48 h and 72 h were measured and compared with the fluorescence spectra of the initial HA. Fluorescence spectra of HA were measured using a fluorescence spectrometer (Hitachi F-7000, Tokyo, Japan) with a 1 cm-path-length quartz cuvette at room temperature. EEM spectra were obtained by subsequently scanning emission (Em) wavelengths from 230 to 600 nm and excitation (Ex) wavelengths from 200 to 450 nm, both stepped by 5 nm intervals. Slit widths were 5 nm for both Ex and Em, and the scanning speed was set at 12000 nm·min^{-1}. The fluorescence index (FI) was calculated as the ratio of the emission intensity at Em 450 nm relative to that at Em 500 nm at Ex 370 nm [29].

2.9. Raman and FTIR Spectroscopy Characterization of CNTs

A micro-Raman imaging spectrometer (DXRxi, Thermo Scientific, Waltham, MA, USA) was used to characterize the CNTs and CNTs-HA. The microscope was equipped with 10× and 50× objectives. During the measurement, a layer of CNTs with a thickness of 2 cm was placed on a glass slide mounted horizontally inside the test chamber. The orientation of the sample stage was adjusted so that the laser spot scanned in parallel. The excitation source had a wavelength of 532 nm. The focused spot size was about 1 μm. The spectral resolution was 1 μm. The optical path length was corrected by the Raman peak intensity at 520 cm^{-1} from a silicon wafer with an excitation power of 2 mW and an exposure time of 10 s. The test results were analyzed and processed with Omnic software with baseline correction and Lorentz peak fitting for peak deconvolution. FTIR spectrometry of CNTs and CNTs-HA was carried out using a Nicolet Magna-IR 750 FTIR spectrometer (Nicolet Magna-IR 750, Nicolet, Madison, WI, USA) using KBr powder as the background. FTIR spectra were recorded from 400 to 4000 cm^{-1} at a resolution of 4 cm^{-1} and averaged over 200 scans.

2.10. Atomic Adsorption Theory Analysis Based on Density Functional Theory (DFT) Calculation

Density function theory calculations were performed using the CP2K package [30]. The PBE functional [31] with Grimme D3 correction [32] was used to describe the system. Unrestricted Kohn–Sham DFT was used as the electronic structure method in the framework of the Gaussian and plane wave methods [33,34]. The Goedecker–Teter–Hutter (GTH) pseudopotentials [35,36] and DZVPMOLOPT-GTH basis sets [33] were utilized to describe the molecules. A plane-wave energy cut-off of 500 Ry was employed.

3. Results and Analysis

3.1. Characterization of CNTs

The TEM images of the MWNTs (Figure 1a–c) and HMWNTs (Figure 1d–f) prior to and after adsorbing HA (Figure 1g–l) are shown in Figure 1. The observed outer tube diameters of the CNTs were consistent with the data given by the manufacturer. Additionally, it can be seen that the MWNTs and HMWNTs exhibited the characteristic of flexible winding, and the winding property of the MWNTs and HMWNTs increased significantly after the adsorption of HA. This result is different from that of nanoparticles, which have better dispersion after HA adsorption [14,26]. Previous research indicated that the tube segments of CNTs are usually closed, and the inner diameter of MWNTs and HMWNTs is too small to allow large molecules such as HA to enter; hence, HA can only be adsorbed on the outer surface of MWNTs and HMWNTs [16,17].

The specific surface areas (BET) after HA adsorption of MWNT-1, MWNT-2, MWNT-3, HMWNT-1, HMWNT-2 and HMWNT-3 were 342.4, 213.8, 129.5, 471.6, 220.4 and 135.1 m^2 g^{-1}, respectively (Table S1). For the CNTs with different surface functional groups, their BET also increased with the decrease in the tube diameter. After the adsorption of HA, the BET of MWNTs and HMWNTs decreased significantly; the reduction rates of MWNTs (19.9–30.9%) were higher than those of HMWNTs (12.0–25.7%), indicating that the adsorption capacity of MWNTs for HA was larger than that of HMWNTs; and HA occupied more adsorption sites of MWNTs. The nitrogen adsorption isotherms for the MWNTs and HMWNTs are shown in Figure S1.

The effect of the solution pH on the changes in the surface charge property and density of the MWNTs and HMWNTs are shown in Figure 2. The MWNTs and HMWNTs had a negative surface charge under different pH conditions, and only MWNT-1 and MWNT-2 had small positive charges at pH < 3.0. With the increase in pH, more OH^- adsorbed on the surface of MWNT-1 and MWNT-2, and the positive charge density decreased, resulting in zero surface charge at pH 3.0. The surface negative charge density of MWNT-1 and MWNT-2 increased gradually with the increase in the solution pH. The zeta potentials of the MWNT-3 and HMWNT surfaces were negative. With the increase in pH, the negative value of the zeta potential increased. The results indicate that the surface functional groups and tube diameters of the CNTs also affected the property and density of the surface charge. For different surface functional groups, the surface negative charge density of the MWNTs and HMWNTs also increased with the increase in the solution pH and tube diameter.

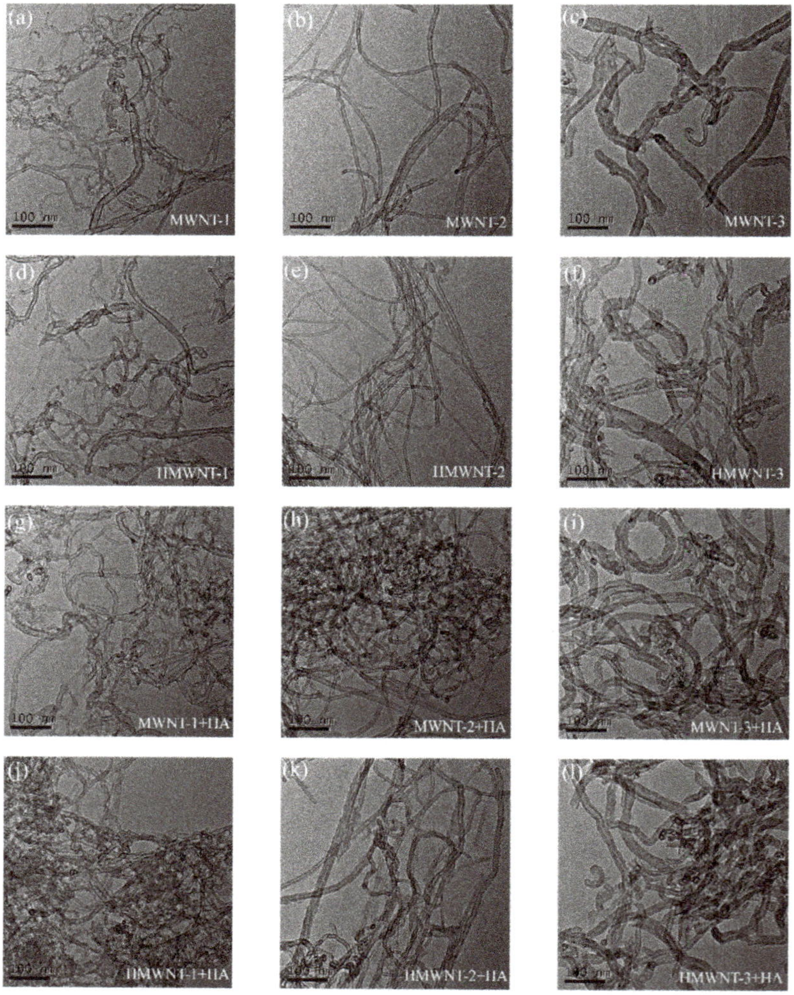

Figure 1. TEM images of MWNTs (**a–c**,**g–i**) and HMWNTs (**d–f**,**j–l**) prior to and after HA adsorption.

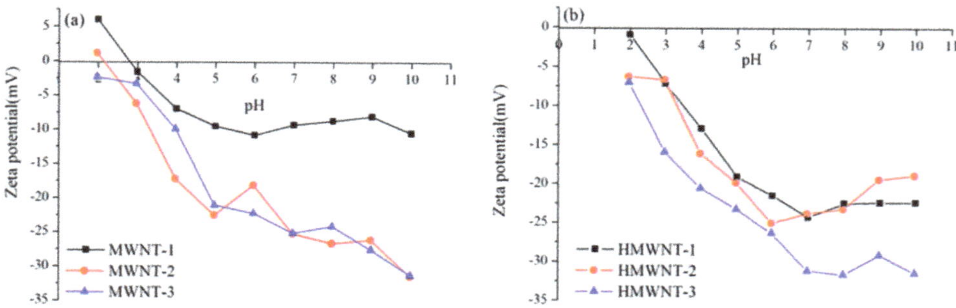

Figure 2. Effect of pH on the surface charge property and density of MWNTs (**a**) and HMWNTs (**b**).

3.2. *Effect of pH on the Adsorption of HA*

The adsorption amount of HA on the surface of the MWNTs and HMWNTs decreased significantly with the increase in the solution pH (Figure 3); especially for the MWNTs and HMWNTs with a larger diameter, HA could hardly be adsorbed on their surface when the solution pH was greater than 7.0, and the maximum adsorption amount of HA was achieved at the lowest pH of 2.0. The adsorption amount of HA on the surface of the MWNTs was slightly higher than that of the HMWNTs, consistent with the above results of BET. For the MWNTs, the maximum adsorption amount of HA decreased with the increase in the tube diameter following the order of MWNT-1 (186.4 mg g^{-1}) > MWNT-2 (181.7 mg g^{-1}) > MWNT-3 (127.4 mg g^{-1}). Similar results were also found for the HMWNTs, with the adsorption amount decreasing in the order of HMWNT-1 (181.1 mg g^{-1}) $\approx$ HMWNT-2 (182.6 mg g^{-1}) > HMWNT-3 (120.8 mg g^{-1}). Therefore, the adsorption amount of HA was greater with smaller tube diameters regardless of the functional groups on the surface of the CNTs. Previous studies demonstrated that the solution pH not only affected the property and density of the surface charge of CNTs, but also the dissociation of HA in the solution [14,16]. Therefore, interactions between the surface functional groups of CNTs and the hydrogen bonds and polar functional groups of HA were inhibited, thus reducing the adsorption amount of HA [14]. The results indicate that the surface functional groups had no significant effect on the adsorption of HA on the CNTs at the low solution pH, and the surface hydroxylation hindered the adsorption of HA at the high solution pH.

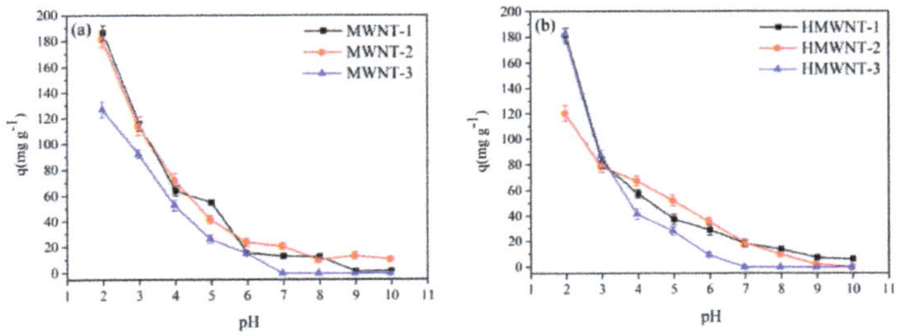

Figure 3. The effect of the solution pH on HA adsorption on the surfaces of MWNTs (**a**) and HMWNTs (**b**).

After the adsorption of HA, the zeta potentials of the MWNTs and HMWNTs decreased first and then increased (Figure 4). This phenomenon can be explained from two aspects: on the one hand, HA was adsorbed on the surface of the CNTs at a low solution pH and then increased the negative charge density of the CNTs due to its negative charge in the solution; on the other hand, under the condition of a high solution pH, HA was difficult to adsorb on the surfaces of the MWNTs and HMWNTs, and the zeta potentials of the MWNTs and HMWNTs were only affected by the solution pH.

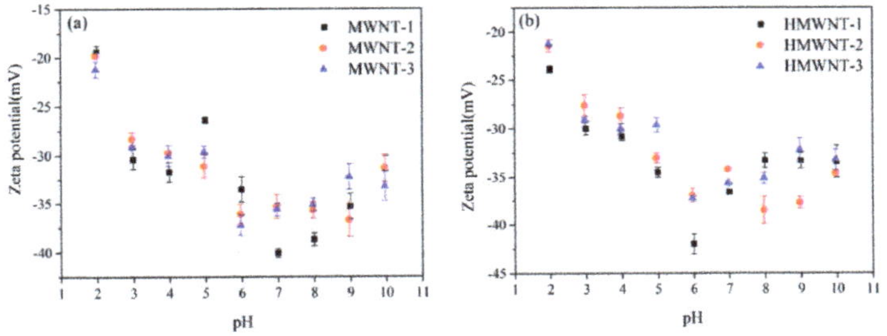

Figure 4. Zeta potential changes after HA adsorption of MWNTs (**a**) and HMWNTs (**b**) at different solution pH values.

3.3. Adsorption Isotherms

The adsorption of HA on the MWNTs and HMWNTs was fitted using both Langmuir (Equation (1)) and Freundlich (Equation (2)) [37,38] adsorption models (Figure 5).

$$q_e = \frac{q_m k_L C_e}{1 + k_L C_e} \quad (1)$$

$$q_e = k_F C_e^{\frac{1}{n}} \quad (2)$$

where q_e is the amount of HA adsorbed at equilibrium (mg g^{-1}); C_e is the concentration of HA in the solution (mg L^{-1}) at equilibrium; q_m (mg g^{-1}) is the maximum adsorption capacity; k_L (L g^{-1}) is the Langmuir equilibrium constant; k_F ($mg^{1-(1/n)}L^{1/n}$ g^{-1}) and n are the Freundlich parameters.

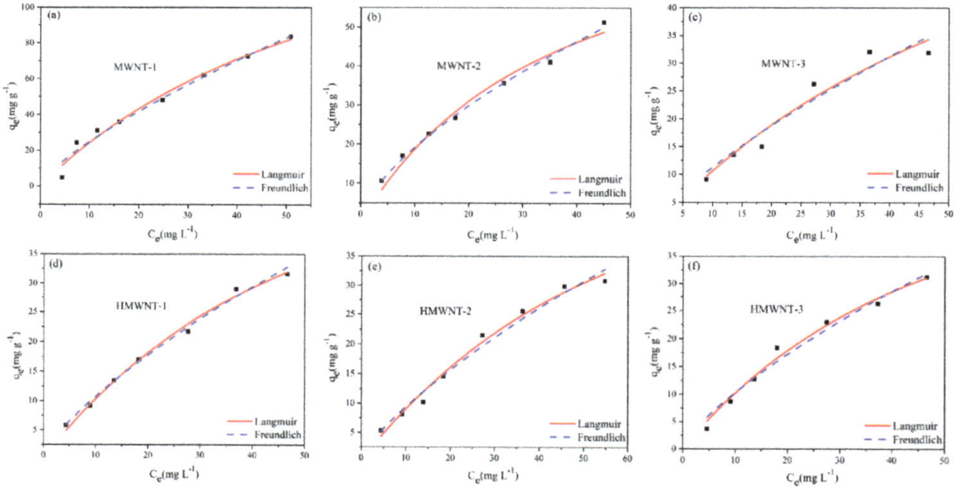

Figure 5. Langmuir and Freundlich adsorption isotherms for HA adsorption on MWNTs (**a**–**c**) and HMWNTs (**d**–**f**).

The Langmuir and Freundlich adsorption isotherms of HA on the MWNT and HMWNT surfaces were obtained by fitting Equations (1) and (2) (Table 2). In general, most of the isotherms achieved a good-quality fitting. Based on the Langmuir isotherm, the maximum adsorption amount of HA on the surface of MWNT-1 was 195.95 mg g^{-1} at pH 6.0. However, under the same conditions, this was reduced to 89.73 and 87.99 mg g^{-1} for MWNT-2 and MWNT-3, respectively. Therefore, the MWNT with the smaller tube diameter and the larger specific surface area had the highest adsorption capacity for HA. The maximum adsorption amounts of HA on the surfaces of HMWNT-1, HMWNT-2 and HMWNT-3 were 74.74, 75.16 and 69.81 mg g^{-1}, respectively. The surface hydroxylation of the HMWNTs greatly reduced the adsorption amount of HA, and the effect of the tube diameter on the adsorption of HA on the HMWNTs was not as significant as that on the MWNTs. The adsorption amount of HA on the MWNTs was larger than that on the HMWNTs, which could be attributed to the fact that the hydroxyl groups in the HMWNTs occupied a part of the adsorption sites, so the surface hydroxylation of the MWNTs may inhibit the adsorption of HA. The constant "*n*" for the strength of adsorption predicted using the Freundlich equation was less than 2.0, indicating that the MWNTs and HMWNTs had a weak adsorption capacity for HA. This is understandable, since at pH 6.0, the surfaces of the MWNTs and HMWNTs were negatively charged, and a part of the HA can be adsorbed on the surfaces of the MWNTs and HMWNTs due to electrostatic repulsion.

Table 2. Langmuir and Freundlich adsorption isotherm constant of HA on MWNTs and HMWNTs.

Samples	Langmuir			Freundlich		
	q_m (mg g^{-1})	K_L (L mg^{-1})	R^2	K_F (mg$^{1-(1/n)}$L$^{1/n}$g^{-1})	n	R^2
MWNT-1	195.95	0.0142	0.975	4.50	1.341	0.974
MWNT-2	89.73	0.0264	0.977	4.48	1.578	0.995
MWNT-3	87.99	0.0137	0.934	2.08	1.362	0.998
HMWNT-1	74.74	0.0161	0.991	2.04	1.383	0.996
HMWNT-2	75.16	0.0136	0.985	1.74	1.362	0.977
HMWNT-3	69.81	0.0172	0.985	1.92	1.368	0.971

To better understand the adsorption behavior of HA on the surfaces of the MWNTs and HMWNTs, Temkin (Equation (S1)) (Figure S2) and Dubinin–Radushkevich (Equations (S2) and (S3)) [37,38] (Figure S3) adsorption isotherms were used to analyze the adsorption of HA. The results of the Temkin and Dubinin–Radushkevich adsorption isotherms are listed in Table S2. Correlation coefficient calculations showed that the Temkin isotherm represents the equilibrium data of 298 K well. The results indicate that the binding energy of HA was evenly distributed at 298 K. Compared with the data in Table 2, the order of the fitting results of each model was $R^2_L > R^2_F > R^2_T > R^2_{D\text{-}R}$. This indicates that the adsorption of HA on the MWNTs and HMWNTs was mainly monolayer adsorption [39].

3.4. Adsorption Kinetics

The tube diameter may influence the adsorption pathway for CNTs, where a smaller tube diameter of CNTs offers a shorter adsorption pathway. Therefore, the study of the adsorption kinetics is helpful for a better understanding of the adsorption of HA on the surfaces of CNTs with different tube diameters and functional groups. As shown in Figure 6, the adsorption amount of HA on the MWNTs was higher than that on the HMWNTs at pH = 6.0, and the adsorption equilibrium times of HA on the surfaces of the MWNTs and HMWNTs were consistent. The adsorption amount of HA increased rapidly in the first 24 h, which slowed down gradually after 24 h until reaching adsorption equilibrium at about 48 h. For the same surface functional groups of the CNTs, the adsorption amount of HA decreased with the increase in the tube diameter, consistent with the result of the adsorption isotherm fitting. Therefore, the CNTs with a larger specific surface area and a shorter adsorption path had a greater adsorption rate of HA.

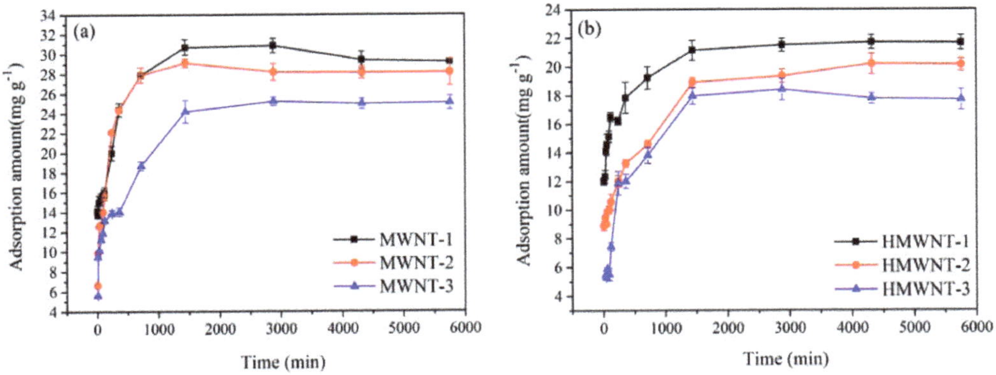

Figure 6. HA adsorption on MWNT (**a**) and HMWNT (**b**) surfaces under different equilibrium times.

In order to quantify the time-dependent variation in the HA adsorption on the MWNT and HMWNT surfaces, pseudo-first-order (Equation (3)) (Figure S4) and pseudo-second-order (Equation (4)) (Figure 7) kinetic models were used for fitting the time-dependent profiles [37,38]. The adsorption kinetic parameters of HA can be calculated by measuring the adsorption rate of HA on MWNTs and HMWNTs.

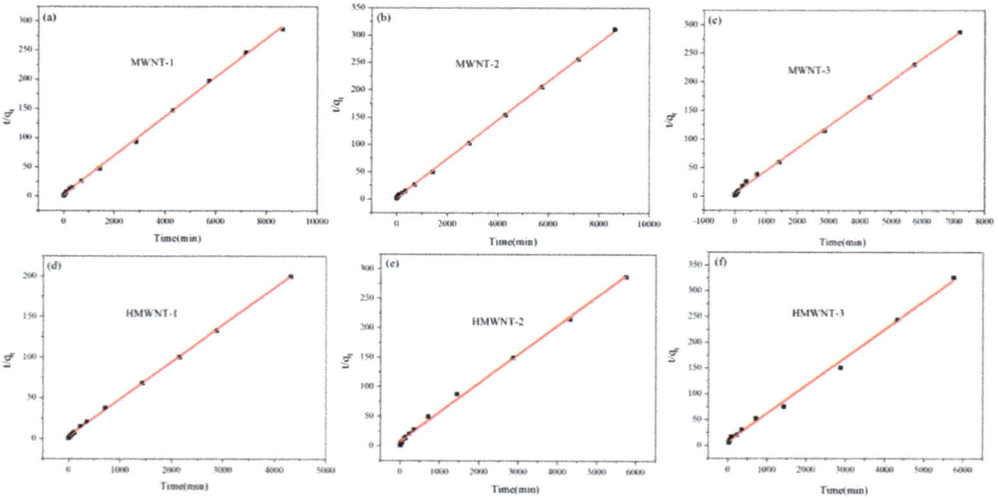

Figure 7. Pseudo-second-order kinetic curves of the adsorption of HA on the surface of MWNTs (**a–c**) and HMWNTs (**d–f**).

Pseudo-first-order model:

$$q_t = q_e\left(1 - e^{-k_1 t}\right) \tag{3}$$

Pseudo-second-order model:

$$\frac{t}{q_t} = \frac{1}{k_2 {q_e}^2} + \frac{1}{q_e}t \tag{4}$$

where k_1 (min^{-1}) and k_2 ($g\ mg^{-1}\ min^{-1}$) are the first-order and second-order adsorption rate constants, q_t ($mg\ g^{-1}$) is the amount of HA adsorbed by the CNTs at time t and q_e ($mg\ g^{-1}$) is the adsorption capacity at adsorption equilibrium. The initial adsorption rate h_0 (mg/g/min) can be defined as follows (Equation (5)):

$$h_0 = k_2 {q_e}^2 (t \to 0) \tag{5}$$

Both k_2 and h_0 can be determined experimentally by plotting t/q_t against t.

The intraparticle diffusion rate was obtained from the plots of q_t versus $t^{1/2}$. The rate parameter for intraparticle diffusion was determined using the following Equation (6) [40]:

$$q_t = k_{int} t^{\frac{1}{2}} + C \tag{6}$$

where C is the intercept, and k_{int} is the intraparticle diffusion rate constant ($mg\ g^{-1}\ min^{-1/2}$).

From Figure S5, the regression of q_t versus $t^{1/2}$ was linear, but none of the straight lines passed through the origin, indicating that the intraparticle diffusion was not the only rate-controlling step.

The adsorption kinetics of HA on the surface of the MWNTs and HMWNTs were in accordance with the second-order kinetic model (R^2 = 0.996–0.999) (Table 3), which indicates that chemical adsorption was the rate-limiting step for HA absorbed on the surface of the MWNTs and HMWNTs. The adsorption kinetic analysis results also demonstrate that the adsorption rates of HA on the MWNT-1 and MWNT-2 surfaces were similar, but much higher than that of MWNT-3. The adsorption rate constant and maximum adsorption amount of HA on HMWNT-1 were also much higher than those of other HMWNTs with a larger tube diameter. By comparing the adsorption rates of HA on the MWNTs with different surface functional groups, it can be seen that surface hydroxylation reduced not only the adsorption amount of HA, but also the adsorption rate. Therefore, the surface functional groups and tube diameters of the CNTs also played a significant role in the adsorption of HA, and the specific surface area, tube diameter and interaction between the surface functional groups and HA synergistically affected the adsorption rate and adsorption capacity of HA.

Table 3. Adsorption kinetic parameters of HA on MWNTs and HMWNTs.

Samples	Pseudo-First-Order Models			Pseudo-Second-Order Models				Intraparticle Diffusion Equation		
	k_1 (min^{-1})	q_e ($mg\ g^{-1}$)	R^2	k_2 ($g\ mg^{-1}\ min^{-1}$)	q_e ($mg\ g^{-1}$)	h_0 ($mg\ g^{-1}\ min^{-1}$)	R^2	k_{int} ($mg\ g^{-1}\ min^{-1/2}$)	C ($mg\ g^{-1}$)	R^2
MWNT-1	0.0111	28.28	0.364	6.54×10^{-4}	30.30	0.612	0.999	0.5329	11.907	0.945
MWNT-2	0.0076	22.65	0.687	6.54×10^{-4}	30.29	0.603	0.999	0.6508	8.476	0.881
MWNT-3	0.0092	27.60	0.831	3.19×10^{-4}	25.64	0.214	0.999	0.4424	7.095	0.944
HMWNT-1	0.1258	18.11	0.210	10.65×10^{-4}	21.88	0.510	0.999	0.2479	12.478	0.929
HMWNT-2	0.0124	16.50	0.199	3.61×10^{-4}	20.45	0.151	0.997	0.2256	8.289	0.973
HMWNTs-3	0.0045	17.60	0.886	4.35×10^{-4}	18.45	0.148	0.996	0.4662	2.203	0.905

The values of C and k_{int} are presented in Table 3. The values of k_{int} for the MWNTs were generally higher than those for the HMWNTs, revealing that the adsorption rate of HA on MWNTs was higher. This result is consistent with the conclusion of the second-order kinetic model. The values of C were proportional to the extent of the boundary layer thickness, that is, the larger the intercept, the greater the boundary layer effect [40]. The values of C decreased with the increasing tube diameters of the MWNTs and HMWNTs, indicating that the CNTs with a smaller outer tube diameter had a larger initial adsorption amount. The C values of MWNT-1 (11.907 mg g^{-1}) and HMWNT-1 (12.478 mg g^{-1}), and MWNT-2 (8.476 mg g^{-1}) and HMWNT-2 (8.289 mg g^{-1}), were close. The results suggest that hydroxylation had little effect on the initial adsorption amount of HA on the CNTs with a smaller outer tube diameter [41]. For the MWNTs with larger outer tube diameters, hydroxylation had a greater effect on the initial adsorption.

3.5. Effect of HA on CNT Suspension/Sedimentation

In the water environment, the suspension and sedimentation behaviors of nanomaterials greatly affect their migration, transformation, fate and ecological effects. Therefore, the study of the suspension and sedimentation behaviors of MWNTs and HMWNTs with different tube diameters and surface functional groups is of great significance for understanding the ecological risks of CNTs. The effects of the HA concentration on the suspension/sedimentation behaviors of the MWNTs and HMWNTs are shown in Figure 8. In the absence of HA, the suspension of the MWNTs was less stable than that of the HMWNTs with the same tube diameter, which was mainly due to the higher surface negative charge density. The effect of the tube diameter on the suspension stability of the MWNTs and HMWNTs was similar, and the CNTs with a smaller tube diameter were more likely to aggregate and settle. The aggregation of the MWNTs and HMWNTs increased their effective hydraulic diameter with reduced surface tension and intensified sedimentation. Thus, the CNTs with a smaller tube diameter were more likely to aggregate and sediment in the same water environment.

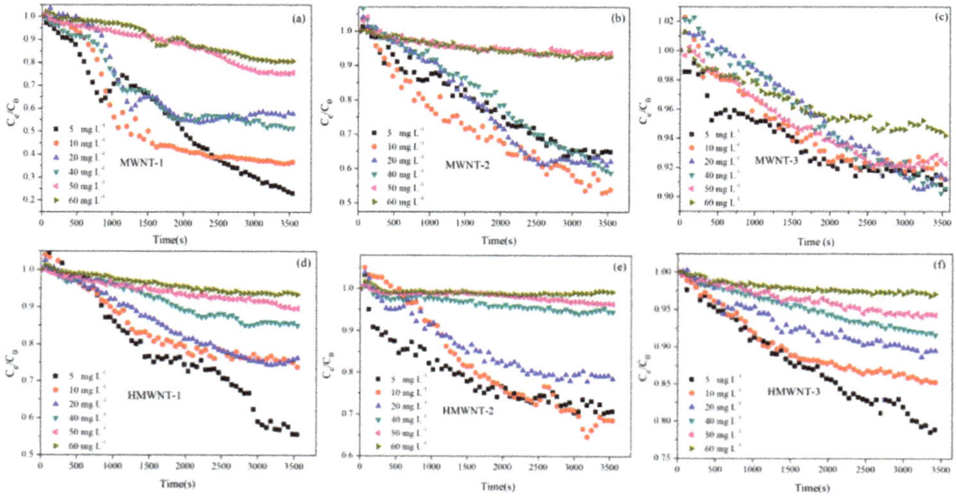

Figure 8. Effect of HA adsorption on sedimentation behaviors of MWNTs (**a–c**) and HMWNTs (**d–f**).

In the presence of HA, the suspension of the CNTs significantly improved with the increase in the HA concentration. Under the condition of pH = 6.0, the surface of the CNTs was negatively charged, and HA was adsorbed on the surface of the CNTs through hydrophobic and π–π interactions [24,26]. The adsorption of HA had two effects on the suspension of the CNTs. On the one hand, HA increased the surface negative charge density of the CNTs, which increased the electrostatic repulsion between the CNTs. On the other hand, the adsorption of HA increased the steric hindrance between the CNTs. The concentration of HA greatly affected the suspension behavior of the CNTs, and the suspension performance of the CNTs with different surface functional groups and tube diameters increased with the increase in the HA concentration. For different surface functional groups, the adsorption of HA had a greater effect on the suspension of the CNTs due to the higher HA adsorption amount on the CNTs. For different tube diameters, the effect of HA on the suspension performance of the CNTs with a smaller tube diameter was higher than that of the CNTs with a larger tube diameter. Therefore, HA significantly promoted the suspension of the MWNTs and HMWNTs, and the adsorption amount of HA was an important factor. CNTs are more easily suspended in the aquatic environment with a higher concentration of natural organic matter, which leads to a greater impact on the carbon cycle and a higher ecological risk.

3.6. Fluorescence Spectral Analysis

The fluorescence spectra of HA prior to and after being adsorbed on the surfaces of the MWNTs and HMWNTs are shown in Figure S6. The characteristic peak of each spectrum at Ex/Em = 275 nm/510 nm (abbreviated as peak A) is very obvious and has a wide spectrum coverage. Peak A of HA significantly reduced after the adsorption, which indicates that HA was successfully adsorbed on the surface of the MWNTs and HMWNTs. The low fluorescence index (FI) values are rich in aromatic moieties [29,42,43]. The FIs of the residual HA after being adsorbed on the CNTs with different surface functional groups and tube diameters were compared (Figure 9). With the increase in the adsorption time, the FI of the residual HA decreased, indicating that the aromatic groups of HA were more difficult adsorb on the surface of the CNTs than other parts, especially for the MWNTs. Therefore, the surface hydroxylation of the CNTs may have increased the adsorption of aromatic compounds of HA. Meanwhile, the FI of the residual HA increased with the increase in the tube diameter of the CNTs, thus the MWNTs and HMWNTs with larger tube diameters were more likely to adsorb the aromatic groups of HA. Therefore, the tube diameters and surface functional groups of CNTs not only affect the adsorption amount of HA, but also the components of HA.

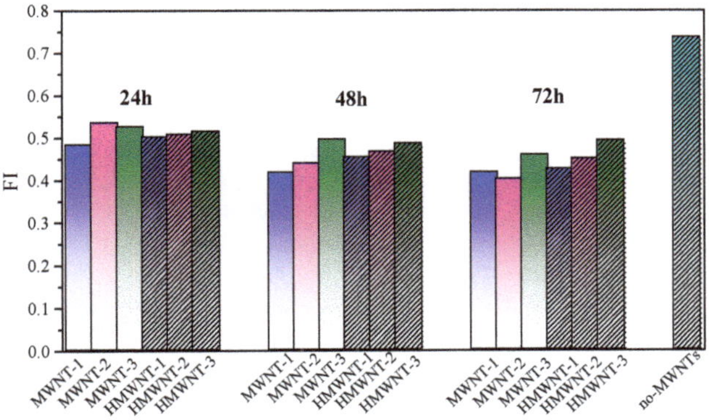

Figure 9. Changes in the florescence index (FI) for HA after adsorption on MWNTs and HMWNTs.

3.7. Raman Spectroscopy and FTIR Study

The Raman spectra of the MWNTs and HMWNTs prior to and after adsorbing HA are shown in Figure 10. After HA was adsorbed on the surfaces of the MWNTs and HMWNTs, there were no significant changes in the positions of the CNTs' characteristic peaks. The peaks at 1570 cm^{-1} (G peak) and 1340 cm^{-1} (D peak) represent ordered and disordered carbon stacking microstructures within the CNTs, respectively. Previous studies indicated that HA also contributed, with Raman peaks at 1379 and 1590 cm^{-1} originating from the symmetric vibration of the carboxyl groups, C=C bond vibration of the carboxyl group in the aromatic group, C-O vibration of the phenolic group and phenol vibration of HA [44]. As the characteristic peaks of the CNTs and HA overlapped to some degree, no significant changes were found in the Raman spectra of the MWNTs and HMWNTs prior to and after HA adsorption. However, it can be seen that the intensities of the characteristic peak significantly increased after HA was adsorbed on the surfaces of the MWNTs and HMWNTs, which indirectly proves the adsorption of HA.

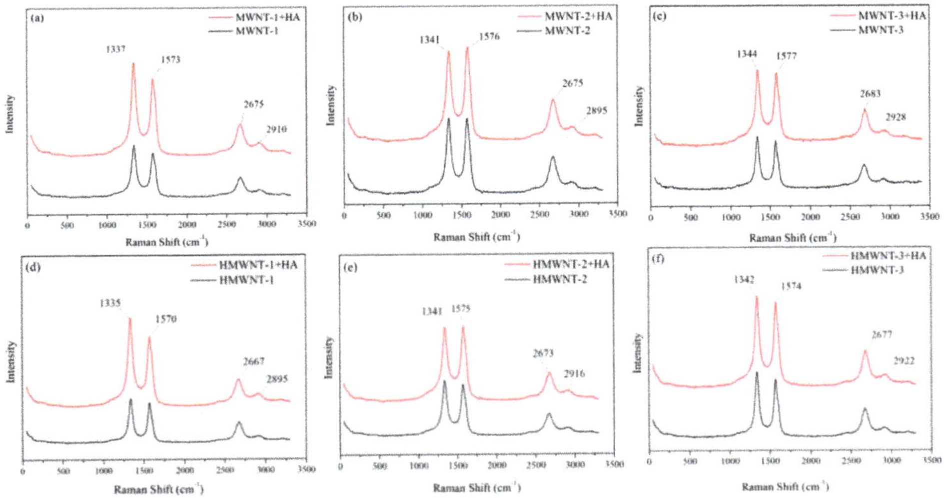

Figure 10. Raman spectra of MWNTs (**a**–**c**) and HMWNTs (**d**–**f**) prior to and after HA adsorption.

The ratio of the D-band to G-band intensities (I_D/I_G) can be used to characterize the degree of defects of CNTs. A higher I_D/I_G value indicates more defects on CNTs [45,46]. The I_D/I_G values of MWNT-1 and HMWNT-1 were 1.17 and 1.10 (Table 4), respectively, indicating that the MWNTs had a more disordered structure, and the corresponding specific surface area was also larger. After the adsorption of HA, the Raman peak intensities of MWNT-1 and HMWNT-1 had the largest increase, accompanied by the increase in I_D/I_G. After the adsorption of HA, the intensity of the characteristic peak in the Raman spectrum increased the most, and the value of I_D/I_G also increased, which indicates that the CNTs with the smallest diameter had the largest adsorption capacity for HA. The reason for this result may be the coincidence of the characteristic peaks of the CNTs and HA. The I_D/I_G values of MWNT-2 and MWNT-3 were smaller than that of MWNT-1, indicating fewer defects and a smaller specific surface area. The I_D/I_G values of HMWNT-2 and HMWNT-3 were very different from that of HMWNT-1, in agreement with the above results of the adsorption isotherms. After HA was adsorbed on the HMWNTs, the value of I_D/I_G decreased due to HA occupying the defect sites. In addition to the G and D peaks, there were two weak Raman peaks, namely, the 2D peak at 2680 cm^{-1} and the D + G peak at 2940 cm^{-1}. The 2D and D + G peaks represent defect-free and disordered carbon accumulation microstructures, respectively. After HA was adsorbed on the MWNTs and HMWNTs, the I_{2D}/I_{D+G} values decreased, indicating that the number of defects increased after the adsorption of HA. This observation is in contrast to the change in the I_D/I_G value prior to and after the adsorption of HA. As the peak heights of the 2D and D + G peaks were much smaller than those of the D and G peaks, the I_D/I_G value should prevail.

Table 4. Raman parameters of MWNTs and HMWNTs prior to and after HA adsorption.

Samples	W_G (cm^{-1})	W_D (cm^{-1})	W_{2D} (cm^{-1})	W_{D+G} (cm^{-1})	I_D/I_G	I_{2D}/I_{D+G}
MWNT-1	1575	1340	2672	2895	1.17	2.24
MWNT-1 + HA	1573	1337	2675	2910	1.20	1.94
MWNT-2	1573	1339	2672	2893	1.01	2.55
MWNT-2 + HA	1576	1341	2675	2895	0.96	2.25
MWNT-3	1568	1340	2673	2922	1.08	2.03
MWNT-3 + HA	1577	1344	2683	2928	1.03	1.72
HMWNT-1	1571	1341	2673	2929	1.10	2.54
HMWNT-1 + HA	1570	1335	2667	2895	1.24	1.72
HMWNT-2	1571	1336	2671	2895	1.07	2.18
HMWNT-2 + HA	1575	1341	2673	2916	0.99	1.97
HMWNT-3	1570	1340	2674	2910	1.12	2.39
HMWNT-3 + HA	1574	1342	2677	2922	1.06	2.08

In order to explain the mechanism of adsorption, Fourier-transform infrared spectroscopy (FTIR) was used to study the surface chemical reactions during adsorption (Figure 11). The spectrograms of each CNT had two obvious characteristic peaks near 1160 and 3430 cm^{-1}, and 3430 cm^{-1} corresponds to the stretching vibration absorption peak of the -OH bond [47]. The absorption peak near 1160 cm^{-1} is the stretching vibration peak of C-O in the carboxyl group [48], and the peak increased significantly after the adsorption of HA. In addition, each CNT had a weak peak near 2915 cm^{-1} and 1400 cm^{-1}, which represent the tensile vibration of the benzene ring and C-C in the carbon nanotube [49]. The peak at 1700 cm^{-1} represents the vibration peak of C=O in the carboxyl group [49,50], indicating that the CNTs were grafted to carboxyl groups on the surface. Peaks at 1625 cm^{-1} were observed for MWNT-2 and HMWNT-2, and the peak strength decreased after HA adsorption, indicating the release of hydroxyl groups [48]. The vibration peak at 1575 cm^{-1} represents the E1u vibration mode of the carbon nanotube wall, which is caused by the stretching vibration of the C=C skeleton in the carbon ring [47], indicating the existence of a graphite structure in the CNTs. The infrared spectra of the CNTs after HA attachment showed an obvious vibration peak at 1575 cm^{-1}, indicating that the overall structure of the CNTs was not damaged to a large extent after HA adsorption.

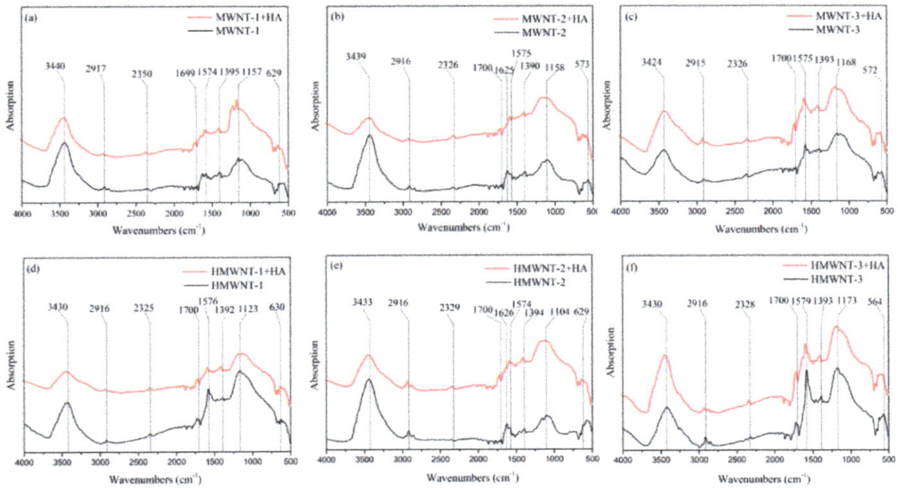

Figure 11. FTIR spectra of MWNTs (**a–c**) and HMWNTs (**d–f**) prior to and after HA adsorption.

3.8. Atomic Adsorption Theory Analysis Based on DFT

The crystal surface model of the carbon nanotubes was based on the first principles of density functional theory (DFT). The simulation was carried out in a cubic box of $20 \times 19.67 \times 36.00$ Angstrom^3. There were 16 -OH groups that were used to modify the carbon nanotube CNT(8,8). Figure 12a shows the atomic arrangement of HA absorbed on the MWNTs, where C-C was neatly arranged in the MWNTs. Figure 12b shows the atomic arrangement of HA absorbed on the HMWNTs, with an O atom bonded to a H atom, which was uniformly attached to the C atom in the HMWNTs. HA molecules were uniformly adsorbed on the surface of the MWNTs and HMWNTs, and HA was parallel to the surface of the MWNTs and HMWNTs. Figure 12c shows the charge difference between the HA molecules and HMWNTs.

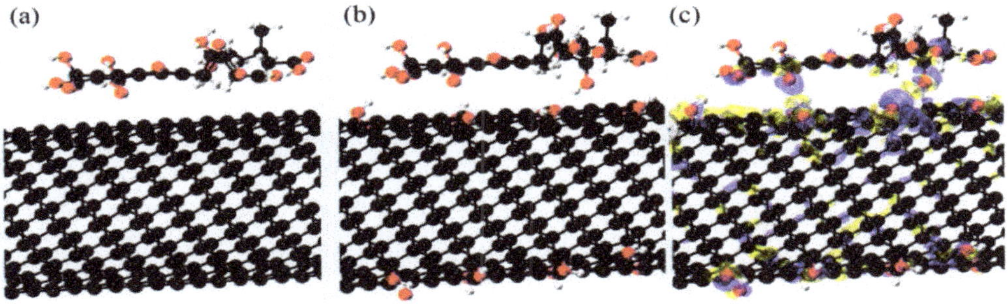

Figure 12. DFT for HA adsorbed on MWNT (**a**) and HMWNT facets (**b**), and the charge difference between the HA molecules and HMWNTs (**c**).

The dark gray, light gray and red balls represent carbon, hydrogen and oxygen atoms, respectively.

The adsorption energy (Ead) value determines the adsorption stability, and the adsorption energy of HA molecules was calculated by Equation (7).

The adsorption energy is defined as

$$E_a = E_{mol/sur} - E_{mol} - \rho_{sur} \quad (7)$$

where $E_{mol/sur}$, E_{mol} and E_{sur} are the DFT energy of the molecule adsorbed on surface, and of the molecule and surface.

The charge density difference is defined as

$$\Delta\rho = \rho_{mol/sur} - \rho_{mol} - \rho_{sur} \quad (8)$$

where $\rho_{mol/sur}$, ρ_{mol} and ρ_{sur} are the electron density of the molecule adsorbed on the surface, and the individual electron densities of the molecule and surface.

The adsorption energy of the molecule on HMWNTs (8,8) was about −2.40 eV, while that on MWNTs (8,8) was about −1.45 eV. This demonstrates that HA was more likely to be adsorbed on the surface of the MWNTs, and that the adsorption was relatively stable, which is also consistent with the results of the adsorption isotherm studies. Therefore, in the natural aquatic environment, MWNTs are more likely to absorb organic pollutants than HMWNTs, resulting in a higher ecological risk.

4. Conclusions

The adsorption amount of HA on both the MWNT and HMWNT surfaces gradually decreased with the increase in the solution pH and tube diameters. At the same tube diameter, the surface hydroxylation prevented the adsorption of HA on the surface of the CNTs. The adsorption rate on the surface of the MWNTs was much higher than that on the surface of the HMWNTs. The suspension of the HMWNTs was higher than that of the MWNTs with the same tube diameter, and the suspension of the CNTs increased with the increase in the tube diameter. The adsorption amount of HA on the surface of the CNTs determined the suspension behavior of the CNTs. Compared with the MWNTs, the HMWNTs could more easily adsorb the aromatic moieties of HA. The adsorption of HA reduced the surface defects of the MWNTs and HMWNTs. Therefore, a larger diameter and surface hydroxylation enhanced the suspension of the CNTs, and HA had a positive effect on the suspension behavior of the CNTs in the water environment and enhanced the migration ability of the CNTs, leading to higher potential ecological risks.

Supplementary Materials: The following are available online at https://www.mdpi.com/article/10.3390/nano12091592/s1, Table S1: Specific surface areas of MWNTs and HMWNTs prior to and after adsorbed HA, Table S2: Parameters of Temkin and Dubinin-Radushkevich adsorption isotherms for HA adsorption on MWNTs and HMWNTs, Figure S1: Nitrogen adsorption isotherms of MWNTs (a–c) and HMWNTs (d–f), Figure S2: Temkin adsorption isotherm of HA adsorption on MWNTs (a–c) and HMWNTs (d–f), Figure S3: Dubinin-Radushkevich adsorption isotherm of HA adsorption on MWNTs (a–c) and HMWNTs (d–f), Figure S4: Pseudo-first-order kinetic curves of adsorption of HA on the surface of MWNTs (a–c) and HMWNTs (d–f), Figure S5: Plots of the intraparticle diffusion kinetic curves of adsorption of HA onto MWNTs (a–c) and HMWNTs (d–f), Figure S6: Fluorescence excitation-emission matrices (EEM) peaks for HA after adsorption of MWNTs and HMWNTs.

Author Contributions: Conceptualization, M.F. and T.Z.; methodology, M.F., T.Z. and Z.T.; software, M.F.; validation, T.Z., Z.T. and J.W.; formal analysis, Z.T.; investigation, X.Z. and S.L.; resources, Z.T. and L.N.; data curation, M.F.; writing—original draft preparation, M.F.; writing—review and editing, M.F. and Z.T.; visualization, Z.T. and T.Z.; supervision, X.Z. and F.W.; project administration, X.Z. and F.W.; funding acquisition, X.Z., Z.T. and T.Z. All authors have read and agreed to the published version of the manuscript.

Funding: This research was funded by Xiaoli Zhao, the National Science Fund for Distinguished Young Scholars, grant number 41925031, and the National Natural Science Foundation of China, grant number 41991315; Zhi Tang, the National Natural Science Foundation of China, grant number 42077349; Tianhui Zhao, the National Natural Science Foundation of China, grant number 42107511.

Institutional Review Board Statement: Not applicable.

Informed Consent Statement: Not applicable.

Data Availability Statement: Not applicable.

Conflicts of Interest: The authors declare no conflict of interest.

Abbreviations

CNTs	carbon nanotubes
MWNTs	multiwall CNTs
HMWNTs	hydroxylated MWNTs
HA	humic acid
MWNT-1	MWNTs with outer tube diameters of 4–6 nm
MWNT-2	MWNTs with outer tube diameters of 5–15 nm
MWNT-3	MWNTs with outer tube diameters of 20–30 nm
HMWNT-1	HMWNTs with outer tube diameters of 4–6 nm
HMWNT-2	HMWNTs with outer tube diameters of 5–15 nm
HMWNT-3	HMWNTs with outer tube diameters of 20–30 nm
Q_1 and Q_2 in Table 1	the maximum charge densities of the two classes of binding sites
$LogK_1$ and $LogK_2$ in Table 1	logK values for proton binding by the two classes of sites
N in Table 1	the number of fitted titration data points
TEM	transmission electron microscopy
C_e in sedimentation test	absorbance at different times in sedimentation test of CNTs
C_0 in sedimentation test	initial absorbance in sedimentation test of CNTs
R^2	correlation coefficient
Em	scanning emission of fluorescence spectral analysis
Ex	scanning excitation of fluorescence spectral analysis
FI	fluorescence index of fluorescence spectral analysis
FTIR	Fourier-transform infrared spectroscopy
DFT	density functional theory
BET ($m^2\ g^{-1}$)	specific surface area
OD (nm)	outer tube diameter
q_e ($mg\ g^{-1}$)	amount of HA adsorbed at equilibrium
C_e ($mg\ L^{-1}$	concentration of HA in solution at equilibrium
q_m ($mg\ g^{-1}$)	maximum adsorption capacity
k_L ($L\ g^{-1}$)	Langmuir equilibrium constant
k_F ($mg^{1-(1/n)}L^{1/n}\ g^{-1}$)	Freundlich parameters
n in Table 2 and Equation (2)	Freundlich parameters
k_1 (min^{-1})	the first-order adsorption rate constant
k_2 ($g\ mg^{-1}\ min^{-1}$) the second-order adsorption rate constant;	
t (min)	adsorption time
q_t ($mg\ g^{-1}$)	the amount of HA adsorbed by CNTs at time t
I_D	D-band intensity in Raman spectroscopy
I_G	G-band intensity in Raman spectroscopy
k($mg^2\ mg^{-2}$)	a constant related to the adsorption amount
a_t	constant of Equation S(1)
b_t	constant of Equation S(1)
R (8.314 J $(mol\ K)^{-1}$) in Equation S(1) and Equation S(3)	the ideal gas constant
T (K)	the thermodynamic temperature
ε	the adsorption potential
C in Table 3 and Equation (6)	the intercept in the intraparticle diffusion rate
k_{int} ($mg\ g^{-1}\ min^{-1/2}$)	the intraparticle diffusion rate constant

References

1. Iijima, S. Helical microtubules of graphitic carbon. *Nature* **1991**, *354*, 56–58. [CrossRef]
2. Liu, Z.; Chen, K.; Davis, C.; Sherlock, S.; Cao, Q.Z.; Chen, X.Y.; Dai, H. Drug delivery with carbon nanotubes for in vivo cancer treatment. *Cancer Res.* **2008**, *68*, 6652–6660. [CrossRef] [PubMed]
3. Mestl, G.; Maksimova, N.I.; Keller, N.; Roddatis, V.; Schlogl, R. Carbon nanofilaments in heterogeneous catalysis: An industrial application for new carbon materials? *Angew. Chem. Int. Ed.* **2001**, *40*, 2066–2068. [CrossRef]
4. Mauter, M.S.; Elimelech, M. Environmental applications of carbon-based nanomaterials. *Environ. Sci. Technol.* **2008**, *42*, 5843–5859. [CrossRef]
5. Khang, D.Y.; Xiao, J.L.; Kocabas, C.; MacLaren, S.; Banks, T.; Jiang, H.Q.; Huang, Y.G.Y.; Rogers, J.A. Molecular scale buckling mechanics in individual aligned single-wall carbon nanotubes on elastomeric substrates. *Nano Lett.* **2008**, *8*, 124–130. [CrossRef] [PubMed]
6. Arvidsson, R.; Sanden, B.A. Carbon nanomaterials as potential substitutes for scarce metals. *J. Clean. Prod.* **2017**, *156*, 253–261. [CrossRef]
7. Alloy, M.M.; Roberts, A.P. Effects of suspended multi-walled carbon nanotubes on daphnid growth and reproduction. *Ecotoxicol. Environ. Saf.* **2011**, *74*, 1839–1843. [CrossRef]
8. Zhu, L.; Chang, D.W.; Dai, L.M.; Hong, Y.L. DNA damage induced by multiwalled carbon nanotubes in mouse embryonic stem cells. *Nano Lett.* **2007**, *7*, 3592–3597. [CrossRef]
9. Lam, C.W. Pulmonary toxicity of single-wall carbon nanotubes in mice 7 and 90 days after intratracheal instillation. *Toxicol. Sci.* **2003**, *77*, 126–134. [CrossRef]
10. Chen, M.; Qin, X.S.; Li, J.; Zeng, G.M. Probing molecular basis of single-walled carbon nanotube degradation and nondegradation by enzymes based on manganese peroxidase and lignin peroxidase. *RSC Adv.* **2016**, *6*, 3592–3599. [CrossRef]
11. Lin, D.H.; Li, T.T.; Yang, K.; Wu, F.C. The relationship between humic acid (HA) adsorption on and stabilizing multiwalled carbon nanotubes (MWNTs) in water: Effects of HA, MWNT and solution properties. *J. Hazard. Mater.* **2012**, *241*, 404–410. [CrossRef] [PubMed]
12. Alpatova, A.L.; Shan, W.Q.; Babica, P.; Upham, B.L.; Rogensues, A.R.; Masten, S.J.; Drown, E.; Mohanty, A.K.; Alocilja, E.C.; Tarabara, V.V. Single-walled carbon nanotubes dispersed in aqueous media via non-covalent functionalization: Effect of dispersant on the stability, cytotoxicity, and epigenetic toxicity of nanotube suspensions. *Water Res.* **2010**, *44*, 505–520. [CrossRef] [PubMed]
13. Lin, D.H.; Liu, N.; Yang, K.; Xing, B.S.; Wu, F.C. Different stabilities of multiwalled carbon nanotubes in fresh surface water samples. *Environ. Pollut.* **2010**, *158*, 1270–1274. [CrossRef] [PubMed]
14. Lin, D.H.; Tian, X.L.; Li, T.T.; Zhang, Z.Y.; He, X.; Xing, B.S. Surface-bound humic acid increased Pb^{2+} sorption on carbon nanotubes. *Environ. Pollut.* **2012**, *167*, 138–147. [CrossRef]
15. Zhang, D.; Pan, B.; Cook, R.L.; Xing, B.S. Multi-walled carbon nanotube dispersion by the adsorbed humic acids with different chemical structures. *Environ. Pollut.* **2015**, *196*, 292–299. [CrossRef] [PubMed]
16. Hyung, H.; Kim, J.H. Natural organic matter (NOM) adsorption to multi-walled carbon nanotubes: Effect of NOM characteristics and water quality parameters. *Environ. Sci. Technol.* **2008**, *42*, 4416–4421. [CrossRef]
17. Yang, K.; Xing, B.S. Adsorption of fulvic acid by carbon nanotubes from water. *Environ. Pollut.* **2009**, *157*, 1095–1100. [CrossRef]
18. Summers, R.S.; Roberts, P.V. Activated carbon adsorption of humic substances: I. Heterodisperse mixtures and desorption. *J. Colloid Interface Sci.* **1988**, *122*, 367–381. [CrossRef]
19. Wang, L.X.; Yang, X.Z.; Wang, Q.; Zeng, Y.X.; Jiang, W. Effects of ionic strength and temperature on the aggregation and deposition of multi-walled carbon nanotubes. *J. Environ. Sci.* **2016**, *51*, 248–255. [CrossRef]
20. Melo, B.A.G.D.; Motta, F.L.; Santana, M.H.A. Humic acids: Structural properties and multiple functionalities for novel technological developments. *Mater. Sci. Eng. C* **2016**, *62*, 967–974. [CrossRef]
21. Polak, J.; Bartoszek, M.; Sułkowski, W.W. Comparison of some spectroscopic and physico-chemical properties of humic acids extracted from sewage sludge and bottom sediments. *J. Mol. Struct.* **2009**, *924*, 309–312. [CrossRef]
22. Chen, G.C.; Shan, X.Q.; Zhou, Y.Q.; Shen, X.E.; Huang, H.L.; Khan, S.U. Adsorption kinetics, isotherms and thermodynamics of atrazine onsurface oxidized multi-walled carbon nanotubes. *J. Hazard. Mater.* **2009**, *169*, 912–918. [CrossRef] [PubMed]
23. Lin, D.H.; Xing, B.S. Adsorption of phenolic compounds by carbon nanotubes: Role of aromaticity and substitution of hydroxyl groups. *Environ. Sci. Technol.* **2008**, *42*, 7254–7259. [CrossRef] [PubMed]
24. Smith, B.; Wepasnick, K.; Schrote, K.E.; Cho, H.; Ball, W.P.; Fairbrother, D.H. Influence of surface oxides on the colloidal stability of multi-walled carbon nanotubes: A structureproperty relationship. *Langmuir* **2009**, *25*, 9767–9776. [CrossRef] [PubMed]
25. Wang, X.L.; Shu, L.; Wang, Y.Q.; Xu, B.B.; Bai, Y.C.; Tao, S.; Xing, B.S. Sorption of peat humic acids to multi-walled carbon nanotubes. *Environ. Sci. Technol.* **2011**, *45*, 9276–9283. [CrossRef] [PubMed]
26. Zhou, X.Z.; Shu, L.; Zhao, H.B.; Guo, X.Y.; Wang, X.L.; Tao, S.; Xing, B.S. Suspending multi-walled carbon nanotubes by humic acids from a peat soil. *Environ. Sci. Technol.* **2012**, *46*, 3891–3897. [CrossRef] [PubMed]
27. International Humic Substances Society. Chemical Properties of IHSS Samples[EB/OL]. Available online: http://www.humicsubstances.org (accessed on 9 August 2019).
28. Tang, Z.; Zhao, X.L.; Zhao, T.H.; Wang, H.; Wang, P.F.; Wu, F.C.; Giesy, J.P. Magnetic nanoparticles interaction with humic acid: In the presence of surfactants. *Environ. Sci. Technol.* **2016**, *50*, 8640–8648. [CrossRef]

29. Ateia, M.; Apul, O.G.; Shimizu, Y.; Muflihah, A.; Yoshimura, C.; Karanfil, T. Elucidating adsorptive fractions of natural organic matter on carbon nanotubes. *Environ. Sci. Technol.* **2017**, *51*, 7101–7110. [CrossRef]
30. Hutter, J.; Iannuzzi, M.; Schiffmann, F.; VandeVondele, J. Cp2k: Atomistic simulations of condensed matter systems. *Wiley Interdiscip. Rev. Comput. Mol. Sci.* **2014**, *4*, 15–25. [CrossRef]
31. Perdew, J.P.; Burke, K.; Ernzerhof, M. Generalized gradient approximation made simple. *Phys. Rev. Lett.* **1996**, *7*, 3865–3868. [CrossRef]
32. Grimme, S. Semiempirical GGA-type density functional constructed with a long-range dispersion correction. *J. Comput. Chem.* **2006**, *27*, 1787–1799. [CrossRef] [PubMed]
33. VandeVondele, J.; Hutter, J. Gaussian basis sets for accurate calculations on molecular systems in gas and condensed phases. *J. Chem. Phys.* **2007**, *127*, 105–114. [CrossRef]
34. VandeVondele, J.; Krack, M.; Mohamed, F.; Parrinello, M.; Chassaing, T.; Hutter, J. Quickstep: Fast and accurate density functional calculations using a mixed Gaussian and plane waves approach. *Comput. Phys. Commun.* **2005**, *167*, 103–128. [CrossRef]
35. Goedecker, S.; Teter, M.; Hutter, J. Separable dual-space Gaussian pseudopotentials. *Phys. Rev. B* **1996**, *54*, 1703–1710. [CrossRef]
36. Hartwigsen, C.; Gœdecker, S.; Hutter, J. Relativistic separable dual-space Gaussian pseudopotentials from H to Rn. *Phys. Rev. B* **1998**, *58*, 3641–3662. [CrossRef]
37. Hassani, A.; Khataee, A.; Karaca, S.; Karaca, M.; Kranan, M. Adsorption of two cationic textile dyes from water with modified nanoclay: A comparative study by using central composite design. *J. Environ. Chem. Eng.* **2015**, *3*, 2738–2749. [CrossRef]
38. Karaca, S.; Gürsesa, A.; Açışlıö, O.; Hassania, A.; Kiransan, M.; Yikilmaz, K. Modeling of adsorption isotherms and kinetics of Remazol Red RB adsorption from aqueous solution by modified clay. *Desalin. Water Treat.* **2013**, *51*, 2726–2739. [CrossRef]
39. Chung, H.K.; Kim, W.H.; Park, J.; Cho, J.; Jeong, T.Y.; Park, P.K. Application of Langmuir and Freundlich isotherms to predict adsorbate removal efficiency or required amount of adsorbent. *J. Ind. Eng. Chem.* **2015**, *28*, 241–246. [CrossRef]
40. Wu, F.C.; Tseng, R.L.; Juang, R.S. Initial behavior of intraparticle diffusion model used in the description of adsorption kinetics. *Chem. Eng. J.* **2009**, *153*, 1–8. [CrossRef]
41. Kumar, B.G.P.; Shivakamy, K.; Miranda, L.R.; Velan, M. Preparation of steam activated carbon from rubberwood sawdust (*Hevea brasiliensis*) and its adsorption kinetics. *J. Hazard. Mater.* **2006**, *136*, 922–929.
42. Ateia, M.; Ran, J.; Fujii, M.; Yoshimura, C. The relationship between molecular composition and fluorescence properties of humic substances. *Int. J. Environ. Sci. Technol.* **2017**, *14*, 867–880. [CrossRef]
43. Shimabuk, K.K.; Kennedy, A.M.; Mulher, R.E.; Summers, R.S. Evaluating activated carbon adsorption of dissolved organic matter and micropollutants using fluorescence spectroscopy. *Environ. Sci. Technol.* **2017**, *51*, 2676–2684. [CrossRef] [PubMed]
44. Roldán, M.L.; Corrado, G.; Francioso, O.; Sanchez-Cortes, S. Interaction of soil humic acids with herbicide paraquat analyzed by surface-enhanced Raman scattering and fluorescence spectroscopy on silver plasmonic nanoparticles. *Anal. Chim. Acta* **2011**, *699*, 87–95. [CrossRef]
45. Martínez, M.T.; Callejas, M.A.; Benito, A.M.; Cochet, M.; Maser, W.K. Sensitivity of single wall carbon nanotubes to oxidative processing: Structural modification, intercalation and functionalisation. *Carbon* **2003**, *41*, 2247–2256. [CrossRef]
46. Qian, W.Z.; Wei, F.; Liu, T.; Wang, Z.W.; Li, Y.D. What causes the carbon nanotubes collapse in a chemical vapor deposition process. *J. Chem. Phys.* **2003**, *118*, 878–882. [CrossRef]
47. Stobinski, L.; Lesiak, B.; Kovér, L.; Tóth, J.; Biniak, S.; Trykowski, G.; Judek, J. Multiwall carbon nanotubes purification and oxidation by nitric acid studied by the FTIR and electron spectroscopy methods. *J. Alloys Compd.* **2010**, *501*, 77–84. [CrossRef]
48. Yang, K.; Lin, D.H.; Xing, B.S. Interactions of humic acid with nanosized inorganic oxides. *Langmuir ACS J. Surf. Colloids* **2009**, *25*, 3571–3576. [CrossRef]
49. Wu, W.; Shan, G.Q.; Xiang, Q.; Zhang, Y.Q.; Yi, S.J.; Zhu, L.Y. Effects of humic acids with different polarities on the photocatalytic activity of nano-TiO_2 at environment relevant concentration. *Water Res.* **2017**, *122*, 78–85. [CrossRef]
50. Chen, H.F.; Li, Q.; Wang, M.X.; Ji, D.B.; Tan, W.F. XPS and two-dimensional FTIR correlation analysis on the binding characteristics of humic acid onto kaolinite surface. *Sci. Total Environ.* **2020**, *724*, 138–154. [CrossRef]

nanomaterials

MDPI

Article

Phytotoxicity and Accumulation of Copper-Based Nanoparticles in *Brassica* under Cadmium Stress

Shiqi Wang [1,2,3], Yutong Fu [1,2,3], Shunan Zheng [4], Yingming Xu [2,3] and Yuebing Sun [2,3,*]

[1] College of Resources and Environment, Northeast Agricultural University, Harbin 150030, China; wwp875822270@163.com (S.W.); 18845141255@163.com (Y.F.)
[2] Key Laboratory of Original Agro-Environmental Pollution Prevention and Control, Agro-Environmental Protection Institute, Ministry of Agriculture and Rural Affairs (MARA), Tianjin 300191, China; ymxu1999@126.com
[3] Tianjin Key Laboratory of Agro-Environment and Agro-Product Safety, Agro-Environmental Protection Institute, Ministry of Agriculture and Rural Affairs (MARA), Tianjin 300191, China
[4] Rural Energy & Environment Agency, Ministry of Agriculture and Rural Affairs (MARA), Beijing 100125, China; zhengshunan1234@163.com
* Correspondence: sunyuebing2008@126.com

Abstract: The widespread use of copper-based nanoparticles expands the possibility that they enter the soil combined with heavy metals, having a toxic effect and posing a threat to the safety of vegetables. In this study, single and combined treatments of 2 mg/L Cd, 20 mg/L Cu NPs and 20 mg/L CuO NPs were added into Hoagland nutrient solution by hydroponics experiments. The experimental results show that copper-based Nanoparticles (NPs) can increase the photosynthetic rate of plants and increase the biomass of *Brassica*. Cu NPs treatment increased the Superoxide Dismutase (SOD), Peroxidase (POD) and catalase (CAT) activities of *Brassica*, and both NPs inhibited ascorbate peroxidase (APX) activity. We observed that Cd + Cu NPs exhibited antagonistic effects on Cd accumulation, inhibiting it by 12.6% in leaf and 38.6% in root, while Cd + CuO NPs increased Cd uptake by 73.1% in leaves and 22.5% in roots of *Brassica*. The Cu content in the shoots was significantly negatively correlated with Cd uptake. The Cd content of each component in plant subcellular is soluble component > cytoplasm > cell wall. Cu NPs + Cd inhibited the uptake of Zn, Ca, Fe, Mg, K and Mn elements, while CuO NPs + Cd promoted the uptake of Mn and Na elements. The results show that copper-based nanoparticles can increase the oxidative damage of plants under cadmium stress and reduce the nutritional value of plants.

Keywords: copper-based nanoparticles; phytotoxicity; bioaccumulation; nutrient element

check for updates

Citation: Wang, S.; Fu, Y.; Zheng, S.; Xu, Y.; Sun, Y. Phytotoxicity and Accumulation of Copper-Based Nanoparticles in *Brassica* under Cadmium Stress. *Nanomaterials* **2022**, *12*, 1497. https://doi.org/10.3390/nano12091497

Academic Editor: Thierry Rabilloud

Received: 28 March 2022
Accepted: 22 April 2022
Published: 28 April 2022

1. Introduction

Cadmium (Cd) is one of the most biologically toxic heavy metals. Cd pollution in farmland not only reduces soil quality and crop yield, but also threatens the health and well-being of animals and humans through the food chain [1]. The long-term consumption of high levels of Cd can lead to hypercalciuria, renal failure, anemia and even death. Cd can be toxic to organisms even at very low concentrations, and can also be toxic to plants when the total concentration exceeds 8 mg/kg [2]. According to the National Soil Pollution Survey Bulletin, the excess rate of Cd in China reached 7.0, showing a trend of gradually increasing from northwest to southeast and from northeast to southwest [3]. A meta-analysis of heavy metals in Chinese farmland and urban soil showed that Cd was the most commonly polluted heavy metal in Chinese soil, accounting for 33.54% and 44.65% of farmland and urban soil pollution, respectively [4].

Cu-based nanoparticles (NPs) have unique properties such as small volume, large specific surface area, high activation energy and many active sites. They have great application potential in industrial, agricultural and commercial fields. There are data to prove that in 2010, the global production of copper-based NPs was 200 tons per year, and it is increasing year by year [5]. Additionally, some studies are exploring the feasibility of using copper-based nanomaterials as nano-fertilizers [6,7], which increases the risk of their entry into the soil environment due to transportation, application, leakage, etc. Copper-based nanoparticles will be absorbed by plants after entering the soil. At low doses, they will have a stimulating effect on plants, which can promote plant growth and development and improve plant tolerance to adverse environmental stress [8], but high doses cause toxic effects on plants [3]. Metal element/metal oxide nanomaterials can also release metal ions, causing cellular oxidative stress and poisoning plants. Studies have shown that the addition of CuO NPs reduced the seed germination rate and the rhizome length of seedlings, decreased root cell viability, and increased the generation of plant reactive oxygen species (ROS) and lipid peroxidation [9,10]. Cu NPs can alter the activity of antioxidant enzymes in plants and activate the antioxidant enzyme defense mechanism against ROS [11]. Copper-based NPs can inhibit the accumulation of nutrient elements, thereby affecting human nutrient intake [12]. Copper-based NPs also affect plant genes; studies have shown that when plants are exposed to different concentrations of CuO NPs, the expressions of the CuZn-superoxide dismutase (CuZnSOD) gene, CAT gene and APX gene in roots are increased to varying degrees [10]. Depending on the plant species, the applied concentration of NPs and the particle size of NPs, different toxic effects will be exhibited.

In addition to causing nanotoxicity, nanoparticles can also adsorb other pollutants, and when they enter the environment, they can act synergistically or antagonistically with heavy metals present in the environment. At present, some studies have investigated the plant performance of other nanomaterials under the combined stress of heavy metal cadmium, but different types of NPs and heavy metals, application ratios, plant species and culture conditions will lead to different toxic effects. Cd^{2+} and TiO_2 NPs exhibited different combined toxicity patterns against *Scenedesmus obliquus* at different combined exposure concentrations. Antagonistic effects were exhibited at low doses, and partial additive and synergistic combined toxicity occurred when the proportion of TiO_2 NPs was increased [13]. ZnO NPs at a concentration of 25 mg/L promoted the growth of *Leucaena Leucocephala* seedlings under cadmium and lead stress, increased the activities of SOD, CAT and other antioxidant enzymes, and significantly decreased the malonaldehyde (MDA) content [14]. After wheat seeds soaked in Fe NPs and ZnO NPs solution were sown in Cd-contaminated soil, the dry weight of wheat was positively correlated with the amount of NP added, which significantly reduced the Cd content of each part of the plant [15]. To date, most studies have focused on the negative effects of copper-based nanomaterials or cadmium alone on plants, and it is crucial to explore the physiological performance and toxicity mechanisms of plants under the combined stress of the two.

Brassica (*Brassica campestris* L. ssp. *chinensis* Makino var. *communis* Tsen et Lee) is widely grown in the north and south of China and has strong enrichment for heavy metal Cd. In this experiment, parameters such as photosynthetic rate, biomass, antioxidant enzyme activity, plant absorption of heavy metals, nutrient element content and plant subcellular element content were determined, in order to explore the combined toxicity of copper-based nanoparticles and Cd in plants. This provided a reference for the in-depth study of the toxicity mechanism of copper-based nanoparticles combined with heavy metals.

2. Materials and Methods

2.1. Cu and CuO NPs Characterization

Nanomaterials were purchased from Zhejiang Ailu Chemical Technology Co., Ltd. (Li Shui City, China). The characterization revealed that the purity of Cu NPs exceeded 99.9%, with a size of 10–30 nm and specific surface area of 6.99 $m^2 \cdot g^{-1}$. CuO NPs purity exceeded 99.5%, particle size was 40 nm and specific surface area was 2.84 $m^2 \cdot g^{-1}$.

2.2. Hydroponic Experiment Design and Exposure Conditions

Seeds of *Brassica* were purchased from Cangzhou Heshuo Agricultural Technology Co., Ltd. (Cangzhou, China). The seeds were washed in deionized water, and the growing medium was vermiculite. After the seedlings were taken out and the roots carefully washed, the *Brassica* seedlings were transferred to hydroponic boxes and incubated with Hoagland nutrient solution for 20 days. The ratio of Hoagland nutrient solution is shown in Table S1. For nanoparticle exposure, according to the effective concentration of previous hydroponic experiments [11,16–18], we set the experimental NPs concentration as 20 mg/L. Five treatments were designed, including Cd (2 mg/L Cd), Cu NPs (20 mg/L Cu NPs), CuO NPs (20 mg/L CuO NPs), Cd + Cu NPs (2 mg/L Cd + 20 mg/L Cu NPs), and Cd + CuO NPs (2 mg/L Cd + 20 mg/L CuO NPs). Each treatment was repeated three times. Each hydroponic box was equipped with an aerator to provide oxygen to the roots while keeping the NPs in suspension. After culturing for 7 days, all the *Brassica* plants were collected to measure various indicators.

2.3. Determination of Physiological Indicators

The photosynthetic rate (Pn) of *Brassica* was measured by a portable photosynthesis measurement system at 10:00 a.m. every day. After 7 days, plants were separated into underground parts and aboveground parts and soaked in 10 $mmol \cdot L^{-1}$ Na_2-EDTA solution for 40 s, to remove metal ions adhering to the root surface. After rinsing with ultrapure water, each part of the fresh sample was weighed with an analytical balance. The fresh samples were refrigerated at −80 °C for later use. The *Brassica* tissues were oven-dried for 7 days at 80 °C and weighed to record the biomass. The oven-dried tissues were ground to powder for subsequent use.

2.4. Determination of Cu, Cd and Other Nutrient Elements

Approximately 0.25 g of the sample was soaked and digested with 8 mL of concentrated HNO_3, and the contents of copper, cadmium and nutrients in *Brassica* were determined by inductively coupled plasma mass spectrometer (ICP-MS, Waltham, MA, USA).

2.5. Determination of Subcellular Cu, Cd and Nutrient Elements

The experimental method refers to Li et al. [19]. A total of 3 g of fresh samples was centrifuged at 2000 rpm at 4 °C for 5 min in the pre-cooled extraction solution (50 mM Tris–HCl, 250 mM sucrose, and 1.0 mM DTE ($C_4H_{10}O_2S_2$), pH 7.5). The precipitate was obtained and defined as the 'cell wall fraction'. The filtrate was transferred to a special tube for a refrigerated centrifuge and centrifuged at 11,900 rpm at 4 °C for 45 min. The deposit was referred to as the 'organelle fraction', and the supernatant solution as the 'soluble fraction'. After the drying and digestion of each component, ICP-MS was used for determination.

2.6. Determination of Antioxidant Enzyme Activity

The enzyme activity of *Brassica* was measured in leaves and root. Peroxidase (POD) activity was determined by the guaiacol method, CAT activity was determined by the potassium permanganate titration method, and superoxide dismutase (SOD) activity was determined by the nitrogen blue tetrazolium photoreduction method [20].

Ascorbate peroxidase (APX) activity was determined as follows: 0.1 g of plant tissue was homogenized in an ice bath and centrifuged at 4 °C for 20 min. To the supernatant was added K_2HPO_4-KH_2PO_4 buffer, 0.3 $mmol \cdot L^{-1}$ ascorbic acid (AsA), 0.1 $mmol \cdot L^{-1}$ EDTA-2NA and 0.06 $mmol \cdot L^{-1}$ H_2O_2. After rapid mixing, the absorbance was measured at 290 nm for 10 and 130 s [21]. The unit of antioxidant enzyme activity is $U \cdot g^{-1}$.

2.7. Statistical Analysis

The results are presented as means ± standard errors of 3 replicates. One-way analysis of variance (one-way ANOVA) was used to determine statistical differences between treatments, followed by an LSD test performed by IBM SPSS Statistics 26. "$p < 0.05$" was used for statistical significance.

3. Results

3.1. Photosynthesis and Plant Growth

The photosynthetic rate was measured from the first to the seventh day of hydroponics of *Brassica* (Table 1). The treatments showed a more obvious change on days 5–7 as the experimental time increased. Consistent with previous research [22,23], compared with CuO NPs treatment, heavy metal Cd significantly inhibited the photosynthetic rate of *Brassica* by 10.02–12.64%. In the measurement of photosynthetic rate in each group within seven days, the photosynthetic rate of CuO NPs treatment was the highest, and the photosynthetic rate of other treatments showed different degrees of decline. Compared with CuO NPs treatment, the photosynthetic rate of the Cu NPs group decreased by 1.07–4.47%. The copper-based nanoparticles alleviated the stress of Cd on the photosynthesis of *Brassica* to a certain extent. Compared with the Cd treatment group, the photosynthetic rate of the Cu NPs + Cd group and the CuO NPs + Cd group both increased, and the results were similar to those of the treatment without heavy metals. In contrast, CuO NPs showed a more significant photosynthesis promotion effect, being increased by 10.2–19.6%. Although the increase in photosynthetic rate was also observed in the Cu NPs group, most changes were not significant.

Table 1. Effect of Cu NPs, CuO NPs and Cd on photosynthetic rate of *Brassica* from 1 to 7 days. The means are averaged from three replicates, and the error bars correspond to the standard deviations of the three values. Different letters above each column indicate a significant difference among treatments in the same group ($p < 0.05$).

	D1	D2	D3	D4	D5	D6	D7
Cd	39.68 ± 1.83 b	38.79 ± 0.65 c	38.80 ± 3.22 c	38.76 ± 3.16 b	37.91 ± 1.48 c	38.59 ± 1.82 c	38.46 ± 3.23 c
Cu NPs	41.37 ± 4.19 b	43.76 ± 2.11 ab	44.26 ± 1.61 ab	44.53 ± 4.84 ab	44.15 ± 2.74 b	44.53 ± 1.28 b	44.60 ± 2.28 ab
CuO NPs	47.30 ± 2.74 a	47.06 ± 3.13 a	47.57 ± 1.42 a	48.14 ± 5.41 a	48.42 ± 1.67 a	49.31 ± 2.61 a	49.86 ± 3.77 a
Cu NPs + Cd	40.87 ± 1.14 b	41.23 ± 1.22 bc	41.54 ± 1.40 bc	42.44 ± 1.88 ab	42.54 ± 0.70 b	42.90 ± 1.80 b	42.76 ± 1.47 bc
CuO NPs + Cd	43.96 ± 3.51 ab	43.48 ± 1.24 ab	43.44 ± 2.56 ab	43.51 ± 2.04 ab	43.42 ± 1.53 b	43.45 ± 2.25 b	45.98 ± 3.34 ab

Figure 1 showed the measurement results of the fresh and dry weights of the leaves and roots of *Brassica*, root fresh weight was not significantly affected by any of the treatments. Cd had a certain inhibitory effect on the biomass of *Brassica*. The fresh and dry weights of *Brassica* under the CuO NPs treatment were the largest among the five treatments (50.65 g and 6.22 g in the shoots, and 3.93 g and 0.37 g in the roots, respectively). Cu NPs and CuO NPs promoted the fresh weight increase of *Brassica* under Cd stress by 41.0% and 44.4%, respectively. Compared with the Cd group, the dry weight of *Brassica* in the Cd + Cu NPs treatment showed a slight increase (2.5% above ground and 3.2% below ground), but these data were not statistically significant. A significant increase in dry weight was observed in the Cd + CuO NPs group (17.4% above ground and 39.8% below ground). This indicates that the two copper-based nanoparticles can alleviate the inhibitory effect of Cd on the biomass of *Brassica* to a certain extent, and the effect of CuO NPs is more obvious.

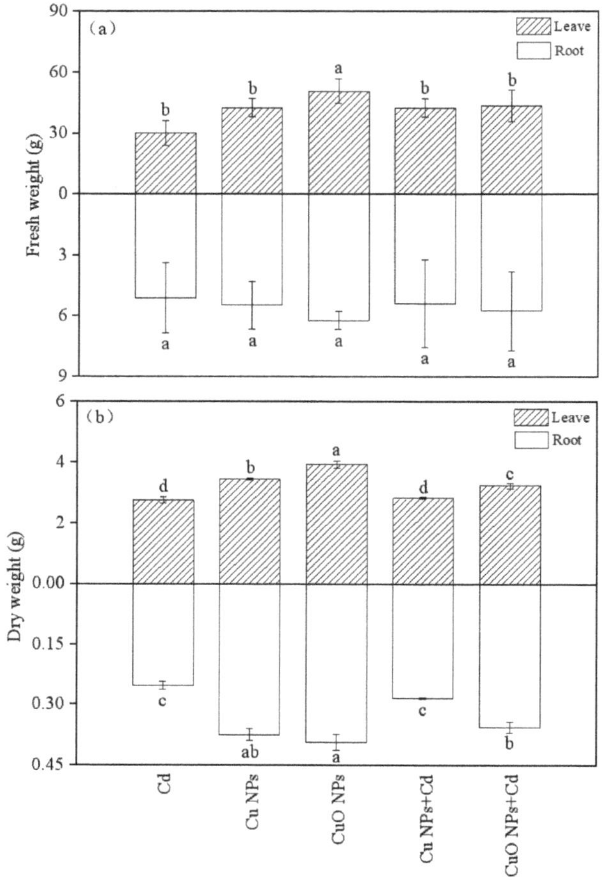

Figure 1. Effect of Cu NPs, CuO NPs and Cd on fresh weight (**a**) and dry weight (**b**) of *Brassica*. The means are averaged from three replicates, and the error bars correspond to the standard deviations of the three values. Different letters above each column indicate a significant difference among treatments in the same group ($p < 0.05$).

3.2. Antioxidase Activity

In this experiment, the activities of SOD, POD and CAT in the underground parts of each treatment did not change significantly (Figure 2). The activities of four antioxidant enzymes in the leaves of Brassica treated with CuO NPs were significantly lower than those of Cu NPs (SOD, POD and CAT activities decreased by 41.8%, 40.0% and 14.58%, respectively). The SOD, POD, and CAT activities of the three treatments contaminated with Cd were Cd + Cu NPs > Cd > Cd + CuO NPs. Compared with the Cd treatment group, Cu NPs promoted the SOD, POD and CAT activities of *Brassica* under Cd stress by 51.3%, 18.2% and 32.65%, respectively. However, the increase in POD activity in the Cu NPs + Cd group was not significant. In contrast to Cu NPs, although CuO NPs did not significantly change the SOD, POD, and CAT activities of Cd-treated *Brassica*, a weak decrease in enzymatic activity was observed. The above-ground and below-ground activities of APX under the treatments of Cu NPs and CuO NPs showed opposite effects. The activity of the APX group in the Cu NPs leaves was much higher (1.68 times) than that in the CuO NPs group, while the APX activity in the roots decreased by 20.6% compared with the CuO NPs treatment. Under cadmium stress, NPs did not significantly change the APX activity in roots. It was observed that compared with Cd treatment, the APX activity of the Cu NPs + Cd and CuO

NPs + Cd groups was significantly inhibited in the shoots, being reduced by 17.26% and 38.49%, respectively.

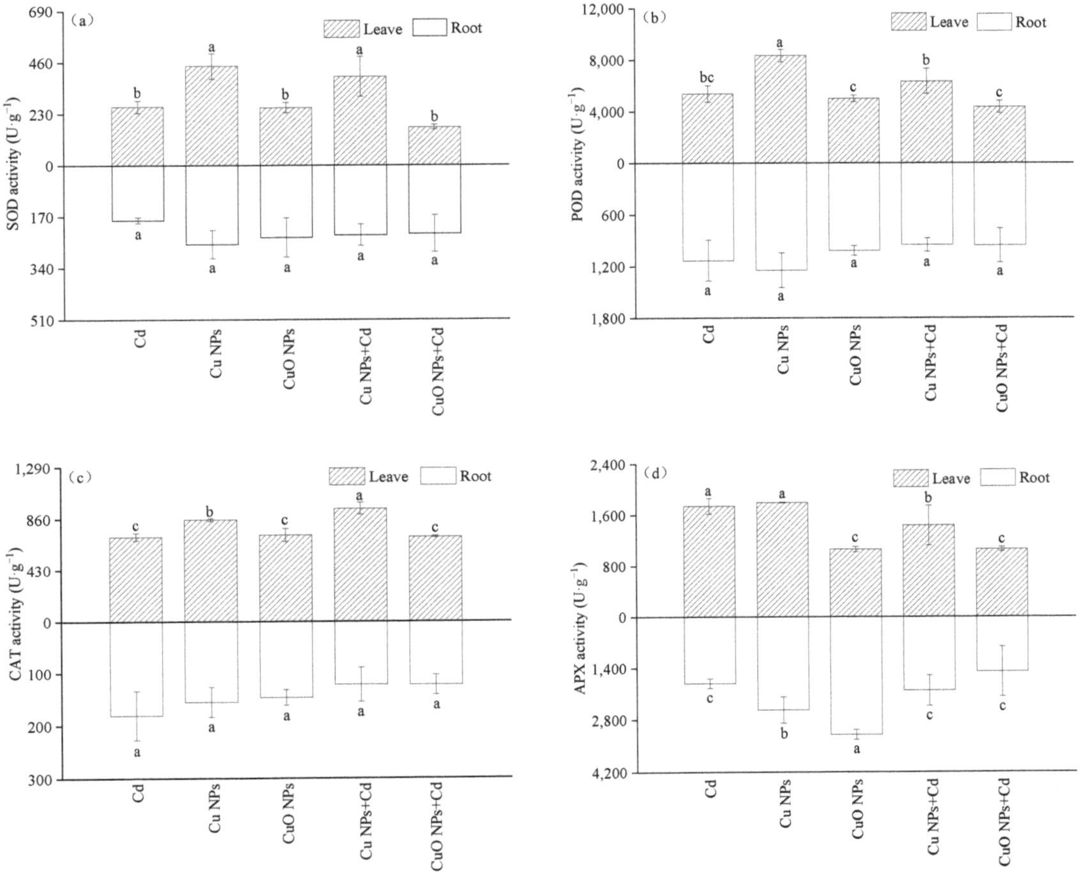

Figure 2. Effect of Cu NPs, CuO NPs and Cd on SOD activity (**a**), POD activity (**b**), CAT activity (**c**) and APX activity (**d**) of *Brassica*. The means are averaged from three replicates, and the error bars correspond to the standard deviations of the three values. Different letters above each column indicate a significant difference among treatments in the same group ($p < 0.05$).

3.3. Uptake of Cu and Cd

The Cu and Cd contents in *Brassica* were determined by the above method (Figure 3). Comparing the two types of nanoparticles, the Cu NPs-treated aboveground Cu accumulation was 2.99 times that of the CuO NPs-treated group, but the underground Cu content (1402.07 $mg \cdot kg^{-1}$) was much lower than that of the CuO NPs group (5594.39 $mg \cdot kg^{-1}$). For the aboveground Cu content of the Cd + Cu NPs and Cd + CuO NPs groups, we observed two opposite influence trends. Compared with Cd treatment, Cu NPs increased by 88.4% while CuO NPs decreased by 15.4%, both trends showing significant changes. Different from the above-ground part, both Cu-containing nanoparticles significantly promoted the accumulation of Cu content in the underground part of *Brassica*, which was 25.58 times and 54.10 times that of the Cd treatment, respectively. The total copper content in *Brassica* in the CuO NPs + Cd group was much higher than that in the Cu NPs + Cd group, but it was mostly accumulated in the roots of *Brassica*, and only a small amount was transported to the shoots.

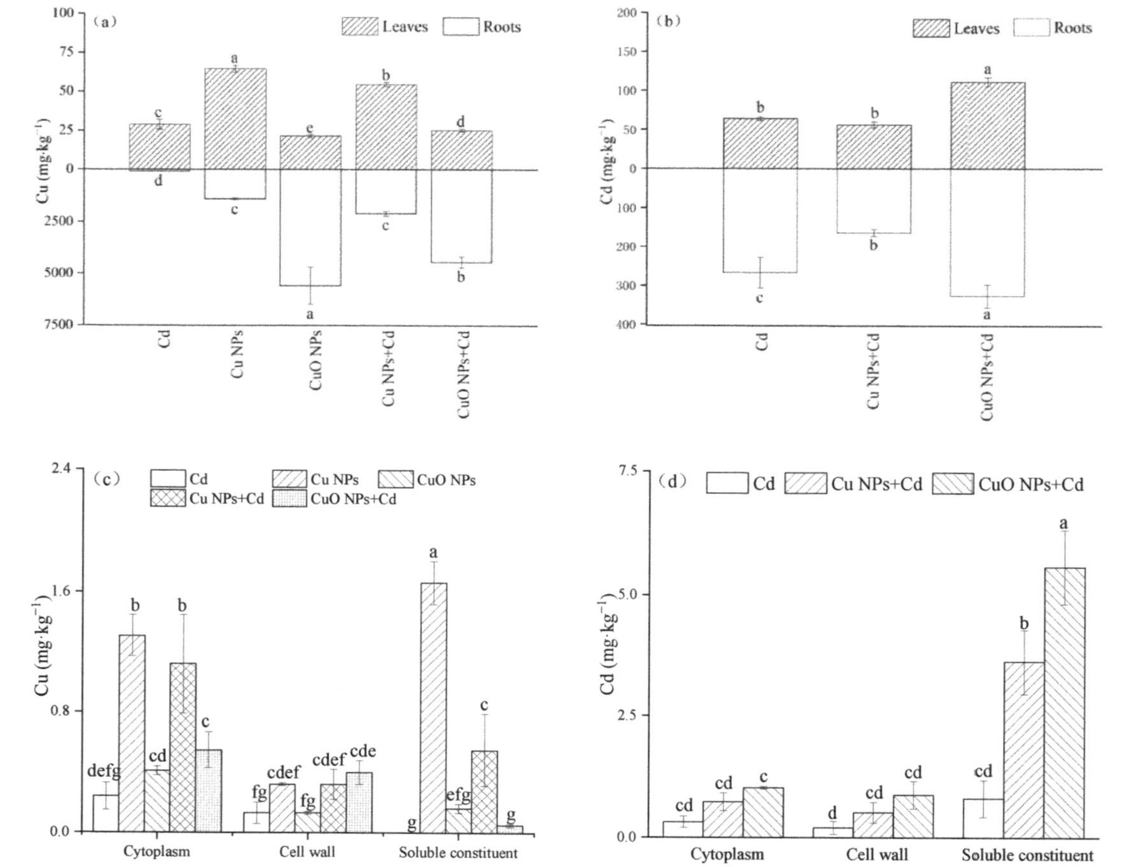

Figure 3. The effects of Cu NPs, CuO NPs and Cd on Cu (**a**), Cd (**b**), subcellular Cu (**c**) and subcellular Cd (**d**) content. The mean is the mean of three replicates and the error bars correspond to the standard deviation of the three values. Different letters above each column indicate significant differences among treatments in the same group ($p < 0.05$).

By comparing the Cd content of *Brassica* after adding two kinds of copper-based nanoparticles, it was found that compared with Cd treatment, Cu NPs showed an inhibitory effect on the absorption of heavy metal Cd in *Brassica*. Cu NPs + Cd decreased by 12.6% and 38.6% in leaves and roots, respectively, but there was no significant difference in the changes in leaves. On the contrary, the addition of CuO NPs increased the accumulation of Cd in the plants. Compared with the Cd group, the Cd content in the leaves and roots of *Brassica* increased significantly, increasing by 73.1% and 22.5%, respectively.

The subcellular Cu and Cd contents were determined using fresh *Brassica* leaf samples from each treatment, and the subcellular Cu content in the shoots is shown in Figure 3c. The subcellular Cu content of *Brassica* after Cu NPs treatment was higher than that of CuO NPs treatment. This is consistent with the results for the total Cu content in the shoots. Under Cd pollution, the copper-containing nanoparticles all caused the Cu content in each component in the subcellular to increase significantly. Compared with the Cd group, the proportion of Cu content in the soluble fraction of the Cu NPs + Cd treatment increased significantly, while the proportion of Cu content in the cell wall in the CuO NPs + Cd group also increased significantly. The Cd content in each subcellular component also showed an upward trend: the Cd content in Cu NPs + Cd and CuO NPs + Cd organelles increased by 130.8% and 221.9%, in the cell wall by 151.4% and 330.3%, and in the soluble fraction by 353.7% and 597.6%, respectively. The subcellular Cd content of *Brassica* mainly accumulated in the soluble fraction of cells.

3.4. Nutrient Element Content

The effects of different treatments on the nutrient content of *Brassica* are shown in Table 2. Compared with the Cu NPs treatment, the contents of nutrient elements in the CuO NPs group were increased in the underground part, while the contents of other elements in the aboveground parts were not statistically significant, except for Mn and Na elements, which decreased significantly. Compared with the Cd group, the contents of nutrient elements in roots measured in Cu NPs + Cd and CuO NPs + Cd treatments all increased to varying degrees. The change in Fe and Ca elements was consistent, and the content of aboveground elements decreased significantly. Fe content in leaves decreased significantly by 46.6% and 43.1%, and Ca content decreased by 22.86% and 17.48%. Cu NPs also significantly inhibited the absorption of Zn, Mg, and K elements in the leaves of *Brassica* under Cd treatment. The contents of Ca in the upper part of the ground decreased after the addition of CuO NPs. Cu NPs reduced the uptake of Mn elements in leaves, whereas CuO NPs showed a promotion effect on the uptake of Mn and Na elements (the contents of Mn and Na elements in leaves were increased by 24.3% and 28.3%, respectively).

Table 2. Effects of Cu NPs, CuO NPs and Cd on nutrient content. The mean is the mean of three replicates and the error bars correspond to the standard deviation of the three values. Different letters above each column indicate significant differences among treatments in the same group ($p < 0.05$).

Nutrition Elements		Cd	Cu NPs	CuO NPs	Cu NPs + Cd	CuO NPs + Cd
Zn	Leaves	90.54 ± 20.71 a	79.28 ± 14.20 ab	71.15 ± 7.23 ab	62.87 ± 6.01 b	69.45 ± 6.52 ab
	Roots	80.30 ± 11.96 b	82.60 ± 3.43 b	98.59 ± 8.42 ab	91.28 ± 8.32 ab	121.82 ± 37.70 a
Ca	Leaves	82.83 ± 0.45 a	74.54 ± 1.86 b	73.57 ± 4.32 b	63.90 ± 4.52 c	68.35 ± 6.98 bc
	Roots	77.28 ± 5.16 b	87.23 ± 2.53 ab	90.09 ± 4.50 a	91.03 ± 8.78 a	89.30 ± 5.32 a
Fe	Leaves	5.77 ± 0.03 a	2.76 ± 0.21 c	2.99 ± 0.18 bc	3.08 ± 0.07 bc	3.27 ± 0.22 b
	Roots	98.04 ± 7.98 b	99.66 ± 5.10 b	147.11 ± 28.01 a	130.31 ± 19.82 ab	114.49 ± 18.89 ab
Mg	Leaves	108.24 ± 4.42 a	88.25 ± 3.10 b	84.02 ± 3.23 b	83.87 ± 1.62 b	110.48 ± 3.55 a
	Roots	55.16 ± 3.31 c	53.72 ± 2.17 c	85.38 ± 10.34 a	80.46 ± 8.76 ab	72.38 ± 1.34 b
K	Leaves	945.79 ± 51.23 ab	911.33 ± 42.53 b	949.09 ± 26.97 ab	745.87 ± 22.82 c	1001.68 ± 46.02 a
	Roots	474.67 ± 43.25 c	493.99 ± 10.67 c	774.41 ± 79.53 a	602.24 ± 67.05 b	697.33 ± 43.84 ab
Mn	Leaves	118.64 ± 1.80 b	116.79 ± 7.25 b	97.33 ± 4.48 d	109.15 ± 2.02 c	147.43 ± 2.11 a
	Roots	40.12 ± 4.71 cd	34.86 ± 0.79 d	64.93 ± 14.16 a	49.12 ± 5.01 bc	54.35 ± 4.94 ab
Na	Leaves	21.22 ± 0.80 b	21.45 ± 0.91 b	18.07 ± 0.56 c	22.13 ± 0.28 b	27.25 ± 1.06 a
	Roots	93.52 ± 6.34 b	101.82 ± 5.99 ab	103.50 ± 12.14 ab	92.38 ± 7.96 b	108.56 ± 1.90 a

Figure 4 is the determination result of the change in nutrient elements in the subcellular components of *Brassica*. Comparing the two NPs, only the Mg content of CuO NPs was significantly increased, by 20.7% compared with Cu NPs. The subcellular contents of other nutrients were not statistically significant. Except for Mn, the addition of NPs all decreased the content of nutrient elements in the subcellular components of *Brassica* under cadmium stress. Compared with the Cd group, Cu NPs + Cd and CuO NPs + Cd increased the content of Mn element in the subcellular plant components by 93.66% and 55.43%, respectively, but the changes in CuO NPs treatment showed no significant effect. The same decreasing trend was observed for the subcellular contents of the three nutrient elements Ca, Fe, and Mg by NPs. CuO NPs significantly reduced the subcellular content of these three elements by

22.84%, 5.76%, and 18.10%, but Cu NPs did not have significant effects. Compared with Cd treatment, copper-based nanoparticles significantly inhibited the absorption of subcellular K and Na elements, which were reduced by 33.83% and 54.91% in the Cu NPs + Cd group, and 30.70% and 40.23% in the CuO NPs + Cd group.

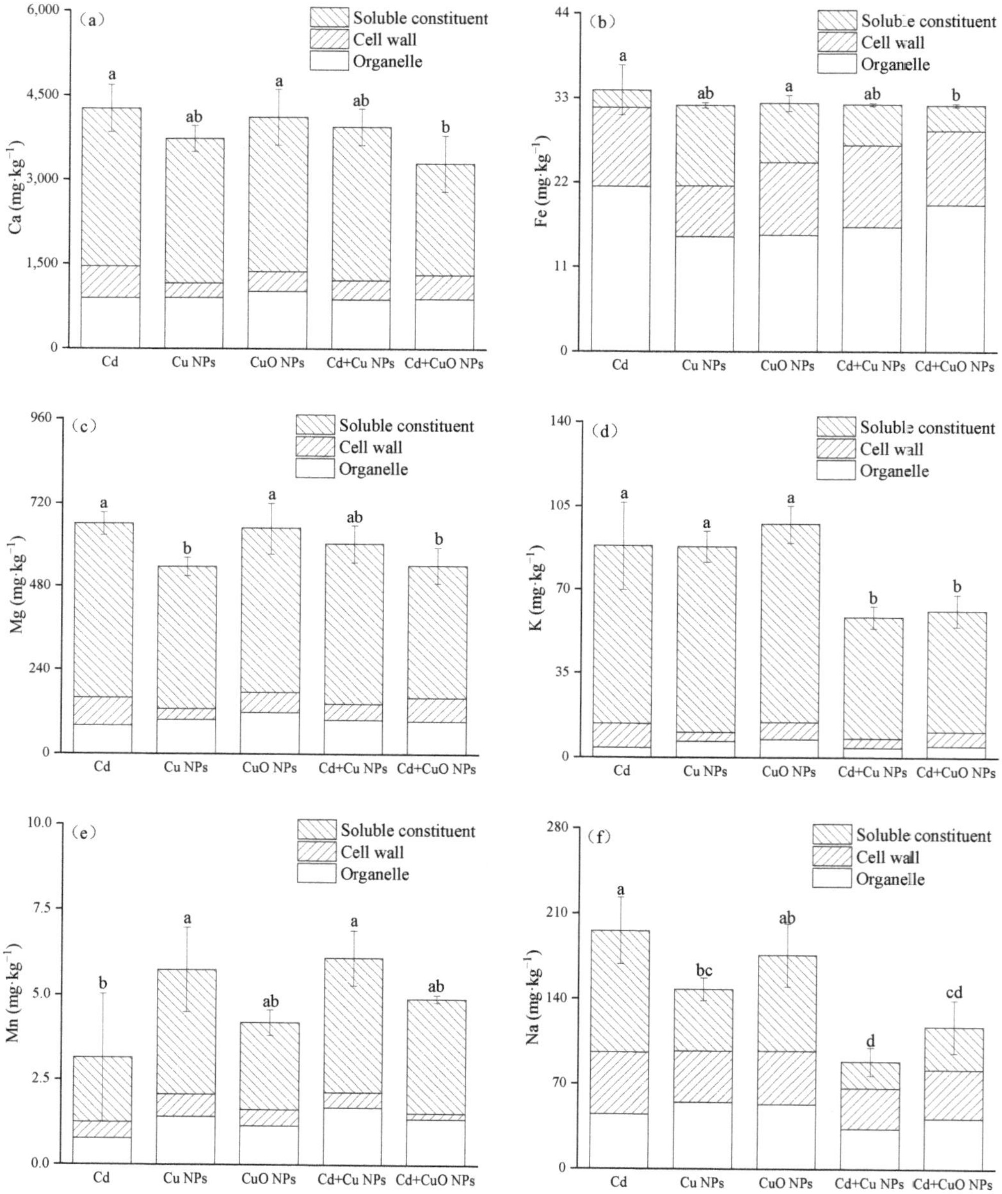

Figure 4. Effects of Cu NPs, CuO NPs and Cd on the subcellular contents of Ca (**a**), Fe (**b**), Mg (**c**), K (**d**), Mn (**e**) and Na (**f**). The mean is the mean of three replicates and the error bars correspond to the standard deviation of the three values. Different letters above each column indicate significant differences between treatments in the same group ($p < 0.05$).

4. Discussion

Cadmium toxicity to plants was observed at the whole plant as well as at the cellular and molecular levels, including energy transfer, photosynthesis, nutritional dysregulation and protein synthesis. Cd can significantly inhibit the photosynthetic rate and growth of plants, and photosynthesis is highly sensitive to Cd and other heavy metal ions [24]. In the present study, copper-based nanoparticles promote the photosynthesis rate of *Brassica* under Cd stress (Table 1). Metal nanoparticles can promote photosynthesis in plants and increase the content of chlorophyll and carotenoids in plants [25]. Govorov et al. believe that metal NPs can improve the efficiency of chemical energy generation in photosynthetic systems [26]. Some studies have shown that copper-based nanoparticles can play a key role in photosynthesis by enhancing chloroplast photosynthetic activity by modulating fluorescence emission, the electron transport chain (ETC), carbon assimilation pathways and photophosphorylation [27]. Plant photosynthesis is very sensitive to Cu [28,29]. Cu plays important roles in mitochondrial respiration, the electron transport chain, photosynthesis, cell wall metabolism and lignin synthesis [30]. Lower concentrations of Cu can promote plant photosynthesis to a certain extent, and the effect is manifested in plant net photosynthetic rate, transpiration rate, relative chlorophyll content and PSII photochemical effect [24]. In the physiological and biochemical study of combined Cu and Cd pollution on *Cinnamomum Camphora*, low-dose Cu promoted the photosynthesis of plants under cadmium stress, and alleviated the damage caused by cadmium to plant photosynthesis to a certain extent [31]. On one hand, the promoting effect of copper-based nanoparticles on *Brassica* biomass is due to the positive effect of photosynthesis on plant growth. On the other hand, lower doses of nanomaterials have a certain stimulatory effect on plants, which can promote plant growth and development and improve plant tolerance to adverse environmental stresses. When the concentration of NPs selected in the experiment is low, the released copper ions can be used as trace elements to stimulate plant growth and increase the biomass of plants [32,33]. Studies have shown that adding a small amount of copper-based nanoparticles can promote plant growth, 10 mg/L CuO NPs increased the root biomass of conventional cotton and transgenic cotton [34]. Low concentrations of nano-copper dioxide particles also had a positive effect on the growth of corn seedlings [8].

Heavy metal stress can negatively affect plant enzyme activity and may cause oxidative damage to cell membranes [35]. Copper-based nanoparticles are phytotoxic by producing excess reactive oxygen species (ROS) or releasing high concentrations of ions [36]. The excessive production of ROS can lead to lipid peroxidation, protein structure destruction, apoptosis and DNA damage [37]. To alleviate the stress caused by ROS, cells have enzymatic mechanisms to eliminate or reduce their damaging effects. Antioxidant enzymes such as catalase (CAT), peroxidase (POD), ascorbate peroxidase (APX) and superoxide dismutase (SOD) play an important role in scavenging ROS to prevent oxidative damage [38]. Antioxidative enzymes are mediators of oxidative damage that help plant biomolecules defend against ROS attack [39]. In plants, CAT is used to remove most of the H_2O_2, POD can sequester the remaining H_2O_2 [40], and SOD catalyzes the conversion of superoxide to oxygen and hydrogen peroxide. In the present study, the CAT, SOD, and POD activities of *Brassica* treated with Cu NPs + Cd were improved (Figure 1), which was similar to the experimental results of Karimi [41] and Kim [42]. Kim et al. found that the enzyme activities of CAT, SOD and POD in cucumbers grown under hydroponic conditions treated with NPs also showed an upward trend. Cu NPs induce oxidative stress in *Brassica*, and promote and stimulate ROS production, and plants enhance the activity of antioxidant enzymes as a defense system in order to resist ROS stress. However, the activity of antioxidant enzymes was decreased by CuO NPs, indicating that CuO NPs produced more ROS and more strongly inhibited the activities of antioxidant enzymes than Cu NPs. Due to their oxidation state, CuO NPs are more toxic than Cu NPs. The CuO NPs destroyed the plant structure and reduced the amounts of nutrients absorbed by the plant. Nutrients cannot support the normal metabolic activities of plants, and cannot synthesize proteins smoothly, thus inhibiting the synthesis of CAT, SOD and POD. APX has the strongest affinity with

H_2O_2, which can quickly remove excess H_2O_2 produced in cells and protect cells from reactive oxygen species poisoning [43]. In this experiment, copper-based nanoparticles both decreased the APX activity in leaves. This result is different from that of increased APX activity in most studies of NP-stressed plants. In an exposure study of Cu NPs and CuO NPs to lettuce (*Lactuca sativa*), the concentration of 10 mg/L also reduced the APX activity of lettuce [44]. We speculate that the result may be related to the applied NPs concentration, but the mechanism is currently unclear and further research is needed.

Consistent with our expected results, the application of copper-based nanoparticles greatly enhanced the copper content in *Brassica*. Copper-based nanometals and nanometal oxides release metal ions which have toxic effects on plants (such as Cu^{2+}). Changes in plant root exudates and culture medium pH can both increase the concentration of soluble metal ions in the rhizosphere solution, so the uptake of metals by plants will also increase [45]. Because of the characteristics of small NPs particles and the large specific surface area, they will adhere to the surface of the root of *Brassica* in large quantities in the culture medium [46]. They can enter plant cells by binding to water channels, carrier proteins, or ion channels. NPs can also increase the permeability of plant cell walls, generate new pores and enter plant cells by endocytosis or by combining with organic chemicals in the environmental medium [47]. The total copper content in Brassica was higher in CuO NPs treatment than in Cu NPs treatment, which is also consistent with the results of Kadri et al. They believed that CuO NPs would release more copper ions than Cu NPs [48]. We observed more Cu accumulated in the roots of *Brassica* than in the leaves. This is due to the plant's repulsion mechanism that can allow metals to accumulate in the roots and prevents their transport to the shoots to maintain the homeostasis of mineral elements in tissues and organs. Since roots are directly exposed to heavy metal pollutants under hydroponic conditions, most of the cadmium uptake by plants is also concentrated in the roots. We believe that the cadmium content in *Brassica* is related to the copper content transported to the shoot. Usually, different metals may share the same transporter on the cell membrane, and plants do not clearly distinguish between important micronutrients and non-essential metals [49]. Studies have shown that plants absorb micronutrients faster than non-essential heavy metals [50]. Since Cd and Cu are metal elements with the same electronic valence, there are competing sites, so Cu has a certain inhibitory effect on the absorption and transfer of Cd elements, which will reduce the Cd content in the shoots of plants. Consistent with the research results of Cvjetko et al., Cu can effectively inhibit the absorption of Cd when copper and cadmium are combined to stress plants [51]. It was observed that the content of Cd absorbed by the roots of *Brassica* in the treatment group of CuO NPs increased. The hydroxyl, carboxyl, amino, sulfhydryl and aldehyde groups of proteins and polysaccharides in plant cell walls can combine with metal cations through complexation, precipitation and ion exchange, limiting the transport of metal cations through the cell membrane [19]. NPs exposure induces ROS generation or other effects that may alter the cell wall structure, such as reducing cell wall thickness, triggering cell wall loosening, or changing cell wall pore size, which may also lead to an upward trend in Cu and Cd content in plants [52,53]. On the other hand, the higher the total metal content in plants, the more severe the oxidative damage to biomolecules. When the content of the two metal elements in the stressed plants is large, the absorption of the two elements by the plants will show a certain additive effect. For example, in the combined lead and cadmium treatment of soybean, as the lead addition increased, the cadmium content in the roots also increased [54]. Additionally, studies have shown that Cu at low concentrations can alleviate Cd-induced damage, while the combined stress of higher concentrations of Cu and Cd exceeds the potential of plants to resist heavy metal stress, and Cd and Cu may interact to increase Cd uptake by plants [55]. Cu NPs + Cd and CuO NPs + Cd used to treat the subcellular elements resulted in increased Cu and Cd content compared with Cd treatment, which verifies the above conjecture that NPs damage the cell membrane and increase the content of Cu and Cd in plants. The cell wall is considered the first barrier to protect protoplasts from toxicity [56]. The Cu content in the subcellular Cu NPs-treated group was much higher than that in the CuO NPs-treated group, which may be because the particle size of Cu NPs we chose is smaller than that of CuO NPs. In the case of agglomeration, nanoparticles with diameters over 20 nm are barely able to penetrate the cell wall [54,57]. The order of Cd content in each component is soluble component > organelle > cell wall. The contents of Cu and Cd in the subcellular soluble fraction of *Brassica* leaves were high, indicating that the soluble fraction in leaf cells was the main part of Cd enrichment. The soluble components of plants are composed of cytoplasm and vacuoles. Cytoplasm is the main location for cell metabolism, and the function of vacuoles is mainly to participate in cellular water metabolism. Various proteins, organic acids, organic bases and other substances contained in vacuoles can interact with Cd combinations. Following the ingestion of metals, plants limit the excess accumulation of metal elements in their roots or store them in tissues or organs that are less sensitive to their toxicity, which may include the cytoplasm and chloroplasts. It is also one of the important mechanisms by which plants resist heavy metal poisoning and participate in heavy metal detoxification [58,59].

The absorption and content of nutrient elements in roots showed an upward trend (Table 2). The effect may be due to the interaction of heavy metals with metal transporters, resulting in altered plant gene expression related to nutrient uptake and changes in plasma membrane permeability when plants are exposed to cadmium and excess copper stress [60,61]. The content of Fe in the roots increased and the content in the shoots decreased significantly. This may be due to the fact that in copper-stressed plants, iron may be complexed in the form of ferritin (iron-binding protein), which is overproduced to protect cells from oxidative damage [62]. However, Fe was more active than Cu, formed insoluble hydroxides, and remained in the roots [11]. NPs reduced the content of most nutrient elements in the leaves of *Brassica*. On the one hand, the surface effect of NPs may lead to the absorption of a large amount of other useful mineral elements on the surface of NPs, thereby reducing the bioavailability of mineral elements by plants [46]. On the other hand, the Cu ions released by NPs will also have negative effects on the nutrient absorption of *Brassica*. Consistent with previous research results, when the Cu concentration caused stress to plants, the contents of Fe, Ca, K, and Mg in the underground parts of plants were significantly higher than those of the control. On the contrary, the contents of Zn, Ca, K, and Mg in the upper part of the ground all decreased after the addition of NPs [63]. Nano-CuO significantly reduced the absorption of manganese, zinc, iron, magnesium, molybdenum and boron by cotton [34]. Cu NPs and CuO NPs also reduced the absorption of manganese, calcium, phosphorus and magnesium by lettuce in Trujillo Reyes et al. [45]. Cu NPs also reduced the uptake of Mn elements in leaves, whereas CuO NPs showed a promotion effect on the uptake of Mn and Na elements (the contents of Mn and Na elements in leaves were increased by 24.3% and 28.3%, respectively). The addition of NPs changed the ability of *Brassica* to absorb and transport certain nutrients and thus their nutritional value. From the experimental results, the negative impact of Cd combined with copper-based NPs on the nutrient absorption of *Brassica* is far more than that of NPs. The results of subcellular nutrient determination were basically consistent with the results in Table 2. Compared with Cd treatment, the content of most nutrient elements in the subcellular components also showed a downward trend after the addition of NPs. The rise of the subcellular Mn element may be due to it being an important part of the Mn-superoxide dismutase (SOD) enzyme [64]. Plants need to increase absorption to synthesize antioxidant enzyme in order to protect plants from active oxygen damage. In general, NPS interferes with the absorption of the upper nutrient elements of the *Brassica*, which reduces the nutritional value of *Brassica*.

The Pearson correlation results between the elements absorbed by the leaves of *Brassica* are shown in Figure 5a. The absorption of Cd and Cu in the shoots showed a significant negative correlation ($\rho = -0.685$), which verifies that the two elements are increasing in the plants. When these two elements are transported in the shoots, there is a competitive relationship. When the concentration is low, the effects of Cd and Cu on plants are antagonistic. However, in the experiment, Cd was significantly positively correlated with K (ρ = 0.733), and extremely significantly positively correlated with Mn (ρ = 0.981) and Na (ρ = 0.931). Previous studies have shown that in some plants, both Cd and K are positively correlated. There may be a certain synergistic effect between the elements, and Na is also not negatively affected by Cd treatment [65]. In this study, Cu showed a negative correlation with the absorption of several elements measured, was moderately correlated with Mn (ρ = −0.768), and showed a very high correlation with K and Mg (ρ = −0.954 and ρ = −0.974, respectively). This may be due to the higher copper content altering membrane permeability and impairing plant nutrient uptake. Among the nutrients absorbed by the leaves of *Brassica*, Zn and Fe (ρ = 0.711), Zn and Ca (ρ = 0.730), Mn and Mg (ρ = 0.738), and Mn and K (ρ = 0. 794) showed a significant positive correlation. The correlations between Fe and Ca (ρ = 0.872), Mg and K (ρ = 0.936), and Mn and Na (ρ = 0. 906) were extremely significant and positive.

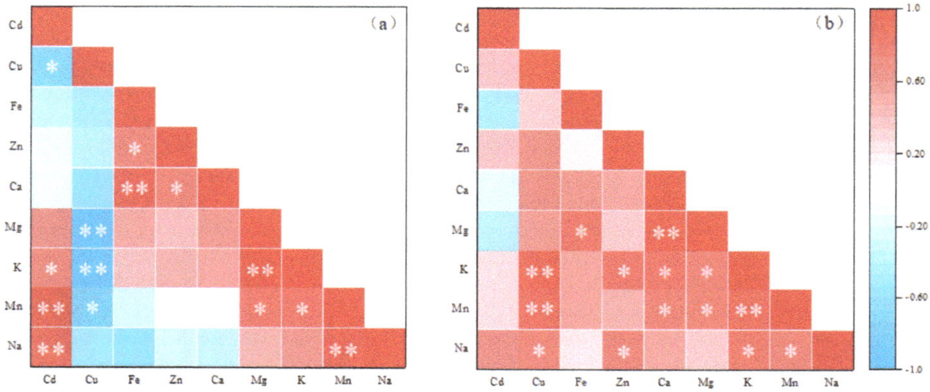

Figure 5. Pearson correlation between antioxidant enzyme activity, heavy metals and mineral elements in *Brassica* leaves (**a**) and roots (**b**). (significant correlation between * and ** at $p \leq 0.05$ and $p \leq 0.01$, respectively).

Figure 5b shows the root element correlation. It is observed that only Cd has a negative correlation with Mg, Ca and Fe, but it is not significant. All other underground elements are positively correlated, among which Cu is extremely significantly positively correlated with Mn and K (ρ = 0.838 and ρ = 0.895, respectively), Ca and Mg (ρ = 0.886), and K and Mn (ρ = 0.812). This may be due to the fact that Cu can increase the permeability of the membrane [66]. NPs can also cause damage to the cells in the root of the plant, which will increase the content of elements entering the *Brassica* [17].

5. Conclusions

Copper-based nanoparticles are currently widely used, and may combine with the existing heavy metal pollution in soil to have a combined effect on plants, causing negative effects on crops and humans. In this study, Cu and CuO NPs showed promoting effects on both the photosynthetic rate and biomass of Cd-stressed plants. The changes in antioxidant enzymes indicated that NPs enhanced the oxidative damage of *Brassica* under Cd treatment, and the oxidative damage caused by CuO NPs was stronger than that of Cu NPs. Both applications of copper-containing nanoparticles increased the Cu content in plants. Cu NPs inhibited Cd uptake in plants, while CuO NPs exhibited a combined effect. NPs caused damage to cells, and both the content of Cu and of Cd in plant subcellular components increased. The combined effect of Cd and copper-based nanoparticles can change the absorption of mineral nutrients by *Brassica* and reduce the nutritional value of crops.

This experiment explored the toxic effects of copper-based nanoparticles on the physiological and biochemical aspects of *Brassica* under Cd stress. However, in this experiment, the experimental method of hydroponics and a lower concentration of NPs were selected. Due to the complex toxicity mechanism of NPs, there may be some differences in the toxic effects caused by the soil environment and the increase in NPs concentration, which warrants further research to prove. The effects of the combination of Cd and NPs on cell and gene damage in plants may be a future research direction.

Supplementary Materials: The following supporting information can be downloaded at: https://www.mdpi.com/article/10.3390/nano12091497/s1. Table S1: The composition and concentration of Hoagland nutrient solution.

Author Contributions: S.W.: conceptualization, methodology, software, investigation, writing—original draft; Y.F.: methodology, investigation; S.Z.: investigation, validation; Y.X.: resources, supervision, formal analysis; Y.S.: investigation, supervision, writing—review and editing. All authors have read and agreed to the published version of the manuscript.

Funding: This work was supported by the National Key Research and Development Program of China (2021YFD2000205) and the National Natural Science Foundation of China (31971525).

Data Availability Statement: Not applicable.

Conflicts of Interest: The authors declare no conflict of interest.

References

1. Xie, L.; Hao, P.; Cheng, Y.; Ahmed, I.M.; Cao, F. Effect of combined application of lead, cadmium, chromium and copper on grain, leaf and stem heavy metal contents at different growth stages in rice. *Ecotoxicol. Environ. Saf.* **2018**, *162*, 71–76. [CrossRef] [PubMed]
2. Singh, J.; Lee, B.K. Influence of nano-TiO_2 particles on the bioaccumulation of Cd in soybean plants (*Glycine max*): A possible mechanism for the removal of Cd from the contaminated soil. *J. Environ. Manag.* **2016**, *170*, 88–96. [CrossRef] [PubMed]
3. Chen, X.; Zhang, C.; Tan, L.; Wang, J. Research progress in toxicity of nanomaterials manufactured on microalgae. *Mar. Sci.* **2017**, *41*, 134–143.
4. Yuan, X.; Xue, N.; Han, Z. A meta-analysis of heavy metals pollution in farmland and urban soils in China over the past 20 years. *J. Environ. Sci.* **2021**, *101*, 217–226. [CrossRef]
5. Keller, A.A.; McFerran, S.; Lazareva, A.; Suh, S. Global life cycle releases of engineered nanomaterials. *J. Nanopart. Res.* **2013**, *15*, 1692. [CrossRef]
6. Kah, M.; Navarro, D.; Kookana, R.S.; Kirby, J.K.; Santra, S.; Ozcan, A.; Kabiri, S. Impact of (nano)formulations on the distribution and wash-off of copper pesticides and fertilisers applied on citrus leaves. *Environ. Chem.* **2019**, *16*, 401–410. [CrossRef]
7. Marmiroli, M.; Pagano, L.; Rossi, R.; De La Torre-Roche, R.; Lepore, G.O.; Ruotolo, R.; Gariani, G.; Bonanni, V.; Pollastri, S.; Puri, A.; et al. Copper Oxide Nanomaterial Fate in Plant Tissue: Nanoscale Impacts on Reproductive Tissues. *Environ. Sci. Technol.* **2021**, *55*, 10769–10783. [CrossRef]
8. Adhikari, T.; Sarkar, D.; Mashayekhi, H.; Xing, B. Growth and enzymatic activity of maize (*Zea mays* L.) plant: Solution culture test for copper dioxide nano particles. *J. Plant Nutr.* **2016**, *39*, 102–118. [CrossRef]
9. Shaw, A.K.; Hossain, Z. Impact of nano-CuO stress on rice (*Oryza sativa* L.) seedlings. *Chemosphere* **2013**, *93*, 906–915. [CrossRef]
10. Nair, P.M.G.; Chung, I.M. The responses of germinating seedlings of green peas to copper oxide nanoparticles. *Biol. Plant.* **2015**, *59*, 591–595. [CrossRef]
11. Hong, J.; Rico, C.M.; Zhao, L.; Adeleye, A.S.; Keller, A.A.; Peralta-Videa, J.R.; Gardea-Torresdey, J.L. Toxic effects of copper-based nanoparticles or compounds to lettuce (*Lactuca sativa*) and alfalfa (*Medicago sativa*). *Environ. Sci.-Process Impacts* **2015**, *17*, 177–185. [CrossRef] [PubMed]
12. Zuverza-Mena, N.; Medina-Velo, I.A.; Barrios, A.C.; Tan, W.; Peralta-Videa, J.R.; Gardea-Torresdey, J.L. Copper nanoparticles/compounds impact agronomic and physiological parameters in cilantro (*Coriandrum sativum*). *Environ. Sci.-Process Impacts* **2015**, *17*, 1783–1793. [CrossRef] [PubMed]
13. Wang, P.; Zhao, L.; Huang, Y.; Qian, W.; Zhu, X.; Wang, Z.; Cai, Z. Combined toxicity of nano-TiO_2 and Cd^{2+} to Scenedesmus obliquus: Effects at different concentration ratios. *J. Hazard. Mater.* **2021**, *418*, 126354. [CrossRef] [PubMed]
14. Venkatachalam, P.; Jayaraj, M.; Manikandan, R.; Geetha, N.; Rene, E.R.; Sharma, N.C.; Sahi, S.V. Zinc oxide nanoparticles (ZnONPs) alleviate heavy metal-induced toxicity in Leucaena leucocephala seedlings: A physiochemical analysis. *Plant Physiol. Biochem.* **2017**, *110*, 59–69. [CrossRef] [PubMed]
15. Rizwan, M.; Ali, S.; Ali, B.; Adrees, M.; Arshad, M.; Hussain, A.; Rehman, M.Z.U.; Waris, A.A. Zinc and iron oxide nanoparticles improved the plant growth and reduced the oxidative stress and cadmium concentration in wheat. *Chemosphere* **2019**, *214*, 269–277. [CrossRef] [PubMed]

16. Wang, Z.Y.; Xu, L.N.; Zhao, J.; Wang, X.K.; White, J.C.; Xing, B.S. CuO Nanoparticle Interaction with Arabidopsis thaliana: Toxicity, Parent-Progeny Transfer, and Gene Expression. *Environ. Sci. Technol.* **2016**, *50*, 6008–6016. [CrossRef]
17. Deng, F.; Wang, S.L.; Xin, H. Toxicity of CuO Nanoparticles to Structure and Metabolic Activity of Allium cepa Root Tips. *Bull. Environ. Contam. Toxicol.* **2016**, *97*, 702–708. [CrossRef]
18. Zhao, L.J.; Huang, Y.X.; Zhou, H.J.; Adeleye, A.S.; Wang, H.T.; Ortiz, C.; Mazer, S.J.; Keller, A.A. GC-TOF-MS based metabolomics and ICP-MS based metallomics of cucumber (*Cucumis sativus*) fruits reveal alteration of metabolites profile and biological pathway disruption induced by nano copper. *Environ.-Sci. Nano* **2016**, *3*, 1114–1123. [CrossRef]
19. Li, H.; Luo, N.; Zhang, L.J.; Zhao, H.M.; Li, Y.W.; Cai, Q.Y.; Wong, M.H.; Mo, C.H. Do arbuscular mycorrhizal fungi affect cadmium uptake kinetics, subcellular distribution and chemical forms in rice? *Sci. Total Environ.* **2016**, *571*, 1183–1190. [CrossRef]
20. Tandy, N.E.; Giulio, R.T.; Richardson, C.J. Assay and Electrophoresis of Superoxide Dismutase from Red Spruce (*Picea rubens* Sarg.), Loblolly Pine (*Pinus taeda* L.), and Scotch Pine (*Pinus sylvestris* L.) A Method for Biomonitoring. *Plant Physiol.* **1989**, *90*, 742–748. [CrossRef]
21. Jurkow, R.; Skara, A.; Pokluda, R.; Smoleń, S.; Kalisz, A. Biochemical response of oakleaf lettuce seedlings to different concentrations of some metal(oid) oxide nanoparticles. *Agronomy* **2020**, *10*, 997. [CrossRef]
22. Shen, S.L.; Xia, B.C.; Xue, F. Cadmium effects on photosynthetic system of sauropus androgynus as a potential cd-accumulator. *Fresenius Environ. Bull.* **2012**, *21*, 3551–3557.
23. Li, P.H.; Lin, Q.; Xu, G.Z. Effects of pb and cd stress on the photosynthetic physiological characters of potato in heavy metal pollution of soil. *Appl. Ecol. Environ. Res.* **2019**, *17*, 12287–12295. [CrossRef]
24. Li, D.; Zhu, Z.; Liu, Y.; Wang, Y. Influence of cadmium on photosynthesis of *Brassica campestris* subsp. *chinensis* L. *J. Zhejiang Univ.* **2005**, *31*, 459–464.
25. Servin, A.; Elmer, W.; Mukherjee, A.; De la Torre-Roche, R.; Hamdi, H.; White, J.C.; Bindraban, P.; Dimkpa, C. A review of the use of engineered nanomaterials to suppress plant disease and enhance crop yield. *J. Nanopart. Res.* **2015**, *17*, 1–21. [CrossRef]
26. Govorov, A.O.; Carmeli, I. Hybrid structures composed of photosynthetic system and metal nanoparticles: Plasmon enhancement effect. *Nano Lett.* **2007**, *7*, 620–625. [CrossRef]
27. Pradhan, S.; Patra, P.; Mitra, S.; Dey, K.K.; Basu, S.; Chandra, S.; Palit, P.; Goswami, A. Copper Nanoparticle (CuNP) Nanochain Arrays with a Reduced Toxicity Response: A Biophysical and Biochemical Outlook on Vigna radiata. *J. Agric. Food Chem.* **2015**, *63*, 2606–2617. [CrossRef]
28. Nuzahat, H.; Zhang, Y.; Asiya, K.; Abdurexit, A. Influence of Different Zinc and Copper Levels on Cadmium Accumulation and Physiological Characterization in *Vetiveria zizanioides* L. *J. Inn. Mong. Univ.* **2012**, *43*, 543–550.
29. Aibibu, N.; Liu, Y.; Zeng, G.; Wang, X.; Chen, B.; Song, H.; Xu, L. Cadmium accumulation in vetiveria zizanioides and its effects on growth, physiological and biochemical characters. *Bioresour. Technol.* **2010**, *101*, 6297–6303. [CrossRef]
30. Mir, A.R.; Pichtel, J.; Hayat, S. Copper: Uptake, toxicity and tolerance in plants and management of Cu-contaminated soil. *Biometals* **2021**, *34*, 737–759. [CrossRef]
31. Zhou, J.; Cheng, K.; Zheng, J.; Liu, Z.; Shen, W.; Fan, H.; Jin, Z. Physiological and Biochemical Characteristics of Cinnamomum camphora in Response to Cu- and Cd-Contaminated Soil. *Water Air Soil Pollut.* **2019**, *230*, 1–11. [CrossRef]
32. Karlsson, H.L.; Gustafsson, J.; Cronholm, P.; Moller, L. Size-dependent toxicity of metal oxide particles-A comparison between nano- and micrometer size. *Toxicol. Lett.* **2009**, *188*, 112–118. [CrossRef] [PubMed]
33. Wierzbicka, M.; Obidzińska, J. The effect of lead on seed imbibition and germination in different plant species. *Plant Sci.* **1998**, *137*, 155–171. [CrossRef]
34. Van, L.; Ma, C.; Shang, J.; Rui, Y.; Liu, S.; Xing, B. Effects of CuO nanoparticles on insecticidal activity and phytotoxicity in conventional and transgenic cotton. *Chemosphere* **2016**, *144*, 661–670. [CrossRef] [PubMed]
35. Santala, K.R.; Ryser, P. Influence of heavy-metal contamination on plant response to water availability in white birch, Betula papyrifera. *Environ. Exp. Bot.* **2009**, *66*, 334–340. [CrossRef]
36. Ahmed, B.; Rizvi, A.; Zaidi, A.; Khan, M.S.; Musarrat, J. Understanding the phyto-interaction of heavy metal oxide bulk and nanoparticles: Evaluation of seed germination, growth, bioaccumulation, and metallothionein production. *RSC Adv.* **2019**, *9*, 4210–4225. [CrossRef]
37. Dietz, K.J.; Herth, S. Plant nanotoxicology (vol 16, pg 582, 2011). *Trends Plant Sci.* **2012**, *17*, 180. [CrossRef]
38. Khatun, S.; Ali, M.B.; Hahn, E.J.; Paek, K.Y.; Botany, E. Copper toxicity in Withania somnifera: Growth and antioxidant enzymes responses of in vitro grown plants. *Environ. Exp. Bot.* **2008**, *64*, 279–285. [CrossRef]
39. Mittler, R. Oxidative stress, antioxidants and stress tolerance. *Trends Plant Sci.* **2002**, *7*, 405–410. [CrossRef]
40. Willekens, H.; Chamnongpol, S.; Davey, M.; Schraudner, M.; Langebartels, C.; VanMontagu, M.; Inze, D.; VanCamp, W. Catalase is a sink for H_2O_2 and is indispensable for stress defence in C-3 plants. *EMBO J.* **1997**, *16*, 4806–4816. [CrossRef]
41. Karimi, P.; Khavari-Nejad, R.A.; Niknam, V.; Ghahremaninejad, F.; Najafi, F. The Effects of Excess Copper on Antioxidative Enzymes, Lipid Peroxidation, Proline, Chlorophyll, and Concentration of Mn, Fe, and Cu in Astragalus neo-mobayenii. *Sci. World J.* **2012**, *2012*, 615670. [CrossRef] [PubMed]
42. Kim, S.; Lee, S.; Lee, I. Alteration of Phytotoxicity and Oxidant Stress Potential by Metal Oxide Nanoparticles in *Cucumis sativus*. *Water Air Soil Pollut.* **2012**, *223*, 2799–2806. [CrossRef]
43. Li, Z.; Li, J.; Zhang, G. Expression regulation of plant ascorbate peroxidase and its tolerance to abiotic stresses. *Hereditas* **2013**, *35*, 45–54. [CrossRef]

44. Trujillo-Reyes, J.; Majumdar, S.; Botez, C.E.; Peralta-Videa, J.R.; Gardea-Torresdey, J.L. Exposure studies of core-shell Fe/Fe_3O_4 and Cu/CuO NPs to lettuce (*Lactuca sativa*) plants: Are they a potential physiological and nutritional hazard? *J. Hazard. Mater.* **2014**, *267*, 255–263. [CrossRef] [PubMed]
45. Franklin, N.M.; Rogers, N.J.; Apte, S.C.; Batley, G.E.; Gadd, G.E.; Casey, P.S. Comparative toxicity of nanoparticulate ZnO, bulk ZnO, and $ZnCl_2$ to a freshwater microalga (*Pseudokirchneriella subcapitata*): The importance of particle solubility. *Environ. Sci. Technol.* **2007**, *41*, 8484–8490. [CrossRef]
46. Shi, J.; Peng, C.; Yang, Y.; Yang, J.; Zhang, H.; Yuan, X.; Chen, Y.; Hu, T. Phytotoxicity and accumulation of copper oxide nanoparticles to the Cu-tolerant plant Elsholtzia splendens. *Nanotoxicology* **2014**, *8*, 179–188. [CrossRef]
47. Miralles, P.; Church, T.L.; Harris, A.T. Toxicity, Uptake, and Translocation of Engineered Nanomaterials in Vascular plants. *Environ. Sci. Technol.* **2012**, *46*, 9224–9239. [CrossRef]
48. Kadri, O.; Karmous, I.; Kharbech, O.; Arfaoui, H.; Chaoui, A. Cu and CuO Nanoparticles Affected the Germination and the Growth of Barley (*Hordeum vulgare* L.) Seedling. *Bull. Environ. Contam. Toxicol.* **2022**, *108*, 585–593. [CrossRef]
49. Verkleij, J.A.C.; Golan-Goldhirsh, A.; Antosiewisz, D.M.; Schwitzguébel, J.; Schröder, P. Dualities in plant tolerance to pollutants and their uptake and translocation to the upper plant parts. *Environ. Exp. Bot.* **2009**, *67*, 10–22. [CrossRef]
50. Mishra, V.K.; Tripathi, B.D. Concurrent removal and accumulation of heavy metals by the three aquatic macrophytes. *Bioresour. Technol.* **2008**, *99*, 7091–7097. [CrossRef]
51. Cvjetko, P.; Tolic, S.; Sikic, S.; Balen, B.; Tkale, M.; Vidakovic-Cifrek, Z.; Pavlica, M. Effect of copper on the toxicity and genotoxicity of cadmium in duckweed (*Lemna minor* L.). *Arh. Hig. Rada. Toksikol.* **2010**, *61*, 287–296. [CrossRef] [PubMed]
52. Kim, J.H.; Lee, Y.; Kim, E.J.; Gu, S.; Sohn, E.J.; Seo, Y.S.; An, H.J.; Chang, Y.S. Exposure of Iron Nanoparticles to Arabidopsis thaliana Enhances Root Elongation by Triggering Cell Wall Loosening. *Environ. Sci. Technol.* **2014**, *48*, 3477–3485. [CrossRef] [PubMed]
53. Ma, C.; White, J.C.; Dhankher, O.P.; Xing, B. Metal-Based Nanotoxicity and Detoxification Pathways in Higher Plants. *Environ. Sci. Technol.* **2015**, *49*, 7109–7122. [CrossRef] [PubMed]
54. Luan, Z.Q.; Cao, H.C.; Yan, B.X. Individual and combined phytotoxic effects of cadmium, lead and arsenic on soybean in Phaeozem. *Plant Soil Environ.* **2008**, *54*, 403–411. [CrossRef]
55. Majsec, K.; Cvjetko, P.; Tolic, S.; Tkalec, M.; Balen, B.; Pavlica, M. Integrative approach gives new insights into combined Cd/Cu exposure in tobacco. *Acta Physiol. Plant.* **2016**, *38*, 1–14. [CrossRef]
56. Weng, B.S.; Xie, X.Y.; Weiss, D.J.; Liu, J.C.; Lu, H.L.; Yan, C.L. *Kandelia obovata* (S., L.) Yong tolerance mechanisms to Cadmium: Subcellular distribution, chemical forms and thiol pools. *Mar. Pollut. Bull.* **2012**, *64*, 2453–2460. [CrossRef]
57. Rondeau-Mouro, C.; Defer, D.; Leboeuf, E.; Lahaye, M. Assessment of cell wall porosity in Arabidopsis thaliana by NMR spectroscopy. *Int. J. Biol. Macromol.* **2008**, *42*, 83–92. [CrossRef]
58. Chen, Y.; Liu, X.; Wang, M.; Wang, J.; Yan, X. Cadmium tolerance, accumulation and relationship with Cd subcellular distribution in *Ricinus communis* L. *Acta Sci. Circumstantiae* **2014**, *34*, 2440–2446.
59. Printz, B.; Lutts, S.; Hausman, J.F.; Sergeant, K. Copper Trafficking in Plants and Its Implication on Cell Wall Dynamics. *Front. Plant Sci.* **2016**, *7*, 601. [CrossRef]
60. Roy, S.K.; Cho, S.W.; Kwon, S.J.; Kamal, A.H.M.; Lee, D.G.; Sarker, K.; Lee, M.S.; Xin, Z.; Woo, S.H. Proteome characterization of copper stress responses in the roots of sorghum. *Biometals* **2017**, *30*, 765–785. [CrossRef]
61. Cambrolle, J.; Garcia, J.L.; Figueroa, M.E.; Cantos, M. Evaluating wild grapevine tolerance to copper toxicity. *Chemosphere* **2015**, *120*, 171–178. [CrossRef] [PubMed]
62. Siddiqui, Z.A.; Khan, M.R.; Abd_Allah, E.F.; Aiman, P.; Parveen, A. Titanium dioxide and zinc oxide nanoparticles affect some bacterial diseases, and growth and physiological changes of beetroot. *J. Veg. Sci.* **2019**, *25*, 409–430. [CrossRef]
63. Zhang, K.; Tong, H.; Huang, S.; Yuan, H. Effect of Cu stress on Cu accumulation and other nutrient element absorption of Iris pseudacaorus and I. lactea var. chinensis. *J. Plant Resour. Environ.* **2007**, *16*, 18–22.
64. DalCorso, G.; Manara, A.; Piasentin, S.; Furini, A. Nutrient metal elements in plants. *Metallomics* **2014**, *6*, 1770–1788. [CrossRef] [PubMed]
65. Moral, R.; Gomez, I.; Navarro-Pedreno, J.; Mataix, J. Effects of cadmium on nutrient distribution, yield, and growth of tomato grown in soilless culture. *J. Plant Nutr.* **1994**, *17*, 953–962. [CrossRef]
66. Gajewska, E.; Glowacki, R.; Mazur, J.; Sklodowska, M. Differential response of wheat roots to Cu, Ni and Cd treatment: Oxidative stress and defense reactions. *Plant Growth Regul.* **2013**, *71*, 13–20. [CrossRef]

nanomaterials

Article

Application of Exogenous Iron Alters the Microbial Community Structure and Reduces the Accumulation of Cadmium and Arsenic in Rice (*Oryza sativa* L.)

Tingting Li [1,2], Jiayuan Li [1], Xin Zhan [1], Xueli Wang [1], Bing He [1], Feishu Cao [3], Changjun Liao [3], Yuefeng Yu [2], Zengyu Zhang [1], Junhui Zhang [2], Bei Li [1], Jiancheng Chen [1], Hong Li [4], Zhiqiang Zhu [5], Yanyan Wei [1,*] and Junming Hu [2,*]

1 State Key Laboratory for Conservation and Utilization of Subtropical Agro-Bioresources, Cultivation Base of Guangxi Key Laboratory for Agro-Environment and Agro-Products Safety, College of Agriculture, Guangxi University, Nanning 530004, China; ltt1210@126.com (T.L.); 1917303005@st.gxu.edu.cn (J.L.); njueczx@163.com (X.Z.); wxl0524@126.com (X.W.); bingh2000@126.com (B.H.); 1817303017@st.gxu.edu.cn (Z.Z.); le_bei@st.gxu.edu.cn (B.L.); 2017392001@st.gxu.edu.cn (J.C.)
2 Agricultural Resources and Environment Research Insititute, Guangxi Academy of Agricultural Sciences, Nanning 530007, China; yuyue202204@163.com (Y.Y.); zjh914zjh914@163.com (J.Z.)
3 Guangxi Bocco Enviromental Protection Technology Co., Ltd., Nanning 530007, China; feishu.cao@hotmail.com (F.C.); lcj19176042408@163.com (C.L.)
4 Key Laboratory of Eco-Environment of Three Gorges Region, Ministry of Education, Chongqing University, Chongqing 400044, China; hongli@cqu.edu.cn
5 College of Tropical Crops, Hainan University, Haikou 570228, China; zqzhu@hainanu.edu.cn
* Correspondence: yanyanwei@gxu.edu.cn or yanyanwei2008@163.com (Y.W.); jmhu06@126.com (J.H.); Tel.: +86-1860-7718-450 (Y.W.)

check for updates

Citation: Li, T.; Li, J.; Zhan, X.; Wang, X.; He, B.; Cao, F.; Liao, C.; Yu, Y.; Zhang, Z.; Zhang, J.; et al. Application of Exogenous Iron Alters the Microbial Community Structure and Reduces the Accumulation of Cadmium and Arsenic in Rice (*Oryza sativa* L.). *Nanomaterials* **2022**, *12*, 1311. https://doi.org/10.3390/nano12081311

Academic Editor: Eleonore Fröhlich

Received: 10 March 2022
Accepted: 6 April 2022
Published: 11 April 2022

Abstract: Cadmium (Cd) and arsenic (As) contamination of soil has been a public concern due to their potential accumulation risk through the food chain. This study was conducted to investigate the performance of ferrous sulfate ($FeSO_4$) and ferric oxide (Fe_2O_3) nanoparticle (Nano-Fe) to stabilize the concentrations of Cd and As in paddy soil. Both Fe treatments led to low extractable Cd and the contents of specifically sorbed As contents, increased ($p < 0.05$) the Shannon index and decreased ($p < 0.05$) the Simpson diversity indices compared with the control. Nano-Fe increased the relative abundances of Firmicutes and Proteobacteria and decreased the abundances of Acidobacteria and Chloroflexi. Moreover, the addition of both forms of Fe promoted the formation of Fe plaque and decreased the translocation factor index (TFs) $_{root/soil}$, TFs $_{shoot/root}$, and TFs $_{grain/shoot}$ of Cd and As. These results suggest that exogenous Fe may modify the microbial community and decrease the soil available Cd and As contents, inhibit the absorption of Cd and As by the roots and decrease the transport of Cd and As in rice grains and the risk intake in humans. These findings demonstrate that soil amendment with exogenous Fe, particularly Nano-Fe, is a potential approach to simultaneously remediate the accumulation of Cd and As from the soil to rice grain systems.

Keywords: cadmium; arsenic; rice; microorganism; ferrous sulfate; ferric oxide nanoparticle

1. Introduction

Cadmium (Cd) and arsenic (As) are common toxic elements in soil and are classified as the number one carcinogens by the International Agency for Research on Cancer [1,2]. Recent nationwide surveys show that the rates of Cd and As in China are 7.0% and 2.7%, respectively, which are higher than China's soil environmental quality limits [3]. Co-pollution with Cd and As in the paddy soil of South China is particularly serious owing to the rapid development of urbanization and mineral production and processing [4]. The excessive intake of Cd can lead to Cd poisoning, which can cause kidney poisoning and chondroplasty [5]. As has a clear causal relationship with the occurrence of skin and lung cancer and is closely related to the occurrence of visceral cancer, such as liver and bladder cancers. Therefore, reducing the amount of Cd and As is of the utmost importance.

Rice is a major grain, which is the main food source of more than half of the global population [6], and is also the main food source of Cd and As in many Asian countries, particularly in China [6]. Cd and As are difficult to control simultaneously because their geochemical behavior and bioavailability differ in the paddy field environment owing to the alternation of soil that is dry and wet [7,8]. The soil redox potential has a substantial influence on the bioavailability of Cd and As. Drainage and flooding cycles in paddy fields can change the soil redox potential, thus, affecting the fate of Cd and As [9]. Owing to the need for water-logging treatment in the process of planting rice under reducing conditions, Cd can form insoluble precipitates with sulfur, which reduces their availability. However, As in the soil transforms into a soluble state, and its bioavailability is enhanced. It is absorbed by the rice plants through their roots. Under oxidation conditions, the solubility of Cd increases owing to the inhibition of sulfur reduction, while the availability of As decreases as the content of Fe(III) oxide increases [6]. Thus, the simultaneous reduction of Cd and As uptake in paddy soil is more challenging than the control of a single pollutant.

To diminish the impact of Cd and As pollution on ecosystems and agriculture, researchers have proposed various biological, physical, and chemical procedures [7], such as soil replacement, leaching, and phytoremediation [10–12]. However, for agricultural land with a large area and a low amount of pollution, these technologies take a long time to repair and improve soil quality, are expensive, easily cause secondary pollution and often destroy the structure of soil microbial community [13,14]. In situ immobilization with a passivator is a widely studied technique because of its high efficiency, low cost, and easy operation [8]. Passivation materials include clay minerals, Fe compounds, and biochar [10], among those that remediate Fe are particularly attractive. The carbon-containing complex with Fe compounds reduced the adsorption of Cd and As in aqueous solutions [15,16]. Some studies used materials modified with Fe in the soil. Namely, a solution of 3.0% (mass ratio) ferrous sulfate-modified nano-silica designated RNS-SFe immobilizes the bioavailable Cd and As in the soil by 85.0% and 80.1%, respectively, by transforming the bioavailable Cd into insoluble mercapto metal compounds (eS-Cd-S-) and bioavailable As into less soluble iron arsenate ($Fe_3[AsO_4])_2$ and $FeAsO_4$), which precipitated on the surface of nano-silica particles [17]. Fe-enriched corncob-eggshell biochar significantly reduced the amounts of Cd and As in brown rice by affecting the rhizosphere soil pH and redox potential (Eh) and causing the formation of Fe-plaque [18]. Some studies also found that biochar modified with Fe could not simultaneously reduce Cd and As. Biochar could reduce the concentrations of Cd in the roots (49–68%) and grains (26–49%), but this results in a simultaneous increase of As concentrations in roots [19]. In contrast to treatment with biochar, 0.5% Fe-biochar decreased the As concentrations in the roots, and also increased the concentrations of Cd. The mobilization of induction of soil Cd by Fe-biochar was probably owing to the acidification of the soil [19]. In addition, as a new environmental-friendly material for heavy metal remediation, iron nanoparticles has attracted great attention [20]. According to the analysis above, the most closely related system described in the literature was the soil-rice system. They mainly used iron solution modified materials to study the morphological transformation and plant cumulative absorption of Cd and As in the soil-rice system. Nevertheless, little information is available on the effect of exogenous Fe materials, particularly Fe oxide nanoparticles (Nano-Fe), on the bioavailability and transportation of Cd and As in the soil-rice system and possible microbial responses.

In this study, a pot experiment was conducted to explore the effects of ferrous sulfate and Nano-Fe applications on (i) rice growth; (ii) Cd and As speciation and the microbial community in soil; (iii) sequestration of Cd and As by root Fe-plaque; and (iv) the uptake and intake risk assessment of Cd and As in paddy rice. The purpose of this experiment was to improve the theoretical basis for the remediation of heavy metal(loid)s and to provide a new way for sustainable development management and food security of rice fields polluted with Cd and As.

2. Materials and Methods

2.1. Site Description

The rice pot experiment was conducted in the greenhouse of the Institute of Agricultural Resources and Environment, Academy of Agricultural Sciences, Nanning City, Guangxi, China (22°51′ N, 108°15′ E). The soil used in the rice pots was collected from the surface soil (0–20 cm) polluted by Cd and As in Jinchengjiang, Hechi, Guangxi, China. Soil properties include a soil pH of 6.80, organic matter content of 30.40 $g \cdot kg^{-1}$, total N content of 1.90 $g \cdot kg^{-1}$, available K content of 327.00 $mg \cdot kg^{-1}$, total Cd content of 1.73 $mg \cdot kg^{-1}$, and total As content of 106.93 $mg \cdot kg^{-1}$. The rice variety was indica type three-line late maturing hybrid rice Yixiang 99E-4.

2.2. Experimental Design and Sampling

The experiment was conducted using a completely randomized design. Three treatments were established in this experiment: (1) control; (2) 0.50 $g \cdot kg^{-1}$ of ferrous sulfate added ($FeSO_4$); and (3) 0.50 $g \cdot kg^{-1}$ of nano iron oxide added (Fe_2O_3) (Nano-Fe). Nano-Fe was shown to possess a negatively charged surface with a zeta potential value of 13.40 ± 0.26 mV. The size and morphology of nano-Fe were studied by scanning electron microscopy (SEM) (TESCAN Inc., Brno, Czech Republic) (Figure S1). X-ray energy spectrum analysis (EDS) (TESCAN Inc., Brno, Czech Republic) and Fourier transform infrared spectroscopy (FTIR) (Thermo Inc., Waltham, MA, USA) of Nano-Fe are shown in Figures S2 and S3, respectively. There were three repetitions for each type of treatment. After natural air drying, the soil was screened for 2 mm to remove large stones and plant stalks. Iron material and base fertilizer were mixed with 8.0 kg soil in a basin (30 cm × 33 cm, d × h). The soil was soaked in tap water for 8 d to ensure that the ions were fully dissolved and mixed. The rice seeds were sterilized with 5.0% sodium hypochlorite for 5 min, soaked in tap water for 8 h, and placed in a dark and humid environment for 48 h. Rice with approximately 1.50 cm high was transferred to the greenhouse of Guangxi University and cultured in the substrate. When the rice had three leaves of the same height, they were transplanted into the basin under ambient air and sunlight conditions from early April to mid-July, within 101 d. Three rice plants were evenly distributed in each basin, and two plants were inserted into each point. The position of the pots was changed every 5 d. The water level in each basin was controlled at approximately 2.0 cm from the soil level. The dosages of N (urea), P (KH_2PO_4), and K (KCl) were 0.1304 $g \cdot kg^{-1}$ (base fertilizer: 0.0391 $g \cdot kg^{-1}$, tiller fertilizer: 0.0391$g \cdot kg^{-1}$, ear fertilizer: 0.0522 $g \cdot kg^{-1}$), 0.4394 $g \cdot kg^{-1}$, and 0.2325 $g \cdot kg^{-1}$, respectively.

Soil and plant samples were collected when the rice had matured on July 15. On the day of sampling, fresh rice rhizosphere soil samples were removed with sterilized glass rods, placed in 5 mL centrifuge tubes, sealed, and placed in a tank of liquid nitrogen. Soil samples were sent to Shanghai Meiji Biomedical Technology Co., Ltd. (Shanghai, China) for high-throughput sequencing to explore the changes in composition of total bacteria (16S rRNA) in different soil treatments. In addition, the soil samples after air drying and screening were analyzed for soil properties, including the contents of different forms of Cd and As. Samples of Fe plaque on the surfaces of rice roots were extracted to determine the concentrations of Fe, Cd, and As on the root surfaces. The plant samples were washed twice with tap water, treated with ultrapure water, dried at 105 °C for 30 min, and then dried to a constant weight at 65 °C. The roots, shoots, and grains yield were measured by weighing after drying, and they were cut and ground. They were screened through a 0.15-mm sieve and bagged to determine the concentrations of Cd and As.

2.3. Soil Chemical Analysis

The soil pH was measured by a pH meter (Leici Inc., Shanghai, China) with a glass electrode using a ratio of 1:2.5 (w/v). The soil organic matter was determined by a potassium dichromate sulfuric acid external heating method. The Kjeldahl method was used to determine the soil total N. A volume of 1 $mol \cdot L^{-1}$ N_4OAc was used as the extractant. The soil was shaken for 0.5 h and filtered, and the contents of soil available K were determined directly by flame photometry (Xinyi Inc., Shanghai, China). The contents of total soil Cd were determined after soil digestion by HF-$HClO_4$-HCl using an atomic absorption spectrometer (Hitachi Inc., Tokyo, Japan) [21]. The total soil As content was determined by acid digestion (HNO_3, HCl and HF) using an atomic fluorescence spectrometer (Titan Inc., Beijing, China) [22].

A four-part continuous extraction method [23] was used to separate the Cd from the samples into four forms, e.g., exchangeable Cd, organically bound Cd, inorganically bound Cd, and residual Cd. The extraction solutions were determined by atomic absorption spectrometry (Hitachi Inc., Tokyo, Japan).

A five-part continuous extraction method [24] was used to separate the As from samples into five forms, e.g., non-specifically sorbed As, specifically sorbed As, amorphous iron oxide As, Fe/Al hydrated oxide As, and residual As. The extraction solutions were determined by atomic fluorescence spectrometry (Titan Inc., Beijing, China).

2.4. Analysis of Fe, Cd, and As on the Roots Surfaces

Fresh roots (10.0 g) were cleaned with deionized water and weighed into plastic bottles to determine the concentrations of Fe, Cd, and As on the roots surfaces. DCB extractant (90 mL of 0.30 mol$\cdot$L^{-1} sodium citrate and 10 mL of 1.0 mol$\cdot$L^{-1} sodium bicarbonate) and approximately 3.0 g of sodium bicarbonate were added in turn, and shaken at 180 rpm for 2 h in a full temperature shaking incubator. After that, the concentrations of Fe, Cd, and As in the extractant were determined by atomic absorption spectrometry (Hitachi Inc., Tokyo, Japan). The extracted rice roots were dried in an oven, and were weighed to determine their dry weight. The concentrations of Fe, Cd, and As on the roots surfaces were calculated (mg$\cdot$kg^{-1} of dry root) [25].

2.5. Analysis of Cd and As in the Roots, Shoots, and Grains

A microwave digestion method was used to measure the concentrations of Cd and As in the roots, shoots, and grains using 0.20 g of plant samples. Each tissue sample was weighed in a digestion tube, and 5.0 mL of concentrated nitric acid was added. A parallel standard sample and a blank were also created, and then microwave digestion (500 power, 150 °C, and 8 min) was conducted. The digested samples were removed and placed on an acid driving rack to drive the acid (120 °C) to the size of a soybean granule. The samples were cooled and transferred to a 25 mL volumetric flask. Finally, the samples were filtered into a plastic bottle. The content of Cd was measured by atomic absorption spectrometry (Hitachi Inc., Tokyo, Japan), and the content of As was measured with an atomic fluorescence spectrometer (Titan Inc., Beijing, China).

2.6. qPCR Amplification and High-Throughput Sequencing of the Soil Bacteria (16S rRNA)

The microbial community genomic DNA in soil was extracted using an E.Z.N.A.® soil DNA Kit (MP Bio-tek, Norcross, GA, USA). The primers 338F (5′-ACTCCTACGGGAGGCAGCAG-3′) were used to amplify the V3-V4 hypervariable region of the 16S rRNA gene. The PCR amplification was performed as follows: initial denaturation at 95 °C for 3 min, 27 denaturing cycles at 95 °C for 30 s, annealing at 55 °C for 30 s, extension at 72 °C for 45 s, single extension at 72 °C for 10 min, and end at 4 °C. As described in the standard protocols of Majorbio Bio-Pharm Technology Co. Ltd. (Shanghai, China), purified amplicons were pooled in equimolar amounts of paired-end sequenced (2 × 300) on an Illumina MiSeq platform (Illumina, San Diego, CA, USA). Operational taxonomic units (OTUs) with a 97% similarity cutoff were clustered by using UPARSE (version 7.1, http://drive5.com/uparse/, accessed on 19 June 2021), and chimeric sequences were identified and removed by using UCHIME. The taxonomy of each OTU representative sequence was analyzed by RDP Classifier (http://rdp.cme.msu.edu/, accessed on 19 June 2021) against the 16S rRNA database (e.g., Silva SSU128) with a confidence threshold of 0.7.

2.7. Data Processing and Analysis

The health risk index (*HRI*) of Cd and As was calculated by estimating the daily intake of Cd (*DIM*) and As (*DIM*) and with an oral reference dose of *RFD* of Cd and As. The *RFD* values of Cd and As were considered to be 0.001 $mg \cdot kg^{-1}$ and 0.0003 $mg \cdot kg^{-1}$, respectively. Bodyweight day^{-1} was determined as previously described [26].

$$HRI = DIM/RFD \tag{1}$$

$$DIM = Cmetal * F * (D\,food\,intake)/(Baverage\,weight) \tag{2}$$

where C_{metal} indicates the accumulation of metal in the grain ($mg \cdot kg^{-1}$); F indicates the conversion factor (0.085); $D_{food\ intake}$ indicates the daily use of grains considered to be 0.40 kg per person, and $B_{average\ weight}$ indicates the body average weight, which was considered to be 70.00 kg per person.

The translocation factor index (TF) was determined as described by Singh et al. [27]. The experimental data were analyzed by SPSS 21.0 (IBM Inc., Armonk, NY, USA) and compared by Duncan's method. GraphPad Prism 7.0 (GraphPad Inc., San Diego, CA, USA) was used for drawing.

3. Results

3.1. Yield of Various Parts of Rice

On the whole, compared with the control, the treatments with exogenous Fe ($FeSO_4$ and Nano-Fe) increased ($p < 0.05$) the yield of shoots and grains, while only Nano-Fe treatment increased ($p < 0.05$) the root yields (Table 1).

Table 1. Effect of exogenous Fe on the yield of various parts of rice.

Treatments	Root Yield (g)	Shoot Yield (g)	Grain Yield (g)
Control	25.66 ± 0.14c	156.07 ± 8.61b	76.33 ± 0.93b
$FeSO_4$	26.18 ± 0.03bc	178.20 ± 0.88a	83.33 ± 0.79a
Nano-Fe	28.40 ± 0.45a	191.00 ± 0.43a	83.46 ± 0.22a

Different lowercase letters in the columns indicate a significant difference among the different treatments at $p < 0.05$.

3.2. Contents of Various Forms of Cd and As in the Soil

Exogenous Fe, in the form of $FeSO_4$ and Nano-Fe, significantly decreased ($p < 0.05$) the contents of exchangeable, organically bound and inorganically bound Cd, which was more pronounced for Nano-Fe in comparison with $FeSO_4$ (Table 2). Moreover, treatments of exogenous Fe increased ($p < 0.05$) the contents of Cd in the residual state (Table 2), decreased ($p < 0.05$) the specifically sorbed As, and increased ($p < 0.05$) the contents of Fe/Al hydrated oxide As (Table 3). However, only Nano-Fe decreased ($p < 0.05$) the amount of amorphous iron oxide As and increased ($p < 0.05$) the amount of residual As.

Table 2. Effect of exogenous Fe applied as $FeSO_4$ and Nano-Fe on the Cd contents in soil.

Treatments	Exchangeable Cd ($\mu g \cdot kg^{-1}$)	Organic Bound Cd ($\mu g \cdot kg^{-1}$)	Inorganic Bound Cd ($\mu g \cdot kg^{-1}$)	Residual Cd ($\mu g \cdot kg^{-1}$)
Control	6.75 ± 0.06a	19.27 ± 0.06a	1119.67 ± 0.98a	583.33 ± 5.44c
$FeSO_4$	6.13 ± 0.01b	13.36 ± 0.18b	1076.67 ± 10.89b	633.33 ± 7.20b
Nano-Fe	3.31 ± 0.09c	6.79 ± 0.26c	965.67 ± 14.03c	753.33 ± 7.20a

Different lowercase letters in the columns indicate a significant difference among the different treatments at $p < 0.05$.

Table 3. Effect of exogenous Fe applied as $FeSO_4$ and Nano-Fe on As contents in soil.

Treatments	Non-Specifically Sorbed As (mg·kg^{-1})	Specifically Sorbed As (mg·kg^{-1})	Amorphous Iron Oxide As (mg·kg^{-1})	Fe/Al Hydrated Oxide As (mg·kg^{-1})	Residual As (mg·kg^{-1})
Control	0.25 ± 0.00a	12.09 ± 0.05a	49.61 ± 0.24a	28.38 ± 0.14c	16.61 ± 0.14b
$FeSO_4$	0.26 ± 0.01a	9.49 ± 0.14b	48.83 ± 0.57ab	30.96 ± 0.25b	17.40 ± 0.62ab
Nano-Fe	0.25 ± 0.01a	8.2 ± 0.13c	48.18 ± 0.31b	32.09 ± 0.36a	18.21 ± 0.12a

Different lowercase letters in the columns indicate a significant difference among the different treatments at $p < 0.05$.

3.3. Concentrations of Fe, Cd, and As on the Roots Surfaces

With the addition of exogenous Fe, the concentration of Fe on the roots surfaces was higher ($p < 0.05$) than that in the control (Figure 1a). Moreover, $FeSO_4$ and Nano-Fe also increased ($p < 0.05$) the concentrations of Cd and As on the roots surfaces, compared with the control (Figure 1b,c).

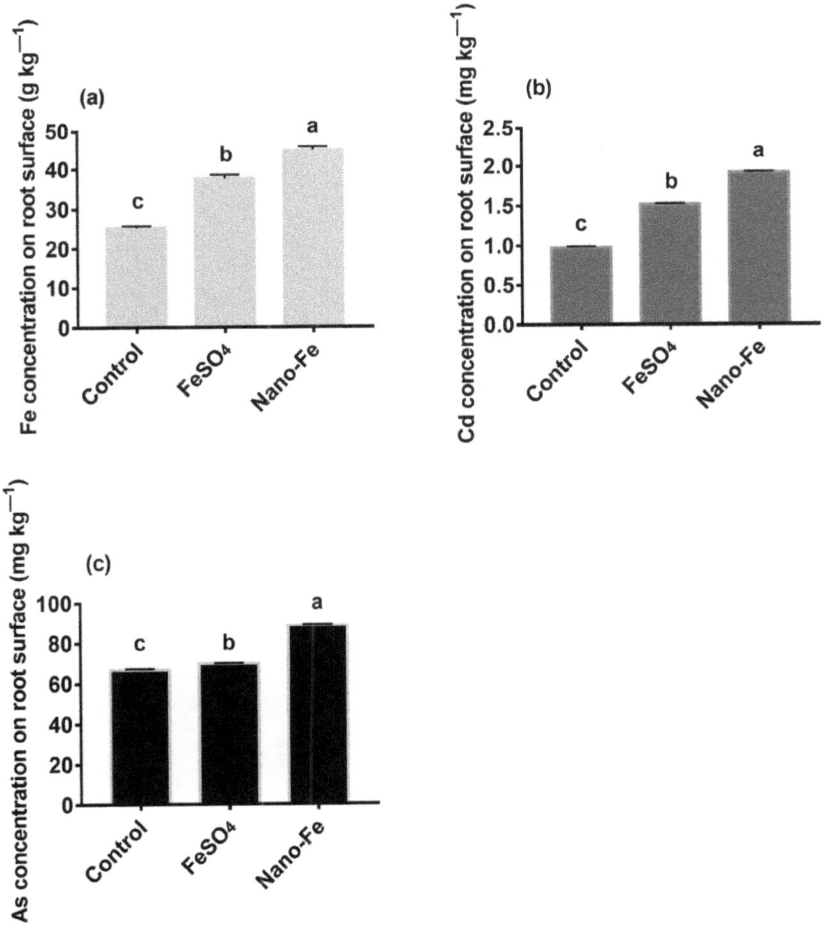

Figure 1. Concentrations of Fe (**a**), Cd (**b**), and As (**c**) on the surfaces of rice roots derived from treatments with exogenous Fe applied as $FeSO_4$ or Nano-Fe. Different lowercase letters in the columns indicate a significant difference among the different treatments at $p < 0.05$.

3.4. Concentrations, Uptake, and TFs of Cd and As in Rice

Exogenous Fe reduced ($p < 0.05$) the Cd concentrations (Figure 2a–c) and its uptake (Figure 2d–f) in the roots, shoots, and grains, reduced ($p < 0.05$) the As concentrations (Figure 3a–c) and its uptake (Figure 3d–f) in the roots, shoots, and grains of rice, compared with the control.

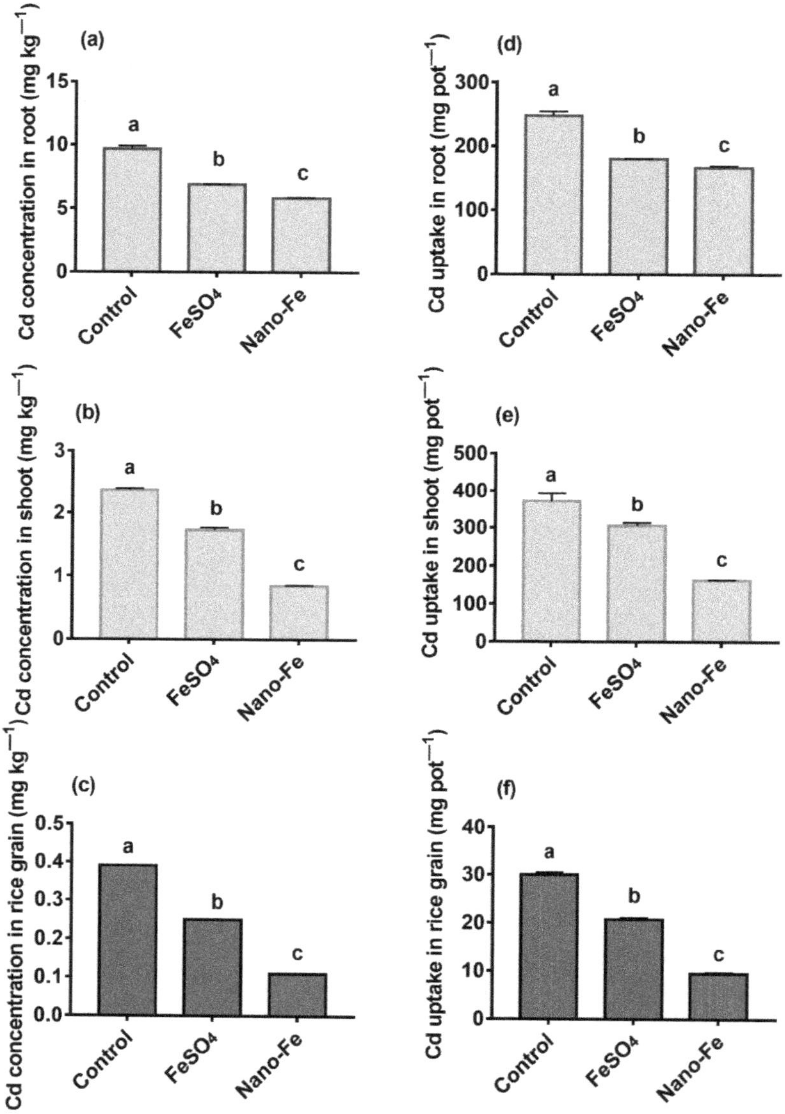

Figure 2. Concentrations of Cd in the roots (**a**), shoots (**b**), grains (**c**) of rice and uptake of Cd in the roots (**d**), shoots (**e**), grains (**f**) of rice derived from treatments with exogenous Fe applied as $FeSO_4$ or Nano-Fe. Different lowercase letters in the columns indicate a significant difference among the different treatments at $p < 0.05$.

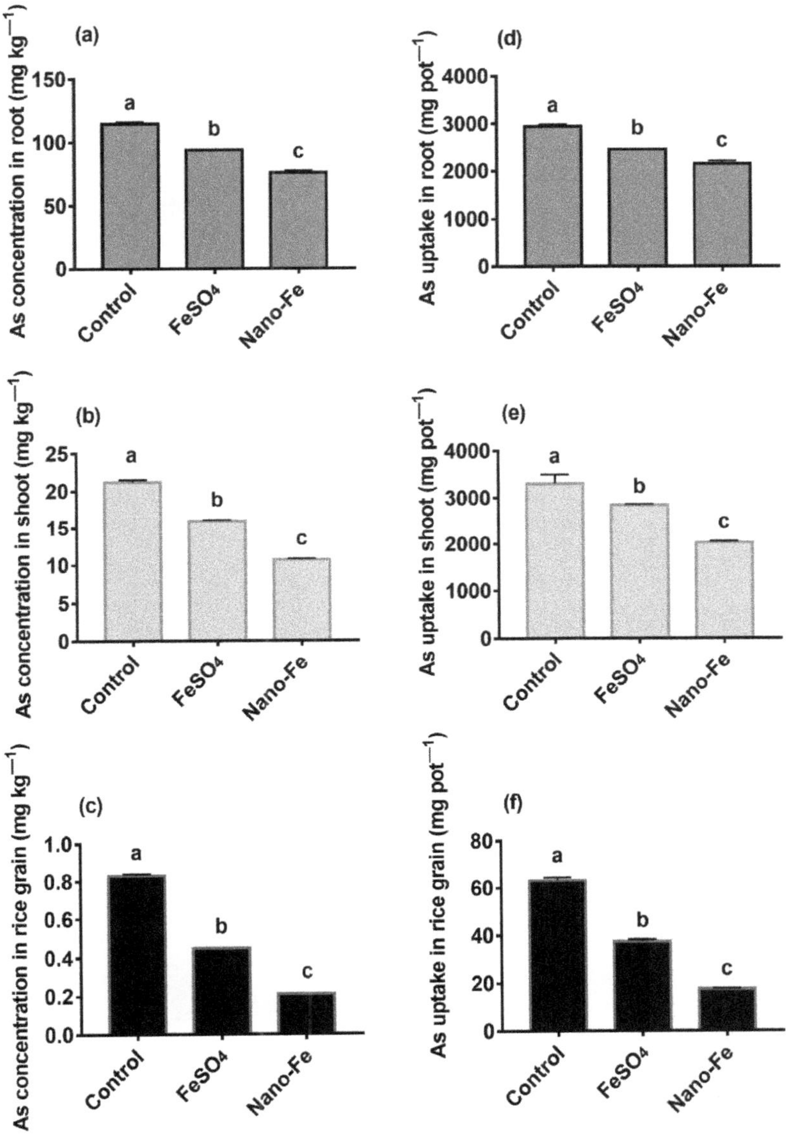

Figure 3. Concentrations of As in the roots (**a**), shoots (**b**), grains (**c**), and uptake of As in the roots (**d**), shoots (**e**), grains (**f**) of rice derived from treatments with exogenous Fe applied as $FeSO_4$ or Nano-Fe. Different lowercase letters in the columns indicate a significant difference among the different treatments at $p < 0.05$.

In terms of Cd, the soil to roots TFs were lower ($p < 0.05$) upon treatments with exogenous Fe, applied in both forms, compared with the control (Figure 4a), although this was more pronounced upon treatment with Nano-Fe. Compared with the control, the lower TFs from roots to shoots were recorded only upon exposure to Nano-Fe (Figure 4b). While both treatments with exogenous Fe equally lowered ($p < 0.05$) TFs values from shoots to grains (Figure 4b,c). In terms of As, the TFs from soil to roots, from roots to shoots and from shoots to grains were lower ($p < 0.05$) upon both treatments with exogenous Fe than those of the control (Figure 4d–f), although this effect was stronger in the Nano-Fe treatment.

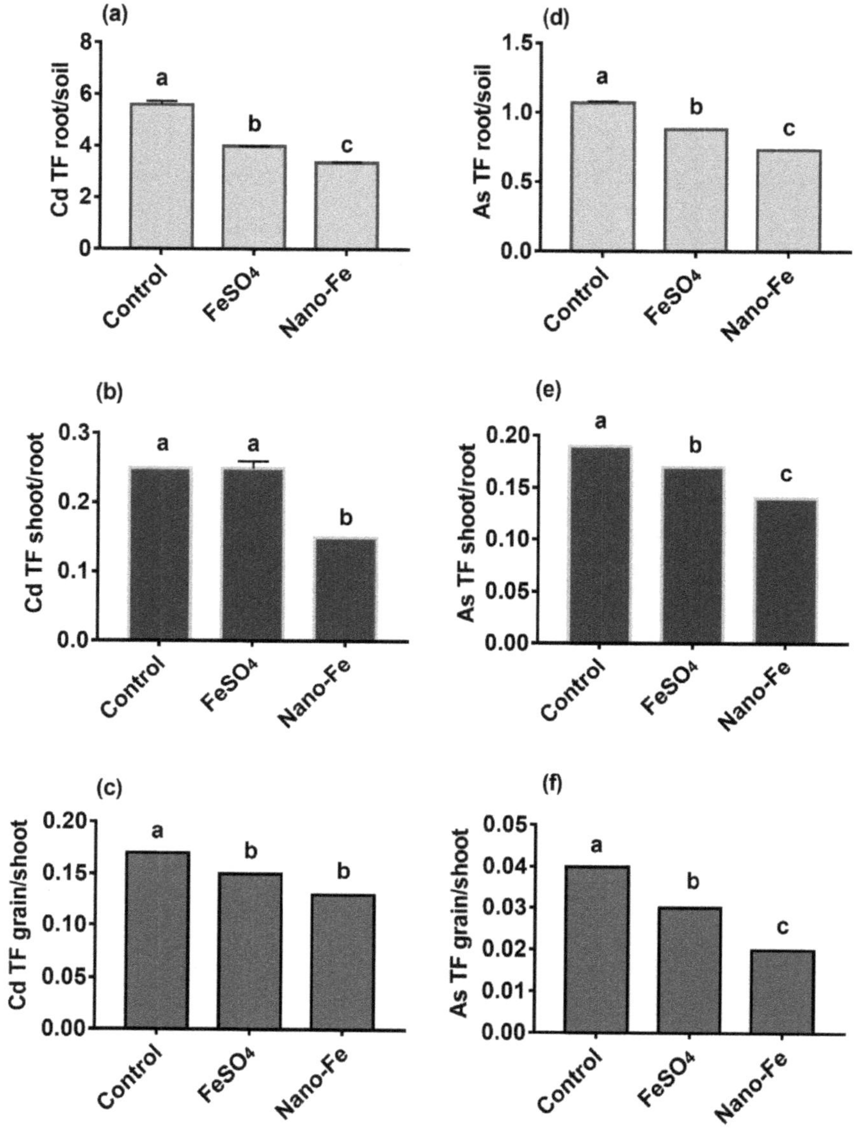

Figure 4. The Cd TF root/soil (**a**), shoot/root (**b**), grain/shoot (**c**), and As TF root/soil (**d**), shoot/root (**e**), grain/shoot (**f**) of rice derived from treatments with exogenous Fe applied as $FeSO_4$ or Nano-Fe. Different lowercase letters in the columns indicate a significant difference among the different treatments at $p < 0.05$.

3.5. Soil Microbial Community

Treatments with exogenous Fe resulted in an increase ($p < 0.05$) in the Shannon diversity indices and a decrease ($p < 0.05$) in the Simpson indices compared with the control (Table 4). Firmicutes, Proteobacteria, Actinobacteria, and Chloroflexi were the dominant phyla and comprised 71.10–72.66% of the bacterial 16S rRNA gene sequences in all the treatments (Figure 5). The relative abundances of Firmicutes, Proteobacteria, Bacteroidetes, Gemmatimonadetes, Ignavibacteriae, unclassified_k__norank, and WS6 in the soil treated with $FeSO_4$ were 12.17%, 7.99%, 10.51%, 21.45%, 33.42%, 46.69%, and 17.27% higher than those of the control, respectively. The relative abundances of Firmicutes, Proteobacteria, Bacteroidetes, Parcubacteria, and Planctomycetes in the soil treated with Nano-Fe were 39.88%, 7.37%, 40.82%, 30.26%, and 36.15% higher than those of the control, respectively. The treatments with exogenous Fe resulted in higher relative abundances of Firmicutes, Proteobacteria, and Bacteroidetes than those of the control (Figure 5).

Table 4. Effect of exogenous Fe applications on Shannon's and Simpson's indices of the bacterial communities.

Treatments	Shannon's Index	Simpson's Index
Control	6.4452 ± 0.0058c	0.00536 ± 0.00004a
$FeSO_4$	6.4920 ± 0.0078b	0.00518 ± 0.00002b
Nano-Fe	6.6208 ± 0.0124a	0.00416 ± 0.00013c

Different lowercase letters in the columns indicate a significant difference among the different treatments at $p < 0.05$.

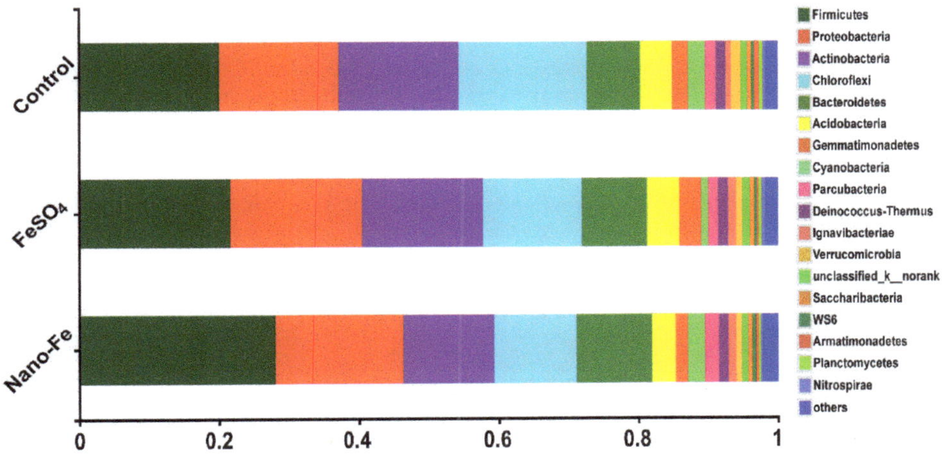

Figure 5. The relative levels of abundance of phyla in treatments with exogenous Fe applied as $FeSO_4$ or Nano-Fe.

3.6. Health Risk Assessment

Based on the *DIM* and *HRQ* formulas of the daily intake of Cd and As, the risk of dietary intake of Cd and As in rice with exogenous Fe was evaluated (Table 5). The results showed that the *DIM* values of Cd and As in rice polluted by Cd and As were 0.00019 and 0.00040 mg·(kg·D)$^{-1}$, respectively. With the application of exogenous Fe, the *DIM* decreased, and the Nano-Fe was the lowest. The data showed that with the exogenous Fe, the *HRI* values of Cd and As were lower than those of the control, which ranged from 0.06 to 0.12 for Cd and from 0.34 to 0.73 for As. The *HRI* of Nano-Fe was the lowest among them. The *HRI* for Cd-As ranged from 0.40 to 0.85. Compared with the control, the *HRI* for Cd-As in rice decreased by 44–74% when exogenous Fe was applied. All of the *HRI* for Cd-As under the exogenous Fe treatments were <1 and met the safety limits.

Table 5. Risk assessment of Cd and As intake via rice grain treated with exogenous Fe.

Treatments	Cd		As		HRI Cd-As
	DIM	HRI	DIM	HRI	
Control	0.00019	0.19	0.00040	1.34	1.53
$FeSO_4$	0.00012	0.12	0.00022	0.73	0.85
Nano-Fe	0.00006	0.06	0.00010	0.34	0.40

4. Discussion

4.1. Speciation of Cd and As in Soil

Heavy metals in the soil can have different toxicities and environmental behavior owing to one or several aspects of the different fractions. The exchangeable component was the most mobile component in soil, which represents the bioavailability of organisms, while the residual phase was the lowest [28]. Based on the results of a continuous extraction experiment, the contents of residual Cd in the soil increased, and the contents of exchangeable Cd decreased after $FeSO_4$ and Nano-Fe were applied to the soil (Table 2), which indicated that they could effectively reduce the migration rate and bioavailability of Cd in the soil. The reason could be that the addition of exogenous Fe to the soil could change the biogeochemical cycle of Fe. The application of $FeSO_4$ can effectively stabilize Cd in the soil by surface precipitation and ion diffusion [29]. The application of Nano-Fe increased the contents of iron oxide/hydroxide, which was the most important factor that affected the mobility of Cd in the range of pH 5.0 to 7.5 [30].

Non-specific sorbtion is usually defined as the outer sphere, which represents the adsorption on the surface of soil minerals owing to the electrostatic attraction [31]. Specifically sorbed As forms a complex with coordinate covalent bonds [11]. Soil available As primarily originates from non-specific and specific sorption [31,32]. In this study, the addition of exogenous Fe did not affect the non-specifically sorbed As, and the contents of specifically sorbed As were significantly lower than that of the control, which was consistent with the results of previous research [32]. The addition of exogenous Fe promoted the increase in amount of Fe/Al hydrated oxide As and the residual As (Table 3). The reason could be that exogenous Fe can promote the diffusion of As from the outer surface of soil minerals to the inner surface and mineral lattice, resulting in the transformation of specifically sorbed As to other components, such as Fe/Al hydrated oxide and the residual fractions [33]. This indicated that exogenous Fe promoted the transformation of As in paddy soil from more mobile fractions to fewer mobile fractions, thus, reducing the uptake of As in the soil by rice.

Exogenous Fe has positive effects on soil microorganisms [34,35], which is the main driving force of the biogeochemical Fe cycle. Heavy metal pollution in soil usually leads to changes in the local microbial community [36]. The addition of exogenous Fe interferes with the effect of microbial community on pollution with Cd and As in the soil. Exogenous Fe altered the bacterial α-diversity and community structure. Nano-Fe resulted in the largest decrease in the contents of soil available Cd and As (Tables 3 and 4). The relative abundances of phyla that help to reduce the phytotoxicity of Cd primarily included Firmicutes, which was consistent with the promotion of Firmicutes with the immobilization of Cd in soil [37]. Acidobacteria can reduce Fe (III) and Fe (II) concentrations [35]. The abundance of Acidobacteria decreased in Nano-Fe, which indicated that the reduction and oxidation of Fe were significantly weakened, which could be the reason why more available Cd was converted into residual Cd. Chloroflexi provides evidence of soil remediation [35]. The relative abundance of Chloroflexi in the Nano-Fe treatment was the lowest among all the treatments. At the phylum level, most of the Fe-oxidizing bacteria are members of Proteobacteria [38], and the increase of these bacteria in the Nano-Fe treatment could indicate that the oxidation of Fe (II) in the soil was enhanced. The change of As fractions in the soil may be due to treatment with Nano-Fe increased the relative abundance of Firmicutes and Proteobacteria in soil. Functional microorganisms can reduce As, dissimilatory As, As oxide, and methylated As, and also participate in nitrate, sulfate, and Fe reduction, which are considered to be resistant to As, as previously described [39,40]. Thus, exogenous $FeSO_4$ and Nano-Fe applications in paddy soil could alter bacterial diversity and community structure, and increase heavy metals resistant bacteria.

In the soil-rice system, iron nanoparticles were applied to the soil, which increased the rice biomass (Table 1) and improved the soil microbial diversity (Table 4), so the iron nanoparticles were not toxic in our study. It is inconsistent with the results of other studies that iron nanoparticles are toxic [41]. The reason may be that the response characteristics of soil with different backgrounds and rice with different varieties to iron nanoparticles are different. In addition, an appropriate amount of iron nanoparticles is the key to affect the accumulation of heavy metals in rice [42,43].

4.2. Absorption of Cd and As in Rice

Rice roots are usually covered with Fe (III) oxide deposits, known as Fe plaque [44]. The structure of the Fe plaque is characterized by a mixture of crystalline and amorphous Fe (III) (oxy hydroxy) oxides, primarily in the form of Fe hydride and goethite [45]. The accumulation of Fe predicted Fe plaque formation. In this study, $FeSO_4$ promoted the formation of Fe plaque (Figure 1a), which was consistent with the findings of Sebastian and Prasad [46]. The reason could be that anions such as SO_4^{2-} attract H^+ ions from the soil solids to soil solutions and form sulfuric acid, which affects the soil pH and forms Fe plaque [47]. The results show that the Fe concentrations in Fe plaque formed by Nano-Fe were higher than those of $FeSO_4$ (Figure 1a). The formation of Fe plaque in treatments with $FeSO_4$ primarily originated from Fe^{2+}, while with Nano-Fe, the formation of Fe plaque primarily originated from the accumulation of nanometer Fe particles [44]. The Nano-Fe treatment, owing to the high level of adhesion and reactivity, resulted in nanoparticles that had a high specific surface area and adhesion to the epithelial root cell wall. Thus, they were more likely to form Fe plaque.

Fe plaque has been confirmed to limit the absorption of Cd and As [48]. The concentrations of Cd and As on the surfaces of rice roots in the exogenous Fe treatments were higher ($p < 0.05$) than those in the control, indicating that Cd and As were enriched in the Fe plaque. However, the Fe plaque that formed on the roots surfaces differed owing to the differences in Fe materials, thus, affecting the absorption of Cd and As by rice. This was also the primary reason why the amount of Cd and As adsorbed on the Fe plaque formed by Nano-Fe was higher than that of $FeSO_4$. Furthermore, Fe oxide can bind heavy metals in the soil [49], and the plants barely absorb the combined portion of Cd or As and Nano-Fe. The concentrations of Cd and As and uptake in rice roots, shoots, and grains of exogenous Fe treatments were reduced ($p < 0.05$), which indicated that the Fe plaque absorbed Cd and As and prevented these compounds from entering the rice roots.

TF represents the plant's ability to translocate the pollutant from the roots to the aerial parts of the plant. The TFs $_{root/soil}$ of Cd and As were lower ($p < 0.05$) than those of the control, which indicated that the application of exogenous Fe reduced the ability of Cd and As to enter into the roots from the soil. There are two factors that could lead to this phenomenon. One could be that the addition of exogenous Fe led to a decrease in the amount of available Cd and As in the soil, whereas the other was that Fe plaque prevented the rice from absorbing Cd and As. The TF $_{shoot/root}$ and TF $_{grain/shoot}$ indicated the translocation, which was affected by xylem loading, intravascular transfer, and the transport of heavy metal through the phloem [50,51]. The Cd TFs $_{shoot/root}$ of the Nano-Fe was lower than that of the control and $FeSO_4$ ($p < 0.05$) (Figure 4b,c). The Cd TFs $_{grain/shoot}$ of the Nano-Fe compared with the $FeSO_4$ treatment, and the Nano-Fe treatment could reduce the ability of Cd to be transferred from the rice roots to shoots, and thus, reduce the accumulation of Cd in rice grains. Compared with the control, exogenous Fe treatments could decrease the TFs $_{shoot/root}$ and TFs $_{grain/shoot}$ of As (Figure 4e,f). The difference in TFs between different treatments could be related to rice genomics or other factors that affect the accumulation of Cd and As in rice, such as translocation in the plant and interactions between heavy metals and other mineral nutrients. Cd was usually absorbed and transported in rice through the transport system of essential cations, such as Ca, Fe, and Mn [52]. These transport systems were regulated by P1B ATPase (heavy metal ATPase), and many genes that encode these transporters are members of the P1B ATPase family [53–55]. The genes that regulate the expression of these proteins could be affected by changing environmental factors, such as exogenous Fe and Fe dosage. The Nano-Fe alleviates As phytotoxicity in rice by improving the uptake of Fe, increasing oxidative stress tolerance and diminishing the accumulation of As [43].

4.3. Health Risk Index

Rice is the key manner of human exposure to heavy metals [56]. It is very important to quantify the health risk assessment of Cd and As in rice. In this study, although the Cd *HRI* value in the control was <1, the risk of long-term consumption of rice produced in this region cannot be ignored [26]. The addition of exogenous Fe reduced the Cd, As, and Cd-As *HRIs*, which indicated that exogenous Fe could effectively reduce their risk. In particular, the reduction of Nano-Fe was the greatest. A health risk index >1 is considered to be a threat to human health. The *HRIs* for Cd-As of the treatments ($FeSO_4$ and Nano-Fe) were <1 and thus, met the safety limits.

5. Conclusions

In this study, $FeSO_4$ and Nano-Fe that were applied at 0.5 g/kg, and were used to treat paddy soil contaminated with Cd-As to examine their influences on the forms of Cd and As in soil and soil microorganisms, the transport and accumulation of Cd and As in rice, and the risk assessment of human intake. The results demonstrated that the application of exogenous Fe as both forms of $FeSO_4$ and Nano-Fe significantly improved the diversity and structure of microbial community, particularly increasing the relative abundances of Cd and As-resistant bacteria, and reducing the contents of available Cd and As in the soil. Moreover, exogeFnous Fe could promote the formation of Fe plaque, reduce the transport and accumulation of Cd and As in rice, and reduce the risk of human intake. The results of our study confirm that supplementation of environmental-friendly Nano-Fe 0.5 $g \cdot kg^{-1}$ to paddy soil is an effective approach to simultaneously reduce Cd and As accumulation in the rice grains grown in co-contaminated soil.

Supplementary Materials: The following supporting information can be downloaded at: https://www.mdpi.com/article/10.3390/nano12081311/s1, Figure S1: Scanning electron microscopy (SEM) of Nano-Fe; Figure S2: X-ray energy spectrum analysis (EDS) of Nano-Fe; Figure S3: Fourier transform infrared spectroscopy (FTIR) of Nano-Fe.

Author Contributions: T.L.: conceptualization, methodology, formal analysis, writing—original draft. J.L.: methodology, formal analysis. X.Z.: formal analysis, writing—original draft. X.W.: formal analysis, investigation. B.H.: methodology, formal analysis, supervision. F.C., C.L.: formal analysis, investigation. Y.Y., Z.Z. (Zengyu Zhang), J.Z., B.L. and J.C.: formal analysis. H.L. and Z.Z. (Zhiqiang Zhu): methodology, formal analysis, supervision. Y.W.: conceptualization, supervision, funding acquisition. J.H.: formal analysis, writing—original draft, supervision, funding acquisition. All authors have read and agreed to the published version of the manuscript.

Funding: This work was supported by the National Natural Science Foundation of China (41401574; 41967045; 41967048); The National Key Research and Development Program of China (2020YFC1808500); Guangxi "Ten Thousand Talents Project" Special Fund of China (2018221); Guangxi Science and Technology Plan Project of China (1598014-4); Science and Technology Development Foundation of Guangxi Academy of Agricultural Sciences (2021YT040; 2021JM57); Pioneer Team of Selenium-rich Agricultural Industry (Guinongkemeng 202214).

Institutional Review Board Statement: Not applicable.

Informed Consent Statement: Not applicable.

Data Availability Statement: The data sets supporting the results of this article are included within the article.

Conflicts of Interest: The authors have no relevant financial or non-financial interests to disclose.

References

1. Humans. *Cadmium and Cadmium Compounds, Beryllium, Cadmium, Mercury, and Exposures in the Glass Manufacturing Industry*; International Agency for Research on Cancer: Lyon, France, 1993.
2. Humans. *Arsenic and Arsenic Compounds, Arsenic, Metals, Fibres and Dusts*; International Agency for Research on Cancer: Lyon, France, 2012.

3. China, M. *The Ministry of Land and Resources Report on the National Soil Contamination Survey*; Ministry of Environmental Protection and Ministry of Land and Resources of the People's Republic of China: Beijing, China, 2014.
4. Abad-Valle, P.; Alvarez-Ayuso, E.; Murciego, A.; Munoz-Centeno, L.M.; Alonso-Rojo, P.; Villar-Alonso, P. Arsenic distribution in a pasture area impacted by past mining activities. *Ecotoxicol. Environ. Saf.* **2018**, *147*, 228–237. [CrossRef] [PubMed]
5. Satarug, S.; Vesey, D.A.; Gobe, G.C. Health Risk Assessment of Dietary Cadmium Intake: Do Current Guidelines Indicate How Much is Safe? *Environ. Health Perspect.* **2017**, *125*, 284–288. [CrossRef] [PubMed]
6. Zhao, F.; Wang, P. Arsenic and cadmium accumulation in rice and mitigation strategies. *Plant Soil* **2020**, *446*, 1–21. [CrossRef]
7. Pan, D.; Liu, C.; Yu, H.; Li, F. A paddy field study of arsenic and cadmium pollution control by using iron-modified biochar and silica sol together. *Environ. Sci. Pollut. Res.* **2019**, *26*, 24979–24987. [CrossRef] [PubMed]
8. Qiao, J.; Liu, T.; Wang, X.; Li, F.; Lv, Y.; Cui, J.; Zeng, X.; Yuan, Y.; Liu, C. Simultaneous alleviation of cadmium and arsenic accumulation in rice by applying zero-valent iron and biochar to contaminated paddy soils. *Chemosphere* **2018**, *195*, 260–271. [CrossRef]
9. Honma, T.; Ohba, H.; Kaneko-Kadokura, A.; Makino, T.; Nakamura, K.; Katou, H. Optimal soil Eh, pH, and water management for simultaneously minimizing arsenic and cadmium concentrations in rice grains. *Environ. Sci. Technol.* **2016**, *50*, 4178–4185. [CrossRef]
10. Hussain, B.; Ashraf, M.N.; Shafeeq-ur-Rahman; Abbas, A.; Li, J.; Farooq, M. Cadmium stress in paddy fields: Effects of soil conditions and remediation strategies. *Sci. Total Environ.* **2021**, *754*, 142188. [CrossRef]
11. Komárek, M.; Vaněk, A.; Ettler, V. Chemical stabilization of metals and arsenic in contaminated soils using oxides—A review. *Environ. Pollut.* **2013**, *172*, 9–22. [CrossRef]
12. Zhang, J.; Martinoia, E.; Lee, Y. Vacuolar transporters for cadmium and arsenic in plants and their applications in phytoremediation and crop development. *Plant Cell Physiol.* **2018**, *59*, 1317–1325. [CrossRef]
13. Cang, L.; Zhou, D.; Wang, Q.; Fan, G. Impact of electrokinetic-assisted phytoremediation of heavy metal contaminated soil on its physicochemical properties, enzymatic and microbial activities. *Electrochim. Acta* **2012**, *86*, 41–48. [CrossRef]
14. Zhang, X.; Wang, H.; He, L.; Lu, K.; Sarmah, A.; Li, J.; Bolan, N.S.; Pei, J.; Huang, H. Using biochar for remediation of soils contaminated with heavy metals and organic pollutants. *Environ. Sci. Pollut. Res.* **2013**, *20*, 8472–8483. [CrossRef] [PubMed]
15. Pawar, R.R.; Kim, M.; Kim, J.G.; Hong, S.M.; Sawant, S.Y.; Lee, S.M. Efficient removal of hazardous lead, cadmium, and arsenic from aqueous environment by iron oxide modified clay-activated carbon composite beads. *Appl. Clay Sci.* **2018**, *162*, 339–350. [CrossRef]
16. Wu, J.; Huang, D.; Liu, X.; Meng, J.; Tang, C.; Xu, J. Remediation of as (III) and Cd (II) co-contamination and its mechanism in aqueous systems by a novel calcium-based magnetic biochar. *J. Hazard. Mater.* **2018**, *348*, 10–19. [CrossRef] [PubMed]
17. Cao, P.; Qiu, K.; Zou, X.; Lian, M.; Liu, P.; Niu, L.; Yu, L.; Li, X.; Zhang, Z. Mercapto propyltrimethoxysilane-and ferrous sulfate-modified nano-silica for immobilization of lead and cadmium as well as arsenic in heavy metal-contaminated soil. *Environ. Pollut.* **2020**, *266*, 115152. [CrossRef]
18. Islam, M.S.; Magid, A.S.I.A.; Chen, Y.; Weng, L.; Ma, J.; Arafat, M.Y.; Khan, Z.H.; Li, Y. Effect of calcium and iron-enriched biochar on arsenic and cadmium accumulation from soil to rice paddy tissues. *Sci. Total Environ.* **2021**, *785*, 147163. [CrossRef]
19. Yin, D.; Wang, X.; Peng, B.; Tan, C.; Ma, L. Effect of biochar and Fe-biochar on Cd and as mobility and transfer in soil-rice system. *Chemosphere* **2017**, *186*, 928–937. [CrossRef]
20. Guha, T.; Barman, S.; Mukherjee, A.; Kundu, R. Nano-scale zero valent iron modulates Fe/Cd transporters and immobilizes soil Cd for production of Cd free rice. *Chemosphere* **2020**, *260*, 127533. [CrossRef]
21. Lambrechts, T.; Gustot, Q.; Couder, E.; Houben, D.; Iserentant, A.; Lutts, S. Comparison of EDTA-enhanced phytoextraction and phytostabilisation strategies with Lolium perenne on a heavy metal contaminated soil. *Chemosphere* **2011**, *85*, 1290–1298. [CrossRef]
22. Larios, R.; Fernández-Martínez, R.; LeHecho, I.; Rucandio, I. A methodological approach to evaluate arsenic speciation and bioaccumulation in different plant species from two highly polluted mining areas. *Sci. Total Environ.* **2012**, *414*, 600–607. [CrossRef]
23. Sposito, G.; Lund, L.; Chang, A. Trace metal chemistry in arid-zone field soils amended with sewage sludge: I. Fractionation of Ni, Cu, Zn, Cd, and Pb in solid phases. *Soil Sci. Soc. Am. J.* **1982**, *46*, 260–264. [CrossRef]
24. Wenzel, W.W.; Kirchbaumer, N.; Prohaska, T.; Stingeder, G.; Lombi, E.; Adriano, D.C. Arsenic fractionation in soils using an improved sequential extraction procedure. *Anal. Chim. Acta* **2001**, *436*, 309–323. [CrossRef]
25. Taylor, G.J.; Crowder, A.A. Use of the DCB technique for extraction of hydrous iron oxides from roots of wetland plants. *Botany* **1983**, *70*, 1254–1257. [CrossRef]
26. Rizwan, M.; Ali, S.; Hussain, A.; Ali, Q.; Shakoor, M.B.; Zia-Ur-Rehman, M.; Farid, M.; Asma, M. Effect of zinc-lysine on growth, yield and cadmium uptake in wheat (*Triticum aestivum* L.) and health risk assessment. *Chemosphere* **2017**, *187*, 35–42. [CrossRef] [PubMed]
27. Singh, A.; Sharma, R.K.; Agrawal, M.; Marshall, F.M. Health risk assessment of heavy metals via dietary intake of foodstuffs from the wastewater irrigated site of a dry tropical area of India. *Food Chem. Toxicol.* **2010**, *48*, 611–619. [CrossRef] [PubMed]
28. Yuan, Y.; Chai, L.; Yang, Z.; Yang, W. Simultaneous immobilization of lead, cadmium, and arsenic in combined contaminated soil with iron hydroxyl phosphate. *J. Soils Sediments* **2017**, *17*, 432–439. [CrossRef]
29. Huang, G.; Ding, C.; Hu, Z.; Cui, C.; Zhang, T.; Wang, X. Topdressing iron fertilizer coupled with pre-immobilization in acidic paddy fields reduced cadmium uptake by rice (*Oryza sativa* L.). *Sci. Total Environ.* **2018**, *636*, 1040–1047. [CrossRef] [PubMed]

30. Muehe, E.M.; Adaktylou, I.J.; Obst, M.; Zeitvogel, F.; Behrens, S.; Planer-Friedrich, B.; Kraemer, U.; Kappler, A. Organic carbon and reducing conditions lead to cadmium immobilization by secondary Fe mineral formation in a pH-neutral soil. *Environ. Sci. Technol.* **2013**, *47*, 13430–13439. [CrossRef]
31. Wang, Y.; Zeng, X.; Lu, Y.; Bai, L.; Su, S.; Wu, C. Dynamic arsenic aging processes and their mechanisms in nine types of Chinese soils. *Chemosphere* **2017**, *187*, 404–412. [CrossRef]
32. Hou, Q.; Han, D.; Zhang, Y.; Han, M.; Huang, G.; Xiao, L. The bioaccessibility and fractionation of arsenic in anoxic soils as a function of stabilization using low-cost Fe/Al-based materials: A long-term experiment. *Ecotoxicol. Environ. Saf.* **2020**, *191*, 110210. [CrossRef]
33. Zhang, Q.; Pei, L.; Liu, C.; Han, M.; Wang, W. Impact of redox condition on fractionation and bioaccessibility of arsenic in arsenic-contaminated soils remediated by iron amendments: A long-term experiment. *Geofluids* **2018**, *2018*, 5243018. [CrossRef]
34. He, S.; Feng, Y.; Ni, J.; Sun, Y.; Xue, L.; Feng, Y.; Yu, Y.; Lin, X.; Yang, L. Different responses of soil microbial metabolic activity to silver and iron oxide nanoparticles. *Chemosphere* **2016**, *147*, 195–202. [CrossRef] [PubMed]
35. Lin, J.; He, F.; Su, B.; Sun, M.; Owens, G.; Chen, Z. The stabilizing mechanism of cadmium in contaminated soil using green synthesized iron oxide nanoparticles under long-term incubation. *J. Hazard. Mater.* **2019**, *379*, 120832. [CrossRef] [PubMed]
36. Guo, H.; Nasir, M.; Lv, J.; Dai, Y.; Gao, J. Understanding the variation of microbial community in heavy metals contaminated soil using high throughput sequencing. *Ecotoxicol. Environ. Saf.* **2017**, *144*, 300–306. [CrossRef] [PubMed]
37. Wang, M.; Li, S.; Chen, S.; Meng, N.; Li, X.; Zheng, H.; Zhao, C.; Wang, D. Manipulation of the rhizosphere bacterial community by biofertilizers is associated with mitigation of cadmium phytotoxicity. *Sci. Total Environ.* **2019**, *649*, 413–421. [CrossRef]
38. Price, A.; Macey, M.C.; Miot, J.; Olsson-Francis, K. Draft genome sequences of the nitrate-dependent iron-oxidizing proteobacteria *Acidovorax* sp. strain BoFeN1 and Paracoccus pantotrophus strain KS1. *Microbiol. Resour. Announc.* **2018**, *7*, e01050-18. [CrossRef]
39. Xu, J.; Han, Y.; Chen, Y.; Zhu, L.; Ma, L. Arsenic transformation and plant growth promotion characteristics of As-resistant endophytic bacteria from As-hyperaccumulator Pteris vittata. *Chemosphere* **2016**, *144*, 1233–1240. [CrossRef]
40. Zhu, L.; Guan, D.; Luo, J.; Rathinasabapathi, B.; Ma, L. Characterization of arsenic-resistant endophytic bacteria from hyperaccumulators *Pteris vittata* and *Pteris multifida*. *Chemosphere* **2014**, *113*, 9–16. [CrossRef]
41. Rashid, M.I.; Shahzad, T.; Shahid, M.; Imran, M.; Dhavamani, J.; Ismail, I.M.I.; Basahi, J.M.; Almeelbi, T. Toxicity of iron oxide nanoparticles to grass litter decomposition in a sandy soil. *Sci. Rep.* **2017**, *7*, 41965. [CrossRef]
42. Khan, S.; Akhtar, N.; Rehman, S.U.; Shujah, S.; Rha, E.S.; Jamil, M. Biosynthesized iron oxide nanoparticles (Fe_3O_4 NPs) mitigate arsenic toxicity in rice seedlings. *Toxics* **2020**, *9*, 2. [CrossRef]
43. Bidi, H.; Fallah, H.; Niknejad, Y.; Tari, D.B. Iron oxide nanoparticles alleviate arsenic phytotoxicity in rice by improving iron uptake, oxidative stress tolerance and diminishing arsenic accumulation. *Plant Physiol. Biochem.* **2021**, *163*, 348–357. [CrossRef]
44. Maisch, M.; Lueder, U.; Kappler, A.; Schmidt, C. From plant to paddy—how rice root iron plaque can affect the paddy field iron cycling. *Soil Syst.* **2020**, *4*, 28. [CrossRef]
45. Chen, C.; Dixon, J.B.; Turner, F.T. Iron coatings on rice roots: Morphology and models of development. *Soil Sci. Soc. Am. J.* **1980**, *44*, 1113–1119. [CrossRef]
46. Sebastian, A.; Prasad, M. Iron plaque decreases cadmium accumulation in *Oryza sativa* L. *and serves as a source of iron. Plant Biol.* **2016**, *18*, 1008–1015.
47. Martínez-Fernández, D.; Komárek, M. Comparative effects of nanoscale zero-valent iron (nZVI) and Fe_2O_3 nanoparticles on root hydraulic conductivity of *Solanum lycopersicum* L. *Environ. Exp. Bot.* **2016**, *131*, 128–136. [CrossRef]
48. Yu, H.; Wang, X.; Li, F.; Li, B.; Liu, C.; Wang, Q.; Lei, J. Arsenic mobility and bioavailability in paddy soil under iron compound amendments at different growth stages of rice. *Environ. Pollut.* **2017**, *224*, 136–147. [CrossRef]
49. Hussain, A.; Ali, S.; Rizwan, M.; ur Rehman, M.Z.; Qayyum, M.F.; Wang, H.; Rinklebe, J. Responses of wheat (*Triticum aestivum*) plants grown in a Cd contaminated soil to the application of iron oxide nanoparticles. *Ecotoxicol. Environ. Saf.* **2019**, *173*, 156–164. [CrossRef] [PubMed]
50. Rasheed, A.; Fahad, S.; Aamer, M.; Hassan, M.U.; Tahir, M.M.; Wu, Z. Role of genetic factors in regulating cadmium uptake, transport and accumulation mechanisms and quantitative trait loci mapping in rice: A review. *Appl. Ecol. Environ. Res.* **2020**, *18*, 4005–4023. [CrossRef]
51. Shinde, A.; Kumar, K. Mechanisms of arsenic transport, accumulation, and distribution in rice. In *Arsenic Toxicity: Challenges and Solutions*; Springer: Singapore, 2021; pp. 279–300.
52. Liu, Y.; Zhang, C.; Zhao, Y.; Sun, S.; Liu, Z. Effects of growing seasons and genotypes on the accumulation of cadmium and mineral nutrients in rice grown in cadmium contaminated soil. *Sci. Total Environ.* **2017**, *579*, 1282–1288. [CrossRef]
53. Takahashi, R.; Bashir, K.; Ishimaru, Y.; Nishizawa, N.K.; Nakanishi, H. The role of heavy-metal ATPases, HMAs, in zinc and cadmium transport in rice. *Plant Signal. Behav.* **2012**, *7*, 1605–1607. [CrossRef]
54. Zhang, H.; Zhang, W.; Huang, S.; Xu, P.; Cao, Z.; Chen, M.; Lin, X. The potential role of plasma membrane proteins in response to Zn stress in rice roots based on iTRAQ and PRM under low Cd condition. *J. Hazard. Mater.* **2022**, *429*, 128324. [CrossRef]
55. Zhiguo, E.; Li, T.; Chen, C.; Wang, L. Genome-wide survey and expression analysis of P1B-ATPases in rice, maize and sorghum. *Rice Sci.* **2018**, *25*, 208–217. [CrossRef]
56. Sharafi, K.; Yunesian, M.; Nodehi, R.N.; Mahvi, A.H.; Pirsaheb, M. A systematic literature review for some toxic metals in widely consumed rice types (domestic and imported) in Iran: Human health risk assessment, uncertainty and sensitivity analysis. *Ecotoxicol. Environ. Saf.* **2019**, *176*, 64–75. [CrossRef] [PubMed]

nanomaterials

Article

Public Perceptions and Willingness-to-Pay for Nanopesticides

Peiyuan Liu [1], Xiaodong Zheng [2], Shuangyue Shangguan [3], Lina Zhao [4], Xiangming Fang [3,*], Yuxiong Huang [1,*] and Slav W. Hermanowicz [1,5]

1 Tsinghua-Berkeley Shenzhen Institute, Shenzhen International Graduate School, Tsinghua University, Shenzhen 518055, China; peiyuan.liu@outlook.com (P.L.); hermanowicz@ceeucb.com (S.W.H.)
2 School of Economics, Zhejiang Gongshang University, Hangzhou 310018, China; zhengxd@cau.edu.cn
3 College of Economics and Management, China Agricultural University, Beijing 100083, China; 18811763751@163.com
4 Institute for Hospital Management, Shenzhen International Graduate School, Tsinghua University, Shenzhen 518055, China; zln20@mails.tsinghua.edu.cn
5 Department of Civil and Environmental Engineering, University of California, Berkeley, CA 94720, USA
* Correspondence: xmfang@cau.edu.cn (X.F.); huang_yuxiong@sz.tsinghua.edu.cn (Y.H.); Tel.: +86-(0755)-3688-1774 (Y.H.)

Abstract: The usage of pesticides is deemed essential to ensure crop production for global food security. Conventional chemical pesticides have significant effects on ecosystems. Nanopesticides are increasingly considered an emerging alternative due to their higher efficiency and lower environmental impacts. However, large knowledge gaps exist in the public perceptions and willingness-to-pay (WTP) for nanopesticides. Thus, we conducted a regional survey of pesticide users and food consumers on perceptions and WTP for nanopesticides across China. We found that 97.4% pesticide users were willing to pay for nanopesticides, with a main price from 25% to 40% higher than for conventional pesticides. Experience with applying pesticides, income, familiarity with and attitude toward nanopesticides, and trust in industries were significant determinants of WTP. Although the public were not familiar with nanopesticides, they had positive attitudes toward their future development and supported labeling nanoscale ingredients on products. Pesticide users presented high trust levels in governments and industries, while 34% of food consumers neutrally or distrusted industries in selling and production. This study highlights the socioeconomic and technological aspects of nanopesticides, which could provide guidance for industries to develop market strategies and for governments to design relevant regulation policies effectively, contributing to crop yield improvement and sustainable agriculture.

Keywords: nanopesticides; public perception; willingness-to-pay; interval regression model; survey

Citation: Liu, P.; Zheng, X.; Shangguan, S.; Zhao, L.; Fang, X.; Huang, Y.; Hermanowicz, S.W. Public Perceptions and Willingness-to-Pay for Nanopesticides. *Nanomaterials* **2022**, *12*, 1292. https://doi.org/10.3390/nano12081292

Academic Editor: Paul M Bertsch

Received: 3 March 2022
Accepted: 7 April 2022
Published: 11 April 2022

1. Introduction

The rapid growth of the global population, which is predicted to reach nearly 9.8 billion people by 2050, requires food production to increase by 50% compared to the levels in 2012 [1]. However, crop production is significantly suppressed by biotic stresses, such as pests, weeds, and diseases caused by fungi, bacteria, and viruses [2,3]. Application of pesticides is a critical way to mitigate these biotic stresses [2]. Although conventional chemical pesticides are effective, they simultaneously pose severe threats to the ecosystem [3,4]. Specifically, due to the low efficiency of conventional pesticides (~90% lost to environment) [5], farmers must increase the application frequency and amount to achieve better crop yields, resulting in 2 million tons per year of conventional pesticides applied worldwide [1]. Such extensive use not only aggravates environmental contamination (e.g., anoxic water bodies, loss of biodiversity, and ecotoxicity) [5,6], but also poses risks to public health directly and indirectly [7,8]. In addition, the long-term use of conventional pesticides induced resistant crop varieties [9] and increased farmers' economic costs [10]. Therefore, a revolution of pesticides is urgently needed to improve crop production and maintain sustainability.

With the rapid development of nanotechnologies, nanopesticides have been increasingly anticipated for the agrochemical sector, including nanoemulsions, nanocapsules (e.g., with polymers), and inorganic engineered nanoparticles (ENPs, for example, metals, metal oxides, and nanoclays) [11]. Composed of nanoscale active ingredients (AIs), nanopesticides exhibited broad-spectrum insecticidal, fungicidal, and herbicidal properties [3,12]. Furthermore, nanopesticides can enhance solubility, control release, increase leaf adhesion, and improve the stability of AIs [13], resulting in elevated efficacy and durability, as well as the reduction in applied AIs [14]. Thus, the application of nanopesticides can maintain or increase crop yields with lower application rates, which would potentially minimize the risks to ecosystems [15]. In addition, the global market for pesticides is estimated to grow from US$75 billion in 2013 to US$90 billion by 2023 [16], and nanopesticides have the potential to result in multibillion-dollar benefits [5]. Nevertheless, only a few commercial products among synthesized nanopesticides have been commercialized (e.g., Kocide 3000 (Dupont), and AZteroid FC (Vive Crop Protection)), as a result of three major barriers to technology readiness and implementation (i.e., efficient delivery at the field scale, regulation and safety concerns, and consumer acceptance) [16].

Currently, research on nanopesticides has been mainly focused on their performance [15], mechanisms [17], environmental fate [2], and ecosystem implications [18,19]. However, huge knowledge gaps exist related to nano-governance and socio-economic aspects, particularly in public perceptions and willingness-to-pay (WTP) for nanopesticides, which are essential to promote the usage and market share of nanopesticides [15]. Specifically, the perception of familiarity with nanotechnology could have a positive impact on risk and benefit perceptions, which would further affect attitudes toward and the acceptance of nanotechnology [20]. In particular, WTP, which is defined as the maximum price that a consumer would accept to purchase one unit of a product or service [21], could shape the direction of the marketing price and enable stakeholders to better match the needs and desires of the public.

Factors that could affect the public perceptions and WTP of nanotechnologies or emerging agri-food technologies depend on complicated mechanisms and multiple aspects. For example, socio-demographic factors (i.e., age, gender, education, and income) were shown to be related to understanding the risks and benefits of nanotechnologies [22,23]. Familiarity with [24] and attitude toward [25] novel agri-food technologies had implications for the willingness to purchase these technologies (e.g., nano-enabled food packaging or ingredients). On the other hand, the safety of nanotechnology applications has been questioned in the past few years, resulting in increased requirements for nanotechnology labeling [26], which could help the public manage risks and benefits while purchasing [27–29]. Another significant aspect of public perceptions and WTP is social trust, which could have an impact on the risk perceptions and communication of new technologies by building social relationships [20]. Multiple studies have shown that the acceptance of nano-enabled products was greatly influenced by trust in industries [30]. Trust in governments to manage risks was also a major concern when assessing nanotechnology applications [31]. There has also been uncertainty regarding the experience of applying pesticides, which could affect farmers' acceptance of new-type pesticides [32]. Users may not want to change their habits if they have used conventional pesticides for a long period time [32]. However, they were more likely to accept new pesticides after realizing the long-term hazards of conventional chemical pesticides [32]. Hence, it is essential to understand the multiple factors influencing public WTP for nanopesticides, which are currently missing in available studies.

To investigate public perceptions of and WTP for nanopesticides, we conducted a regional survey of pesticide users and food consumers across China, with almost 400 valid responses. China, as a large agricultural country with 23.6% of its workforce employed in the agricultural sector in 2020 [33], is the largest consumer of pesticides in the world [34]. China has made impressive progress, taking 9% of the planet's arable land to feed 22% of the worlds' population [35]. However, as a result of its ongoing urbanization, population growth, and severe environmental impacts, China is facing new challenges to sustainable agriculture [35]. As nano-enabled agriculture has exhibited potential to address these challenges, it is worthwhile to investigate pesticide users and food consumers in China as a representative case study on nanopesticides to explore future agricultural development. The objectives of this study were to (1) examine the price ranges of public WTP for nanopesticides; (2) identify the factors influencing pesticide users' WTP for nanopesticides; (3) estimate the WTP for nanopesticides under different pesticide user profiles; and (4) explore general public perspectives on nanopesticides. In this study, we combined the advantages of multiple models (i.e., the Heckman model, interval regression model, ordinary square least model, and ordered logistic model) to estimate statistical outcomes. Our findings could narrow the gaps among academia, the public, industries, and governments, thereby helping to assess the market potential, facilitate research and development, and design regulation policies for nanopesticides. This research further aimed at meeting the increasing demands in food production and making agriculture more sustainable.

2. Materials and Methods

2.1. Data Collection

The survey was conducted between 24 July 2020 and 5 August 2020 via face-to-face interviews with a questionnaire. We stratified the survey sites into the western, middle, and eastern parts of China. The locations included the countryside located in two municipalities directly under the central government (i.e., Chongqing and Shanghai), eight cities of two autonomous regions (i.e., Guangxi and Tibet) and 56 cities in 11 provinces (details in Figure A1 of Appendix A). The respondents were randomly chosen and we collected 395 fully completed surveys. The survey included 232 pesticide users (i.e., farmers using pesticides) and 163 food consumers (i.e., people from aquaculture and animal husbandry not using pesticides). Eighteen surveys were not completed and were discarded in the following analysis.

More specifically, the appropriate sample size was estimated before investigation by calculating the equation of simple random sampling with substitution [36,37]:

$$N = \frac{4(Z_{crit})^2 p(1-p)}{D^2} \tag{1}$$

where N is the sample size; Z_{crit} is the standard normal deviation corresponding to the selected confidence level (CI); D is the minimum expected difference, which is specified here subjectively to reflect the difference between the upper and lower limit of an expected CI (i.e., the total width of the expected CI); and p is a pre-study estimate of the proportion to be measured. We set the CI at 95 percent that yielded Z_{crit} = 1.960, assumed D = 10% (0.1), and estimated p = 0.9 (using the proportion from a preliminary survey on pesticide users' willingness-to-pay for nanopesticides; approximately 90% replied "yes"). Based on these assumptions, Equation (1) yielded a sample size of N = 138. Therefore, considering possible invalid responses, we expanded the survey scale and the final 232 valid responses of pesticide users met the requirements of the sample size.

2.2. Questionnaire and Measurements

2.2.1. Variable Selection

Ten independent variables were selected to evaluate public acceptance of emerging technologies. Factors related to socio-demographic information (i.e., gender, age, education, and income) and public perceptions of nanopesticides (i.e., familiarity with and attitude toward nanopesticides, labeling preference, and trust in governments and industries) were investigated. In addition, we included experience of applying pesticides and the associated quadratic term to examine the possible incremental or diminishing effects.

The dependent variables included (a) the decision to spend money on nanopesticides when the price was lower than that of conventional pesticides. If respondents indicated unwillingness, they would end the questionnaire. Otherwise, respondents were asked specific follow-up questions about (b) the price ranges of WTP for nanopesticides (Figure A2 in Appendix A).

2.2.2. Questionnaire Design

As illustrated in Figure 1, the questionnaire incorporated 12 questions that were divided into three sections; the complete questionnaire is presented in Appendix B. The first section included four socio-demographic questions (i.e., gender, age, education, and income), followed by five questions relevant to the perceptions of nanopesticides (i.e., familiarity with and attitude toward nanopesticides, labeling preference, and trust in governments and industries) in the second section. Respondents were then asked whether they planted crops needing pesticides. The food consumers (i.e., people from aquaculture and animal husbandry) would not need to purchase pesticides and quitted the survey. Only pesticide users (i.e., farmers) continued with the third-section questions, including the experience of applying pesticides and the WTP for nanopesticides (Figure 1).

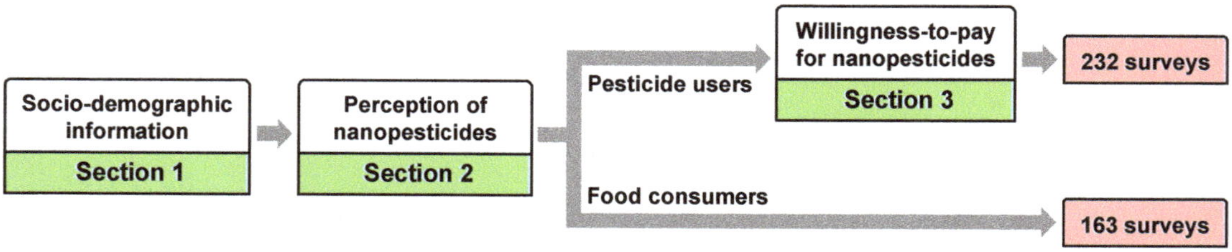

Figure 1. Flowchart for the questionnaire survey.

The WTP questions were designed using the contingent valuation method (CVM), a mature tool used to estimate public WTP for environmental goods and services in the marketplace [21,38], and widely applied in the sectors of foods and pesticides [39–42]. The CVM-based questionnaire is typically framed as an open-ended question, such as "how much money you would be willing to pay for the target goods or services?" or as a "yes/no" question that determines whether or not the respondent would be willing to pay $X for the target goods or services [43]. We chose the doubled-bounded CVM to ask a series of questions to progressively narrow down each respondent's bounds on WTP, resulting in nine intervals (in %) that consumers would be willing to pay for nanopesticides per kilogram over that for conventional pesticides per kilogram: −100–0%, 0–10%, 10–25%, 25–40%, 40–50%, 50–75%, 75–100%, 100–130%, and ≥130% (Figure A2 in Appendix A). Compared with point data by asking a single open-ended or yes/no question, intervals can generate more efficient estimations and be closer to reality by avoiding the randomness of respondent answers [44,45].

2.3. *Data Analysis*

The survey data were analyzed using the Stata programming software. A descriptive statistical analysis of 232 pesticide users was conducted. In order to explore the influencing factors of WTP for nanopesticides, the Heckman model was firstly used to test whether there would be sample selection bias [46] if we excluded the six samples who missed the specific price ranges of WTP (i.e., they would not like to spend money on nanopesticides even at a lower price, as Quit 1 shown in Figure A2 in Appendix A).

2.3.1. Theory of the Heckman Model

Sample selection bias may arise when values of dependent variables are missing or unobserved caused by another process (e.g., self-selection by individuals or data units investigated, sample selection decisions by analysts or data processers) [46,47]. For example, if the appearance of outcome variable y_i depends on a selection variable z_i, such incidental truncation may result in a missing data problem of y_i and biased coefficient estimation using standard regression techniques (e.g., OLS). In order to resolve this potential bias, the Heckman model was introduced and assumed a two-stage relationship (Equations (A1)–(A3) in Appendix A). The first step in this model is to determine whether an observation in an overall population appears in the final representative samples, and the second step is to model the relationship between the dependent and independent variables in the final selected samples [46]. With the maximum likelihood estimation in the Heckman model, rho (ρ; the correlation between error terms in the selection and outcome equations) could be examined to indicate whether or not sample selection bias exists [46]. If rho is significant, traditional techniques (e.g., OLS) would report biased β estimation. In this situation, the results of the Heckman model can provide consistent and asymptotically efficient estimates by correcting selection bias [48]. Otherwise, traditional regression methods could generate efficient estimates by using selected samples. More details on the Heckman model were provided in Appendix A.

2.3.2. Interval Regression Model

To further examine the significance levels of different independent variables for 226 samples, the interval regression model was used as a preferred method when the outcome was measured as interval data, left-censored data, or right-censored data [48,49]. Other models (i.e., the ordinary least squares (OLS) model and ordered logistic model) were not chosen due to limitations. Specifically, the OLS model would use the interval medians as a dependent variable's values and use the upper or lower limit values for left-censored data or right-censored data, which neglects the uncertainty distribution of the dependent variable and reduces the accuracy of the results [50]. In addition, the ordered logistic model would order intervals sequentially as dependent variable's values, which does not take the threshold values into account and results in a loss of information within the dependent variable [49].

By using the interval regression model, we assumed that each respondent i had a WTP for nanopesticides Y_i^* that was related to independent variables X_i in the following way:

$$Y_i^* = X_i\beta + \varepsilon_i \quad (2)$$

where ε_i was assumed to be a normally distributed term with zero mean [48].

We did not observe Y_i^* directly, but we knew it fell within some interval $[Y_{i1}, Y_{i2}]$ based on the responses from a series of double-bounded CVM questions (Figure A2 in Appendix A). Therefore, the likelihood contribution of respondent i was $\Pr(Y_{i1} \leq Y_i^* \leq Y_{i2})$ or $\Pr(Y_{i1} \leq X_i\beta + \varepsilon_i \leq Y_{i2})$. For left-censored data (the unobserved Y_i^* was less than or equal to a fixed upper endpoint) and right-censored data (the unobserved Y_i^* was greater than or equal to a fixed lower endpoint), the likelihood contributions were $\Pr(X_i\beta + \varepsilon_i \leq Y_{i2})$ and $\Pr(Y_{i1} \leq X_i\beta + \varepsilon_i)$, respectively. The maximum likelihood function was estimated using the command *intreg* in Stata, and the specific Equations (A4)–(A8) were illustrated in Appendix A.

2.3.3. Robustness Test

The ordinary least squares (OLS) and ordered logistic models were used to identify the robustness and credibility of the interval regression model. Specifically, we converted the interval data, left-censored data, and right-censored data of WTP into point data of the interval median, upper limit value, and lower limit value, respectively, to estimate the OLS regression model. Meanwhile, nine price ranges of WTP (Figure A2 in Appendix A) were converted into ordinal numbers 1–9 sequentially for the ordered logistic model using maximum likelihood estimation. The independent variables in the OLS and ordered logistic models remained unchanged with that in the interval regression model.

In addition, the relative influence importance of different variables was compared using standard beta coefficients. The plots of the public's perspectives on nanopesticides were created using the online OmicShare Tools [51].

3. Results and Discussion

3.1. Descriptive Statistics of Variables

Among all 395 samples, 163 food consumers answered the survey, except for questions regarding the experience of applying pesticides and the WTP for nanopesticides. A total of 232 pesticide users responded to all the questions. The descriptive statistics of the pesticide users were summarized in Table 1.

As shown in Table 1, the 232 pesticide users were between 25 and 75 years old (median = 46), and 17.2% were female while 82.8% were male. Overall, the participants were educated and had an average 11-year education level (mean = 11.1, median = 12). There was a significant range in annual incomes (standard deviation = 15.7), with the median level at 130,000 RMB (approximately 20,000 USD). The participants' average experience in applying pesticides was more than 15 years, and the maximum was 52 years. Although the current level of familiarity with nanopesticides was low (mean = 2.6), pesticide users had relatively supportive attitudes toward the future development of nanopesticides (mean = 4, median = 4). For labeling indications, the participants generally preferred to be informed that the product contains nano-components (mean = 4.2, minimum = 3). Pesticide users strongly trusted governments and industries regarding supervision, production, and selling (medians = 4).

Not surprisingly, based on the above positive attitudes, most pesticide users (97.41%) were willing to spend money on nanopesticides. Only six pesticide users (2.59%) would not like to spend any money on nanopesticides, even if the price was lower than that of conventional pesticides (Table 1 and Figure 2). The high proportion of WTP for nanopesticides was much higher than that of WTP for other nano-enabled food products. For example, almost 50% of consumers refused to purchase foods (e.g., canola oil) with nano-packaging, nanodrop, and nano-sensor attributes [52]. The distinct proportions of WTP for different nanoproducts could result from various survey subjects; particularly, the more directly consumers were in contact with nanoproducts, the less likely they were willing to use nanoproducts [53]. Compared to nanopesticide users, fewer food consumers were willing to purchase foods engaged with nanotechnology. This high public purchase intention of nanopesticides could motivate academia, industries, and governments to advance the research and development of nanopesticides, rather than being impeded by worries and uncertainties about public rejection. Moreover, as illustrated in Figure 2, 2.16% of pesticide users would be willing to purchase nanopesticides only if the price was lower than that of conventional pesticides. The main price range that respondents were willing to pay for nanopesticides was 25–40% higher than that of conventional pesticides (Figure 2), guiding industries to improve market strategies and price nanopesticides more appropriately in the future.

Table 1. Overview and measurements of the variables and descriptive statistics of 232 pesticide users.

Dependent Variable	Description and Measurement	Mean	Median	Standard Deviation	Min	Max
Decision to spend money on nanopesticides at a lower price	No = 0, Yes = 1	0.97	1	0.16	0	1
Price ranges of willingness-to-pay	The percentage that consumers were willing to pay higher than conventional pesticides for nanopesticides: (−100%, 0) = 1, [0, 10%) = 2, [10%, 25%) = 3, [25%, 40%) = 4, [40%, 50%) = 5, [50%, 75%) = 6, [75%, 100%) = 7, [100%, 130%) = 8, ≥130% = 9	4.95	4	2.56	1	9
Independent Variable	**Description and Measurement**	**Mean**	**Median**	**Standard Deviation**	**Min**	**Max**
Gender	Female = 1, Male = 0	0.17	0	0.38	0	1

Table 1. *Cont.*

Dependent Variable	Description and Measurement	Mean	Median	Standard Deviation	Min	Max
Years of education	Seven categories: uneducated = 0, primary school = 6, middle school = 9, high school/professional high school/technical school/secondary school = 12, junior college = 15, undergraduate education = 16, postgraduate education = 19 Unit: years	11.08	12	2.72	6	16
Total household income in 2019	Unit: 100,000 RMB (approximately 15,385 USD)	4.06	1.3	15.69	0.07	222
Experience of applying pesticides	Unit: years	15.57	12.5	10.89	0	52
Familiarity with nanopesticides	Completely unfamiliar = 1, A little unfamiliar = 2, General = 3, Quite familiar = 4, Very familiar = 5	2.64	3	1.07	1	5
Attitude toward the future development of nanopesticides	Very opposed = 1, A little opposed = 2, Neutral = 3, Quite supportive = 4, Very supportive = 5	4.00	4	0.69	2	5
Labeling preference	Do you agree that the product label of nanopesticides must indicate that it contains nano-components? Completely disagree = 1, A little disagree = 2, Neutral = 3, Quite agree = 4, Strongly agree = 5	4.22	4	0.68	3	5
Social trust	Completely distrust = 1, A little distrust = 2, General = 3, Quite trust = 4, Strongly trust = 5					
Trust in governments	Do you trust that governments could supervise the safety risks of nanopesticides?	4.13	4	0.78	1	5
Trust in industries	Do you trust that manufactures and retailers could produce and sell nanopesticides legally?	3.86	4	0.79	1	5

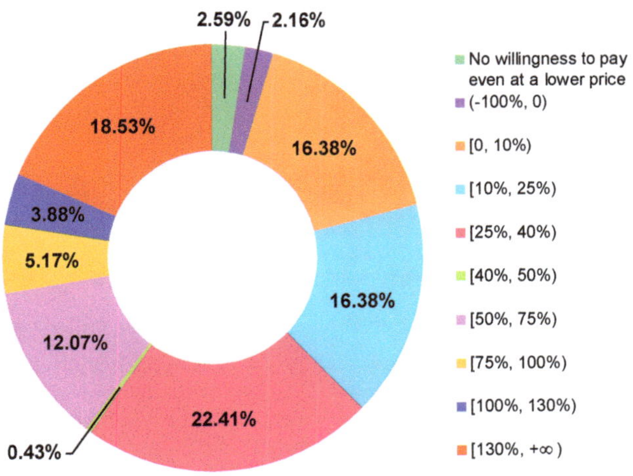

Figure 2. The distribution of the price ranges (% of WTP for nanopesticides over that for conventional pesticides).

3.2. Sample Selection Bias and Model Robustness Evaluation

As shown in Figure 2, six pesticide users were not willing to pay for nanopesticides even if the price was lower than that of conventional pesticides, which led to the missing data for the price range. The Heckman model was used to evaluate the sampling selection bias, and we confirmed that these six samples could be excluded as the value of rho (Table A1 in Appendix A) in the Heckman model was not significant [46]. Therefore, the interval regression model on 226 samples was subsequently utilized to evaluate the factors influencing WTP for nanopesticides (Table 2), with detailed discussion in the following section. In order to enhance the credibility of the results, OLS and ordered logistic models were used to verify the robustness of the interval regression model (Table 2). The coefficients of the OLS model were aligned well with those of the interval regression model (Table 2). No obvious differences in the significance levels of variables were found among the OLS, ordered logistic, and interval regression models (Table 2), indicating good reliability of the results obtained by the interval regression model.

Table 2. Interval regression model for evaluating factors influencing pesticide users' WTP for nanopesticides, and OLS and ordered logistic models for testing the robustness of the interval regression model.

Variable	Interval Regression Model		OLS Model		Ordered Logistic Model	
	Coefficient	Robust Standard Error	Coefficient	Robust Standard Error	Coefficient	Robust Standard Error
Gender	−3.61	10.99	−3.56	9.00	−0.41	0.41
Age	−0.10	0.54	−0.08	0.47	−0.02	0.02
Years of education	1.97	1.38	1.70	1.18	0.12 *	0.05
Experience of applying pesticides	−2.77 *	1.08	−2.27 *	0.89	−0.08 *	0.04
Quadratic term of experience of applying pesticides	0.05 *	0.02	0.04 *	0.02	0.00 *	0.00
Income	1.39 *	0.57	1.15 **	0.39	0.05 **	0.02
Familiarity with nanopesticides	11.08 **	3.39	8.55 **	2.76	0.46 **	0.12
Attitude toward nanopesticides	13.38 *	6.65	13.70 *	5.87	0.61 *	0.26
Trust in governments	−7.64	5.60	−6.37	4.84	−0.27	0.22
Trust in industries	13.83 *	5.52	10.52 *	4.59	0.29	0.19
Labeling preference	0.04	6.26	−0.39	5.44	0.20	0.27
Constant	−43.22	38.93	−36.62	33.08		
Wald test	Chi-square = 57.34; $p = 0.00$					
VIF †	Mean = 1.58					
Numbers of observations	226		226		226	

Notes: ** and * indicate significance at $p < 0.01$ and $p < 0.05$ levels, respectively. The unit of the coefficients is percentage points. † The mean value of VIF (variance inflation factor) was smaller than 2, indicating no multicollinearity between the independent variables in the regression model.

3.3. Determinants of Willingness-to-Pay for Nanopesticides

As shown in Table 2, both the experience of applying pesticides and the associated quadratic term were statistically significant. There were diminishing and incremental trends before and after 27-year experience, which was a relatively intermediate-level of experience for pesticide users. Early career and richer-experience pesticide users reported higher WTP price ranges than intermediate-experience pesticide users. Specifically, compared with early career pesticide users, intermediate-experience pesticide users would not like to change their habits to adapt to the new routine as they have already formed usage patterns with the conventional pesticides [32]. Meanwhile, the rich-experience pesticide users would try emerging alternatives with higher efficiency and better sustainability after realizing the long-term hazards of conventional chemical pesticides [32].

On the other hand, pesticide users with higher income would be willing to pay more for nanopesticides compared to individuals with lower income. For every 100,000 RMB (approximately 15,385 USD) increase in annual income, the WTP price range would increase by 1.39% (Table 2), which was attributed to greater purchasing power [32]. Furthermore, income was positively correlated with risk preference [54], and risk-takers were more prone to invest in emerging alternative technologies [55,56]. Therefore, in the early stage of promoting nanopesticides, industries could target main markets to regions with better economic situations, and pesticide users with junior and rich experience.

Familiarity with and attitude toward nanopesticides were found to positively and significantly influence the price range of WTP for nanopesticides (Table 2). A sense of familiarity can be created by a generally positive framing of nanotechnology in the media, which could mitigate consumers' negative responses to risky content, while positive beliefs may confirm benefit information [57]. Moreover, people with low familiarity may be initially less interested in emerging alternatives [57]. It would be important to use media exposure (e.g., science-related news and education programs) and interpersonal communication in an elaborative manner (e.g., lectures) to strengthen perceived familiarity [58]. Increased familiarity would lead people to have a more favorable attitude toward technology [59], further facilitating the acceptance of nanopesticides.

In addition, the price range of WTP for nanopesticides could be notably improved with the rise of the trust level in industries (Table 2). General social trust in the food industry can evoke the preference for emerging nanotechnology products [60,61]. Therefore, maintaining a good corporate reputation would play a key role in marketing nanopesticides. Industries are supposed to operate in accordance with laws and regulations and to carry out effective public-relations strategies simultaneously (e.g., media advertisements, posters, and proactive dialog between different stakeholders through workshops or forums). While trust in governments did not significantly influence pesticide users' WTP for nanopesticides (Table 2), governments have responsibilities in transferring relevant knowledge to the public (e.g., through training programs for farmers to learn practice techniques and the benefits and risks of nanopesticides) and developing regulations (e.g., registration of nanopesticides, legality of industries, and use and recycling management).

Gender, age, and education level were not significant determinants of WTP. Although labeling preference was also not a significant influencing factor of pesticide users' WTP for nanopesticides (Table 2), the participants generally agreed that product labels must indicate the usage of nanocomponents (Table 1). Such labeling would not only increase public familiarity with nanotechnology, but also be beneficial for consumers who want to avoid risks, in addition to those who aim to benefit from nanotechnology [57]. However, labeling alone is insufficient to educate the public [28], and comprehensive knowledge of nanopesticides should also be provided.

The standardized beta coefficients in the regression model were further calculated to examine which variables contributed most to the interval regression model (Table A2 in Appendix A), with a higher absolute value indicating a stronger influencing effect of the corresponding independent variable [62]. The experience of applying pesticides was found to have a greater influence on public WTP for nanopesticides than the other variables. Familiarity, trust in industries, attitude, and income had similar influencing importance. These results were based on statistical regressions, which may be different in reality. Overall, these significant determinants of WTP for nanopesticides could provide a direction for industries about which group of pesticide users would be the target customers (e.g., people with high income and high familiarity) and also indicate to policy-makers how they can influence the public acceptance of nanopesticides (e.g., by improving public familiarity and strengthening regulations to increase trust levels in industries).

3.4. Estimations of Willingness-to-Pay for Distinct Consumer Profiles

Based on the results of the interval regression model, we estimated the actual WTP for distinct pesticide users' profiles (Table A3 in Appendix A). For example, pesticide users with 23-year experience in applying pesticides who were a little unfamiliar with nanopesticides and completely distrusted industries, would be willing to pay prices 1% lower for nanopesticides than that for conventional pesticides. In contrast, pesticide users with 13-year experience alongside general familiarity with nanopesticides and a neutral trust level in industries, would be willing to pay 47% more for nanopesticides. Table A3 also illustrates that pesticide users with 5-year experience in applying pesticides, who were very familiar with nanopesticides and strongly trusted industries, would be willing to pay 112% more for nanopesticides.

3.5. General Public Perspectives on Nanopesticides

As discussed above, 163 food consumers (i.e., those in aquaculture and animal husbandry who did not use pesticides) also participated in the survey. We analyzed all responders' (i.e., 163 food consumers and 232 pesticide users) perspectives on nanopesticides to assess the overall perceptions (Figure 3).

In general, nearly half of the survey participants were not familiar with nanopesticides (see light blue and dark boxes in Figure 3a), which was consistent with the results of various surveys that indicated the knowledge of food-relevant nanotechnologies in the general population was low [63]. Food consumers had a lower familiarity level with nanopesticides than pesticide users (Figure 3a). Nevertheless, few people in both groups opposed the future development of nanopesticides (Figure 3b). It implied that there are significant expectations regarding nanopesticides, which have the potential to be highly accepted in the market. In addition, most of the public had neutral positions or agreed that products should have labeling indications for nanocomponents (Figure 3c). The requirement for labeling indications should be incorporated into regulations by governments. Although the general public highly trusted governments, 6% of food consumers distrusted industries, and 28% of food consumers had general trust levels in industries (Figure 3d). Industries producing and selling nanopesticides need to put effort into communicating not only with pesticide users but also with food consumers to enhance general social trust. Otherwise, food consumers would not purchase the nanopesticide-engaged foods, which would further negatively influence pesticide users' WTP for nanopesticides. It is, therefore, critical to understand current public perspectives on nanopesticides among both pesticide users and food consumers, thereby helping industries and governments assess the development trends of nanopesticides and make relevant strategies for production and regulations in the next stage.

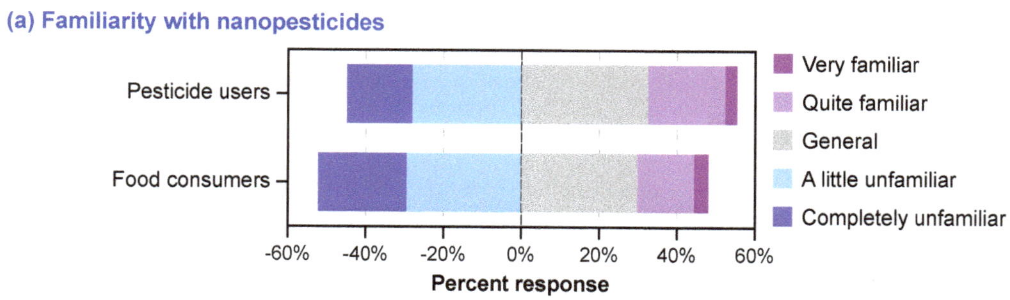

Figure 3. *Cont.*

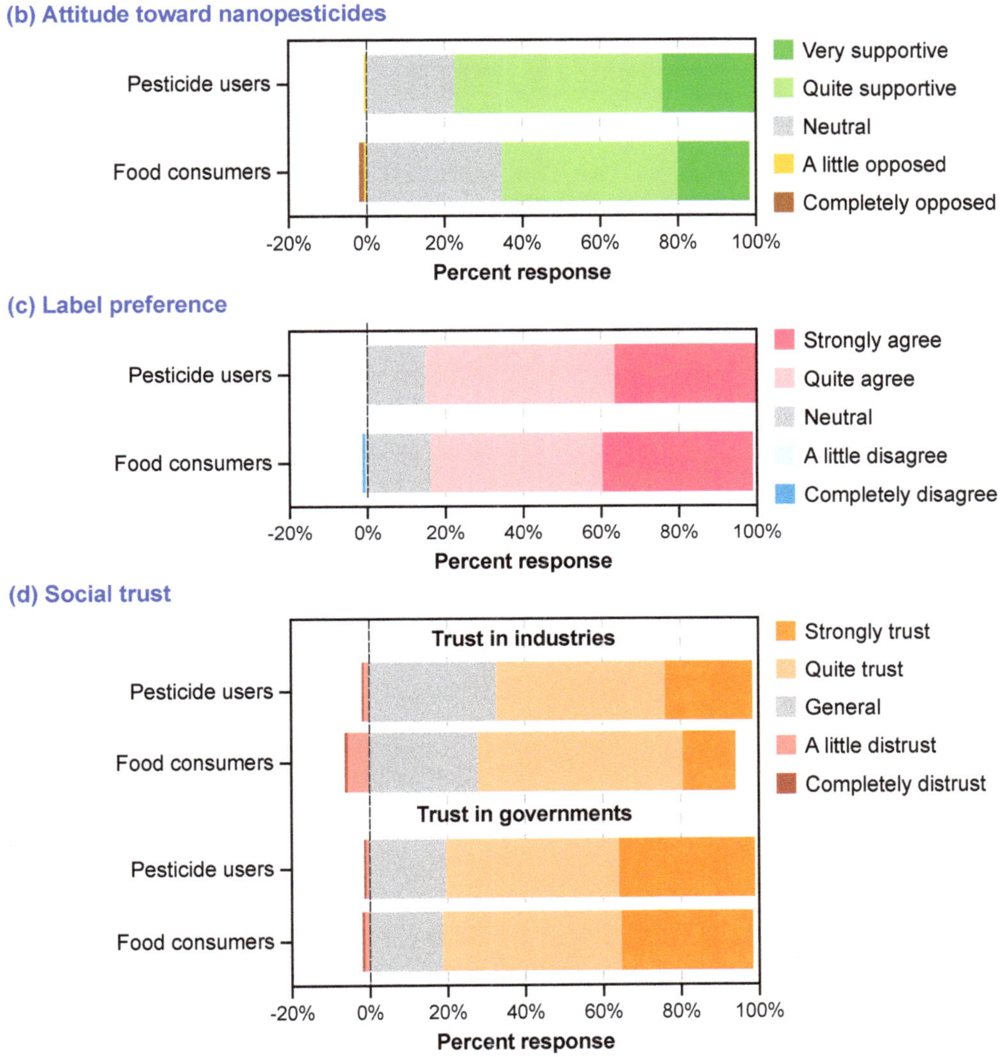

Figure 3. Comparison of public's perspectives on nanopesticides between pesticide users (i.e., farmers) and food consumers (i.e., people from aquaculture and animal husbandry). Public responses when asked (**a**) "How familiar are you with nanopesticides?"; (**b**) "What is your attitude toward the future development of nanopesticides?"; (**c**) "Do you agree that the product label of nanopesticides must indicate that it contains nano-components?"; (**d**) "Do you trust that industries (manufactures and retailers) could produce and sell nanopesticides legally?" and "Do you trust that governments could supervise the safety risks of nanopesticides?".

4. Conclusions

This study combined socioeconomic and technological aspects to evaluate factors that affect public willingness-to-pay (WTP) for nanopesticides and public perceptions from both pesticide users and food consumers perspectives. The findings provide key information for industries and governments to improve marketing strategies and regulations for the large-scale future application of nanopesticides, thus ensuring crop production for global food security and maintaining agricultural sustainability.

As this study demonstrated, nanopesticides were highly accepted by pesticide users, and 97.4% were willing to spend money on them. The main price range (%) pesticide users were willing to pay for nanopesticides was 25–40% higher than that for conventional pesticides. The experience of applying pesticides had a greater influence on the WTP for nanopesticides than the other variables. Familiarity, trust in industries, attitude, and income were also positive and significant determinants of WTP for nanopesticides. The general public's familiarity level with nanopesticides was low. Nevertheless, both pesticide users and food consumers supported the future development of nanopesticides quite strongly. Most of the participants agreed that nanopesticides must include labels indicating that the product contains nanocomponents. Pesticide users generally trusted governments and industries, while a few food consumers had neutral or distrust levels in industries for selling and production.

Based on our findings, we suggest that governments should take label requirements into account when developing regulations. The related knowledge of nanopesticides should also be provided to the public via media, lectures, and training programs. In addition, governments should take responsibilities for optimizing relevant regulatory frameworks, such as the standard code of nanopesticides for entering markets, the legality of industries, and the use and recycling of nanopesticides.

The current study is also subject to certain limitations, as survey results were based on Chinese samples. It would be important to conduct local studies in different countries with larger sample sizes, since public responses may vary with cultures and traditions [64]. Moreover, although we measured public WTP for nanopesticides, a divergence may exist between intentions and actual purchasing behaviors [65]; hypothetical WTP values were typically higher than the real WTP values [66]. Compared with the survey scenarios, people may be more frugal in real life as a result of budget constraints, policy implications, etc. [32]. Furthermore, the current study only investigated the factors influencing pesticide users' WTP for nanopesticides. It would also be essential to identify the factors influencing food consumers' WTP for nanopesticide-engaged foods. The social acceptance and successful application of nanoproducts depend on complex aspects [63]. The associated considerations of nanopesticides, such as cost assessment, environmental impact, risks to human health, and ethical issues, still need to be addressed more comprehensively in future research.

Author Contributions: Conceptualization, P.L.; methodology, P.L., X.Z., X.F. and L.Z.; software, P.L.; formal analysis, P.L.; investigation, X.Z., S.S. and X.F.; data curation, P.L.; writing—original draft preparation, P.L.; writing—review and editing, Y.H. and S.W.H.; visualization, P.L.; supervision, Y.H. and X.F.; project administration, Y.H.; funding acquisition, Y.H. All authors have read and agreed to the published version of the manuscript.

Funding: This research was funded by the National Natural Science Foundation of China [grant numbers 42077293 and 22006088], Natural Science Foundation of Guangdong Province, China [grant numbers 2019A1515011692 and 2019QN01L797], and Tsinghua Shenzhen International Graduate School [grant numbers QD2021010N and HW2020002].

Data Availability Statement: The data presented in this study are available on request from the corresponding author. The data are not publicly available due to privacy.

Acknowledgments: The authors would like to thank all respondents for completing the questionnaires voluntarily; their identities and personal data were protected.

Conflicts of Interest: The authors declare no conflict of interest.

Appendix A

Appendix A contains the distribution of sampling sites in the western, middle, and eastern parts of China (Figure A1), and the sequence of questions about the willingness-to-pay for nanopesticides (Figure A2). The theory underlying the Heckman model and formulas of interval regression model are also provided in Appendix A. Table A1 presents the Heckman model for testing sample selection bias. Tables A2 and A3 provide the standardized beta coefficients of influencing factors, and different pesticide user profiles' willingness-to-pay for nanopesticides, respectively.

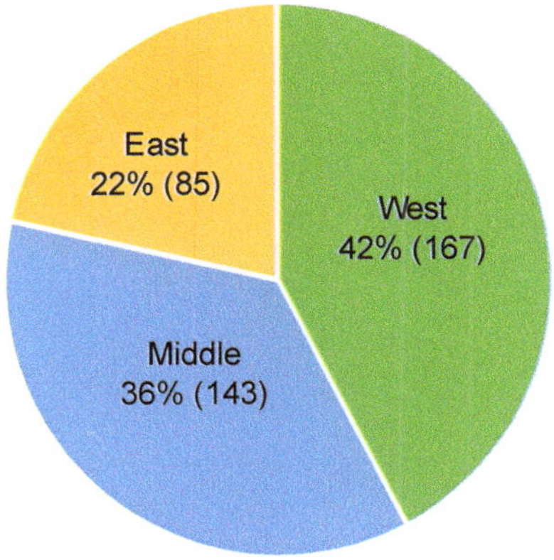

Figure A1. The distribution of sampling sites in the western, middle, and eastern parts of China. Numbers in the parentheses represent the sample size of regions. Eastern part: Fujian province (16), Jiangsu province (24), Zhejiang province (11), Hainan province (18), Shanghai (16). Middle part: Hubei province (42), Hunan province (19), Anhui province (32), Jiangxi province (50). Western part: Yunnan province (24), Sichuan province (17), Guizhou province (27), Chongqing (59), Tibet autonomous region (15), Guangxi autonomous region (25).

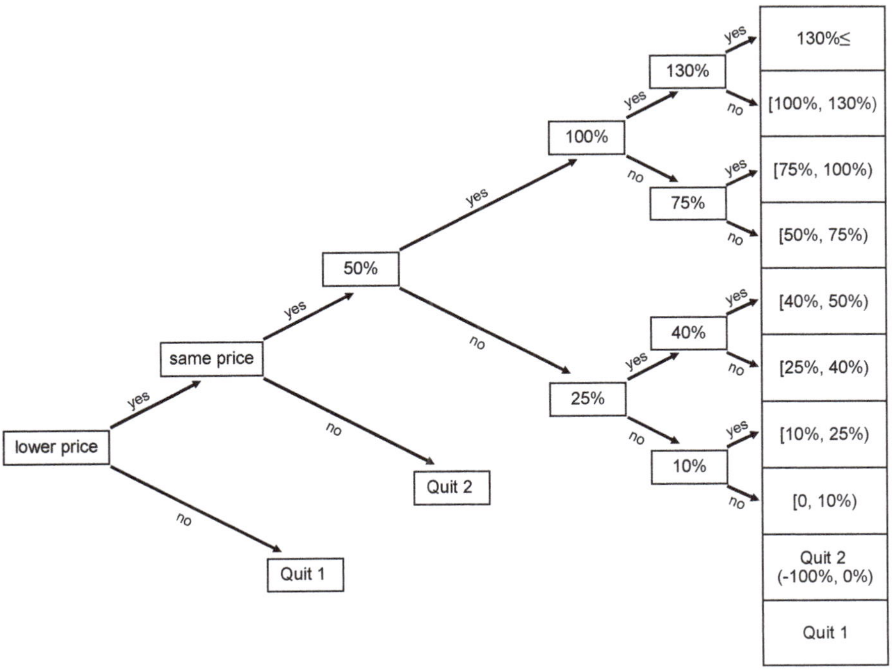

Figure A2. Sequence of questions about the willingness-to-pay for nanopesticides. At each node, respondents were asked whether they would be willing to purchase nanopesticides if the price was lower, the same as, or X% higher than conventional pesticides. The right column shows bounds on the percentages that respondents were willing to pay for nanopesticides above that for conventional pesticides.

Appendix A.1. Theory of Heckman Model

Sample selection bias may arise when values of dependent variables are missing or unobserved, caused by another process (e.g., self-selection by individuals or data units investigated, sample selection decisions by analysts or data processers) [46,47]. For example, if the appearance of outcome variable y_i depends on a selection variable z_i, such incidental truncation may result in a missing data problem of y_i and biased coefficient estimation using standard regression techniques (e.g., OLS). In order to resolve this potential bias, the Heckman model was introduced and assumed the underlying two-stage relationship [48]:

The selection equation is shown below:

$$z_i^* = \alpha_i\gamma + u_i z_i = \begin{cases} 1 & if\ z_i^* > 0 \\ 0 & if\ z_i^* \leq 0 \end{cases} \tag{A1}$$

The outcome equation is as follows:

$$y_i = \begin{cases} x_i\beta + \varepsilon_i & if\ z_i = 1 \\ unobserved & if\ z_i = 0 \end{cases} \tag{A2}$$

where x_i are covariates that affect the outcome and α_i are covariates that affect selection, $\varepsilon_i \sim N(0, \sigma^2)$, $u_i \sim N(0,1)$, $corr(\varepsilon_i, u_i) = \rho$.

The log likelihood for observation i is $\ln L_i$:

$$\ln L_i = \begin{cases} \omega_i \ln\Phi\left\{\frac{\alpha_i\gamma + (y_i - x_i\beta)\rho/\sigma}{\sqrt{1-\rho^2}}\right\} - \frac{\omega_i}{2}\left(\frac{y_i - x_i\beta}{\sigma}\right)^2 - \omega_i \ln\left(\sqrt{2\pi}\sigma\right) & y_i observed \\ \omega_i \ln\Phi(-\alpha_i\gamma) & y_i unobserved \end{cases} \tag{A3}$$

where Φ is the standard cumulative normal distribution and ω_i is an optional weight for observation i.

The first stage is to determine whether an observation in an overall population appears in the final representative samples, and the second stage is to model the relationship between the dependent and independent variables in the final selected samples [46]. With maximum likelihood estimation in the Heckman model, rho (ρ; the correlation between error terms in the selection and outcome equations) could be examined to indicate whether or not sample selection bias exists [46]. If ε_i and u_i are correlated, traditional techniques (e.g., OLS) would report biased β estimation. In this situation, the results of the Heckman model could be consistent and asymptotically efficient estimates by correcting selection bias [48]. Otherwise, traditional regression methods could generate efficient estimates by using selected samples. Since the formulas above are appropriate for general-type data, the formulas extended for the interval data of our study could be found in the command *eintreg* of Stata manual Release 15 [67].

Appendix A.2. Formulas of Interval Regression Model

The equation of the interval regression model [48] is as follows:

$$Y_i^* = X_i\beta + \varepsilon_i \tag{A4}$$

where Y_i^* is a continuous outcome for the ith observation with covariates X_i and corresponding coefficients β. ε is the error term that is assumed to be mean zero and normally distributed; $\varepsilon \sim N(0, \sigma^2)$.

If observation $i \in C$ (not censored), we observe Y_i^* as the point data. If observation $i \in L$ (left-censored), the unobserved Y_i^* is in the interval $(-\infty, Y_{i2}]$. The likelihood contribution is as follows:

$$\Pr(Y_i^* \leq Y_{i2}) = \Pr(X_i\beta + \varepsilon_i \leq Y_{i2}) \tag{A5}$$

If $i \in I$ (interval-censored), the unobserved Y_i^* is in the interval $[Y_{i1}, Y_{i2}]$. The likelihood contribution is as follows:

$$\Pr(Y_{i1} \le Y_i^* \le Y_{i2}) = \Pr(Y_{i1} \le X_i\beta + \varepsilon_i \le Y_{i2}) \quad \text{(A6)}$$

If $i \in R$ (right-censored), the unobserved Y_i^* is in the interval $[Y_{i1}, +\infty)$. The likelihood contribution is shown below:

$$\Pr(Y_{i1} \le Y_i^*) = \Pr(Y_{i1} \le X_i\beta + \varepsilon_i) \quad \text{(A7)}$$

The total loglikelihood function is given as:

$$\begin{aligned} \ln L = -\tfrac{1}{2} \sum_{i \in C} \omega_i \left\{ \left(\tfrac{Y_i^* - X_i\beta}{\sigma} \right)^2 + \log 2\pi\sigma^2 \right\} \\ + \sum_{i \in L} \omega_i \log \Phi\left(\tfrac{Y_{i2} - X_i\beta}{\sigma} \right) \\ + \sum_{i \in I} \omega_i \log\left\{ \Phi\left(\tfrac{Y_{i2} - X_i\beta}{\sigma} \right) - \Phi\left(\tfrac{Y_{i1} - X_i\beta}{\sigma} \right) \right\} \\ + \sum_{i \in R} \omega_i \log\left\{ 1 - \Phi\left(\tfrac{Y_{i1} - X_i\beta}{\sigma} \right) \right\} \end{aligned} \quad \text{(A8)}$$

Note that Φ is the cumulative standard normal distribution and ω_i is the weight for the ith observation [48]. The coefficients could be estimated by maximizing the value of the loglikelihood function $\ln L$.

Table A1. Heckman model for testing sample selection bias.

Variable	Heckman Model	
	Coefficient	**Robust Standard Error**
Gender	−4.46	10.94
Age	−0.12	0.54
Years of education	2.15	1.39
Experience of applying pesticides	−2.88 **	1.09
Quadratic term of experience of applying pesticides	0.05 *	0.02
Income	1.42 *	0.57
Familiarity with nanopesticides	11.19 **	3.40
Attitude toward nanopesticides	13.79 *	6.67
Trust in governments	−7.96	5.66
Trust in industries	13.57 *	5.53
Labeling preference	−0.65	6.29
Constant	−39.45	38.97
rho	−0.42	0.33
Wald test	Chi-square = 57.95; p = 0.00	
VIF	Mean = 1.59	
Numbers of observations	232	

Notes: ** and * indicate significance at the $p < 0.01$ and $p < 0.05$ levels, respectively. The unit of the coefficients is percentage points.

Table A2. Standardized beta coefficients of influencing factors.

Variable	Standardized Beta Coefficient
Experience of applying pesticides	−0.58 *
Familiarity with nanopesticides	0.23 **
Trust in industries	0.21 *
Attitude toward nanopesticides	0.18 *
Income	0.17 *
Trust in governments	−0.12
Years of education	0.10
Gender	−0.03
Age	−0.02
Labeling preference	0.00

Note: ** and * indicate significance at the $p < 0.01$ and $p < 0.05$ levels, respectively.

Table A3. Different pesticide user profiles' willingness-to-pay for nanopesticides.

Experience of Applying Pesticides	Familiarity with Nanopesticides	Trust in Industries	The Percentage That Pesticide Users Were Willing to Pay Higher for Nanopesticides than That for Conventional Pesticides
23	2	1	−1.00%
13	2	1	8.27%
23	1	3	15.57%
5	2	1	23.04%
13	1	3	24.84%
23	2	3	26.66%
13	2	3	35.92%
23	3	3	37.74%
23	2	4	40.48%
13	3	3	47.00%
13	2	4	49.75%
5	2	3	50.69%
23	3	4	51.56%
13	3	4	60.83%
5	3	3	61.77%
5	2	4	64.52%
5	3	4	75.60%
5	4	5	100.51%
5	5	5	111.59%

Appendix B Questionnaire of Public Perceptions and Willingness-to-Pay for Nanopesticides

1. What is your gender? 1 = Female, 0 = Male
2. What is your full year of age?
3. What is your educational level? 1 = uneducated, 2 = primary school, 3 = middle school, 4 = high school, 5 = professional high school/technical school, 6 = secondary school, 7 = junior college, 8 = undergraduate education, 9 = postgraduate education
4. What is your total household income in 2019?
5. How are you familiar with nanopesticides? 1 = Completely unfamiliar, 2 = A little unfamiliar, 3 = General, 4 = Quite familiar, 5 = Very familiar

Nanopesticides are composed of nanomaterials (1–100 nm, 1 nm = $1/10^9$ m) with broad-spectrum insecticidal, fungicidal, or herbicidal properties [15]. Compared with conventional chemical pesticides, nanopesticides could improve efficacy 24~150% [15], prolong effective time period [68], lower application rates [12], and increase yields [15]. In addition, nanopesticides could have less residuals and decrease environmental burden [68].

The potential negative impacts on human health and environment are low, but there is still a lack of comprehensive evaluation [69].

6. What is your attitude toward the future development of nanopesticides? 1 = Completely opposed, 2 = A little opposed, 3 = Neutral, 4 = Quite supportive, 5 = Very supportive
7. Do you agree that the product label of nanopesticides must indicate it contains nanocomponents? 1 = Completely disagree, 2 = A little disagree, 3 = Neutral, 4 = Quite agree, 5 = Strongly agree
8. Do you trust that governments could supervise the safety risks of nanopesticides? 1 = Completely distrust, 2 = A little distrust, 3 = General, 4 = Quite trust, 5 = Strongly trust
9. Do you trust that industries (manufactures and retailers) could produce and sell nanopesticides legally? 1 = Completely distrust, 2 = A little distrust, 3 = General, 4 = Quite trust, 5 = Strongly trust
10. Do you plant grain, vegetables, and fruit that need pesticides? 1 = Yes, 0 = No (If yes, please answer the following questions; If no, please quit the survey)
11. You have ___ years of experience in applying pesticides.
12. The price ranges of willingness-to-pay for nanopesticides:

a. Would you be willing to purchase nanopesticides if the price is lower than conventional pesticides?	1 = Yes (continue with Question b); 2 = No (stop answering and quit the survey)
b. Would you be willing to purchase nanopesticides if the price is as the same as conventional pesticides?	1 = Yes (continue with Question c); 2 = No (stop answering and quit the survey)
c. Would you be willing to purchase nanopesticides if the price is 50% higher than conventional pesticides?	1 = Yes (skip to Question d); 2 = No (skip to Question e)
d. Would you be willing to purchase nanopesticides if the price is 100% higher than conventional pesticides?	1 = Yes (skip to Question d-1); 2 = No (skip to Question d-2)
d-1. Would you be willing to purchase nanopesticides if the price is 130% higher than conventional pesticides?	1 = Yes; 2 = No
d-2. Would you be willing to purchase nanopesticides if the price is 75% higher than conventional pesticides?	1 = Yes; 2 = No
e. Would you be willing to purchase nanopesticides if the price is 25% higher than conventional pesticides?	1 = Yes (skip to Question e-1); 2 = No (skip to Question e-2)
e-1. Would you be willing to purchase nanopesticides if the price is 40% higher than conventional pesticides?	1 = Yes; 2 = No
e-2. Would you be willing to purchase nanopesticides if the price is 10% higher than conventional pesticides?	1 = Yes; 2 = No

References

1. Kah, M.; Tufenkji, N.; White, J.C. Nano-enabled strategies to enhance crop nutrition and protection. *Nat. Nanotechnol.* **2019**, *14*, 532–540. [CrossRef] [PubMed]
2. Su, Y.; Ashworth, V.; Kim, C.; Adeleye, A.S.; Rolshausen, P.; Roper, C.; White, J.; Jassby, D. Delivery, uptake, fate, and transport of engineered nanoparticles in plants: A critical review and data analysis. *Environ. Sci. Nano* **2019**, *6*, 2311–2331. [CrossRef]
3. Zhao, L.; Lu, L.; Wang, A.; Zhang, H.; Huang, M.; Wu, H.; Xing, B.; Wang, Z.; Ji, R. Nano-Biotechnology in Agriculture: Use of Nanomaterials to Promote Plant Growth and Stress Tolerance. *J. Agric. Food Chem.* **2020**, *68*, 1935–1947. [CrossRef] [PubMed]
4. Kyriakopoulos, G.; Doulia, D.; Anagnostopoulos, E. Adsorption of pesticides on porous polymeric adsorbents. *Chem. Eng. Sci.* **2005**, *60*, 1177–1186. [CrossRef]
5. Gilbertson, L.M.; Pourzahedi, L.; Laughton, S.; Gao, X.; Zimmerman, J.B.; Theis, T.L.; Westerhoff, P.; Lowry, G.V. Guiding the design space for nanotechnology to advance sustainable crop production. *Nat. Nanotechnol.* **2020**, *15*, 801–810. [CrossRef]
6. Kyriakopoulos, G.; Doulia, D. Adsorption of Pesticides on Carbonaceous and Polymeric Materials from Aqueous Solutions: A Review. *Sep. Purif. Rev.* **2006**, *35*, 97–191. [CrossRef]
7. Fernandez-Cornejo, J.; Nehring, R.F.; Osteen, C.; Wechsler, S.; Martin, A.; Vialou, A. Pesticide Use in U.S. Agriculture: 21 Selected Crops, 1960–2008. *SSRN Electron. J.* **2014**, *124*, 1–80. [CrossRef]
8. Kyriakopoulos, G.; Xiarchos, I.; Doulia, D. Treatment of contaminated water with pesticides via adsorption. *Int. J. Environ. Technol. Manag.* **2006**, *6*, 515–524. [CrossRef]
9. Li, Y.; Yang, D.; Cui, J. Graphene oxide loaded with copper oxide nanoparticles as an antibacterial agent against Pseudomonas syringae pv. tomato. *RSC Adv.* **2017**, *7*, 38853–38860. [CrossRef]
10. Hayles, J.; Johnson, L.; Worthley, C.; Losic, D. Nanopesticides: A review of current research and perspectives. In *New Pesticides and Soil Sensors*; Elsevier: Amsterdam, The Netherlands, 2017; pp. 193–225.
11. Kookana, R.S.; Boxall, A.B.A.; Reeves, P.T.; Ashauer, R.; Beulke, S.; Chaudhry, Q.; Cornelis, G.; Fernandes, T.F.; Gan, J.; Kah, M.; et al. Nanopesticides: Guiding Principles for Regulatory Evaluation of Environmental Risks. *J. Agric. Food Chem.* **2014**, *62*, 4227–4240. [CrossRef]
12. Walker, G.W.; Kookana, R.S.; Smith, N.E.; Kah, M.; Doolette, C.L.; Reeves, P.T.; Lovell, W.; Anderson, D.J.; Turney, T.W.; Navarro, D.A. Ecological Risk Assessment of Nano-enabled Pesticides: A Perspective on Problem Formulation. *J. Agric. Food Chem.* **2018**, *66*, 6480–6486. [CrossRef] [PubMed]
13. Kah, M.; Beulke, S.; Tiede, K.; Hofmann, T. Nanopesticides: State of Knowledge, Environmental Fate, and Exposure Modeling. *Crit. Rev. Environ. Sci. Technol.* **2013**, *43*, 1823–1867. [CrossRef]
14. Pérez-de-Luque, A.; Rubiales, D. Nanotechnology for parasitic plant control. *Pest Manag. Sci.* **2009**, *65*, 540–545. [CrossRef] [PubMed]
15. Kah, M.; Kookana, R.S.; Gogos, A.; Bucheli, T.D. A critical evaluation of nanopesticides and nanofertilizers against their conventional analogues. *Nat. Nanotechnol.* **2018**, *13*, 677–684. [CrossRef]
16. Hofmann, T.; Lowry, G.V.; Ghoshal, S.; Tufenkji, N.; Brambilla, D.; Dutcher, J.R.; Gilbertson, L.M.; Giraldo, J.P.; Kinsella, J.M.; Landry, M.P.; et al. Technology readiness and overcoming barriers to sustainably implement nanotechnology-enabled plant agriculture. *Nat. Food* **2020**, *1*, 416–425. [CrossRef]
17. Chatterjee, A.K.; Chakraborty, R.; Basu, T. Mechanism of antibacterial activity of copper nanoparticles. *Nanotechnology* **2014**, *25*, 135101. [CrossRef]
18. Klaine, S.J.; Koelmans, A.A.; Horne, N.; Carley, S.; Handy, R.D.; Kapustka, L.; Nowack, B.; von der Kammer, F. Paradigms to assess the environmental impact of manufactured nanomaterials. *Environ. Toxicol. Chem.* **2012**, *31*, 3–14. [CrossRef]
19. Pandey, S.; Giri, K.; Kumar, R.; Mishra, G.; Raja Rishi, R. Nanopesticides: Opportunities in Crop Protection and Associated Environmental Risks. *Proc. Natl. Acad. Sci. India Sect. B Biol. Sci.* **2018**, *88*, 1287–1308. [CrossRef]
20. Ganesh Pillai, R.; Bezbaruah, A.N. Perceptions and attitude effects on nanotechnology acceptance: An exploratory framework. *J. Nanopart. Res.* **2017**, *19*, 41. [CrossRef]
21. Vo, N.X.; Huyen Nguyen, T.T.; Van Nguyen, P.; Tran, Q.V.; Vo, T.Q. Using Contingent Valuation Method to Estimate Adults' Willingness to Pay for a Future Coronavirus 2019 Vaccination. *Value Health Reg. Issues* **2021**, *24*, 240–246. [CrossRef]
22. George, S.; Kaptan, G.; Lee, J.; Frewer, L. Awareness on adverse effects of nanotechnology increases negative perception among public: Survey study from Singapore. *J. Nanopart. Res.* **2014**, *16*, 2751. [CrossRef]
23. Lee, C.J.; Scheufele, D.A.; Lewenstein, B.V. Public attitudes toward emerging technologies: Examining the interactive effects of cognitions and affect on public attitudes toward nanotechnology. *Sci. Commun.* **2005**, *27*, 240–267. [CrossRef]
24. Frewer, L.J.; Bergmann, K.; Brennan, M.; Lion, R.; Meertens, R.; Rowe, G.; Siegrist, M.; Vereijken, C. Consumer response to novel agri-food technologies: Implications for predicting consumer acceptance of emerging food technologies. *Trends Food Sci. Technol.* **2011**, *22*, 442–456. [CrossRef]
25. Yue, C.; Zhao, S.; Cummings, C.; Kuzma, J. Investigating factors influencing consumer willingness to buy GM food and nano-food. *J. Nanopart. Res.* **2015**, *17*, 283. [CrossRef]
26. Amenta, V.; Aschberger, K.; Arena, M.; Bouwmeester, H.; Botelho Moniz, F.; Brandhoff, P.; Gottardo, S.; Marvin, H.J.P.; Mech, A.; Quiros Pesudo, L.; et al. Regulatory aspects of nanotechnology in the agri/feed/food sector in EU and non-EU countries. *Regul. Toxicol. Pharmacol.* **2015**, *73*, 463–476. [CrossRef] [PubMed]

27. D'Silva, J.; Bowman, D.M. To Label or Not to Label?—It's More than a Nano-sized Question. *Eur. J. Risk Regul.* **2010**, *1*, 420–427. [CrossRef]
28. Brown, J.; Kuzma, J. Hungry for Information: Public Attitudes Toward Food Nanotechnology and Labeling. *Rev. Policy Res.* **2013**, *30*, 512–548. [CrossRef]
29. Stokes, E. Regulating nanotechnologies: Sizing up the options. *Leg. Stud.* **2009**, *29*, 281–304. [CrossRef]
30. Sodano, V.; Gorgitano, M.T.; Verneau, F.; Vitale, C.D. Consumer acceptance of food nanotechnology in Italy. *Br. Food J.* **2016**, *118*, 714–733. [CrossRef]
31. Macoubrie, J. Nanotechnology: Public concerns, reasoning and trust in government. *Public Underst. Sci.* **2006**, *15*, 221–241. [CrossRef]
32. Yang, M. *The Study on Pesticide Use Behavior of Greenhouse Vegetable Growers: Taking Shandong as an Example*; China Agricultural University: Beijing, China, 2019.
33. Statista Distribution of the Workforce Across Economic Sectors in China from 2010 to 2020. Available online: https://www.statista.com/statistics/270327/distribution-of-the-workforce-across-economic-sectors-in-china/ (accessed on 1 November 2021).
34. Wang, W.; Jin, J.; He, R.; Gong, H.; Tian, Y. Farmers' Willingness to Pay for Health Risk Reductions of Pesticide Use in China: A Contingent Valuation Study. *Int. J. Environ. Res. Public Health* **2018**, *15*, 625. [CrossRef] [PubMed]
35. Yu, J.; Wu, J. The Sustainability of Agricultural Development in China: The Agriculture–Environment Nexus. *Sustainability* **2018**, *10*, 1776. [CrossRef]
36. Eng, J. Sample Size Estimation: How Many Individuals Should Be Studied? *Radiology* **2003**, *227*, 309–313. [CrossRef] [PubMed]
37. Skordoulis, M.; Ntanos, S.; Arabatzis, G. Socioeconomic evaluation of green energy investments. *Int. J. Energy Sect. Manag.* **2020**, *14*, 871–890. [CrossRef]
38. Wang, Z.; Gong, Y.; Mao, X. Exploring the value of overseas biodiversity to Chinese netizens based on willingness to pay for the African elephants' protection. *Sci. Total Environ.* **2018**, *637–638*, 600–608. [CrossRef]
39. Karina Gallardo, R.; Wang, Q. Willingness to pay for pesticides' environmental features and social desirability bias: The case of apple and pear growers. *J. Agric. Resour. Econ.* **2013**, *38*, 124–139. [CrossRef]
40. Khan, J.; Jan, A.; Lim, K.H.; Shah, S.A.; Khanal, A.R.; Ali, G. Household's Perception and their Willingness to Pay for Pesticides-Free Fruits in Khyber Pakhtunkhwa (Kp) Province of Pakistan: A Double-Bounded Dichotomous Choice Contingent Valuation Study. *Sarhad J. Agric.* **2019**, *35*, 1266–1271. [CrossRef]
41. Petrescu-Mag, R.M.; Banatean-Dunea, I.; Vesa, S.C.; Copacinschi, S.; Petrescu, D.C. What Do Romanian Farmers Think about the Effects of Pesticides? Perceptions and Willingness to Pay for Bio-Pesticides. *Sustainability* **2019**, *11*, 3628. [CrossRef]
42. Vidogbéna, F.; Adégbidi, A.; Tossou, R.; Assogba-Komlan, F.; Martin, T.; Ngouajio, M.; Simon, S.; Parrot, L.; Zander, K. Consumers' Willingness to Pay for Cabbage with Minimized Pesticide Residues in Southern Benin. *Environments* **2015**, *2*, 449–470. [CrossRef]
43. Carson, R.T.; Hanemann, W.M. Chapter 17 Contingent Valuation. In *Handbook of Environmental Economics*; Elsevier: Amsterdam, The Netherlands, 2005; Volume 2, pp. 821–936. ISBN 9780444511454.
44. Cawley, J. Contingent valuation analysis of willingness to pay to reduce childhood obesity. *Econ. Hum. Biol.* **2008**, *6*, 281–292. [CrossRef]
45. Juster, F.T.; Smith, J.P. Improving the Quality of Economic Data: Lessons from the HRS and AHEAD. *J. Am. Stat. Assoc.* **1997**, *92*, 1268–1278. [CrossRef]
46. Certo, S.T.; Busenbark, J.R.; Woo, H.; Semadeni, M. Sample selection bias and Heckman models in strategic management research. *Strateg. Manag. J.* **2016**, *37*, 2639–2657. [CrossRef]
47. Heckman, J.J. Sample Selection Bias as a Specification Error. *Econometrica* **1979**, *47*, 153–161. [CrossRef]
48. StataCorp Stata. *Version: Stata Base Reference Manual: Release*; StataCorp LLC: College Station, TA, USA, 2019.
49. Scheerder, J.; Vos, S.; Taks, M. Expenditures on Sport Apparel: Creating Consumer Profiles through Interval Regression Modelling. *Eur. Sport Manag. Q.* **2011**, *11*, 251–274. [CrossRef]
50. Ming, J.; Zhang, J. The Homebound Intention, Migration Costs and the Remittance to Home—An Analysis of Interval Regression Model. *South China Popul.* **2011**, *26*, 48–56.
51. Gene Denovo OmicShare Tools. Available online: https://www.omicshare.com/tools (accessed on 29 May 2021).
52. Zhou, G.; Hu, W. Public acceptance of and willingness-to-pay for nanofoods in the U.S. *Food Control* **2018**, *89*, 219–226. [CrossRef]
53. Kidd, J.; Westerhoff, P.; Maynard, A. Survey of industrial perceptions for the use of nanomaterials for in-home drinking water purification devices. *NanoImpact* **2021**, *22*, 100320. [CrossRef]
54. Sulaiman, E.K. An Empirical Analysis of Financial Risk Tolerance and Demographic Features of Individual Investors. *Procedia Econ. Financ.* **2012**, *2*, 109–115. [CrossRef]
55. Dong, Y.-C. Risk Preference Theory and Family Portfolio—The Evidence from Chinese Household Finance Survey. In Proceedings of the 2018 5th International Conference on Management Science and Management Innovation (MSMI 2018), Wuhan, China, 20–22 April 2018; Atlantis Press: Paris, France, 2018; Volume 54, pp. 93–98.
56. Khanna, M.; Miao, R. Inducing the adoption of emerging technologies for sustainable intensification of food and renewable energy production: Insights from applied economics. *Aust. J. Agric. Resour. Econ.* **2022**, *66*, 1–23. [CrossRef]
57. Bieberstein, A.; Roosen, J.; Marette, S.; Blanchemanche, S.; Vandermoere, F. Consumer choices for nano-food and nano-packaging in France and Germany. *Eur. Rev. Agric. Econ.* **2013**, *40*, 73–94. [CrossRef]

58. Lee, E.W.J.; Ho, S.S. The perceived familiarity gap hypothesis: Examining how media attention and reflective integration relate to perceived familiarity with nanotechnology in Singapore. *J. Nanopart. Res.* **2015**, *17*, 228. [CrossRef]
59. Liu, H.; Priest, S. Understanding public support for stem cell research: Media communication, interpersonal communication and trust in key actors. *Public Underst. Sci.* **2009**, *18*, 704–718. [CrossRef]
60. Siegrist, M.; Cousin, M.E.; Kastenholz, H.; Wiek, A. Public acceptance of nanotechnology foods and food packaging: The influence of affect and trust. *Appetite* **2007**, *49*, 459–466. [CrossRef] [PubMed]
61. Siegrist, M.; Keller, C.; Kastenholz, H.; Frey, S.; Wiek, A. Laypeople's and Experts' Perception of Nanotechnology Hazards. *Risk Anal.* **2007**, *27*, 59–69. [CrossRef]
62. Siegel, A.F. Multiple Regression. In *Practical Business Statistics*; Elsevier: Amsterdam, The Netherlands, 2016; pp. 355–418. ISBN 978-0-12-804250-2.
63. Lombi, E.; Donner, E.; Dusinska, M.; Wickson, F. A One Health approach to managing the applications and implications of nanotechnologies in agriculture. *Nat. Nanotechnol.* **2019**, *14*, 523–531. [CrossRef]
64. Brunsø, K.; Grunert, K.G. Cross-Cultural Similarities and Differences in Shopping for Food. *J. Bus. Res.* **1998**, *42*, 145–150. [CrossRef]
65. Zahedi, S.; Batista-Foguet, J.M.; van Wunnik, L. Exploring the public's willingness to reduce air pollution and greenhouse gas emissions from private road transport in Catalonia. *Sci. Total Environ.* **2019**, *646*, 850–861. [CrossRef]
66. Venkatachalam, L. The contingent valuation method: A review. *Environ. Impact Assess. Rev.* **2004**, *24*, 89–124. [CrossRef]
67. StataCorp Stata. *Version: Stata Extended Regression Models Reference Manual: Release 15*; StataCorp LLC: College Station, TA, USA, 2017.
68. Chhipa, H.; Joshi, P. Nanofertilisers, Nanopesticides and Nanosensors in Agriculture. In *Nanoscience in Food and Agriculture 1*; Springer: Cham, Switzerland, 2016; pp. 247–282. ISBN 9783319393032.
69. Agathokleous, E.; Feng, Z.; Iavicoli, I.; Calabrese, E.J. Nano-pesticides: A great challenge for biodiversity? The need for a broader perspective. *Nano Today* **2020**, *30*, 100808. [CrossRef]

nanomaterials

MDPI

Review

Environmental Fate and Toxicity of Sunscreen-Derived Inorganic Ultraviolet Filters in Aquatic Environments: A Review

Shengwu Yuan [1,2], Jingying Huang [2], Xia Jiang [1], Yuxiong Huang [2], Xiaoshan Zhu [2,3,4,*] and Zhonghua Cai [2]

[1] National Engineering Laboratory for Lake Pollution Control and Ecological Restoration, State Environment Protection Key Laboratory for Lake Pollution Control, Chinese Research Academy of Environmental Science, Beijing 100012, China; dirk_yuan@163.com (S.Y.); jiangxia@craes.org.cn (X.J.)
[2] Shenzhen International Graduate School, Tsinghua University, Shenzhen 518055, China; yuan.shengwu@craes.org.cn (J.H.); huang_yuxiong@sz.tsinghua.edu.cn (Y.H.); caizh@sz.tsinghua.edu.cn (Z.C.)
[3] School of Environment, Tsinghua University, Beijing 100084, China
[4] Southern Laboratory of Ocean Science and Engineering (Guangdong, Zhuhai), Zhuhai 519000, China
* Correspondence: zhu.xiaoshan@sz.tsinghua.edu.cn

Abstract: An increasing number of inorganic ultraviolet filters (UVFs), such as nanosized zinc oxide (nZnO) and titanium dioxide ($nTiO_2$), are formulated in sunscreens because of their broad UV spectrum sunlight protection and because they limit skin damage. However, sunscreen-derived inorganic UVFs are considered to be emerging contaminants; in particular, nZnO and $nTiO_2$ UVFs have been shown to undergo absorption and bioaccumulation, release metal ions, and generate reactive oxygen species, which cause negative effects on aquatic organisms. We comprehensively reviewed the current study status of the environmental sources, occurrences, behaviors, and impacts of sunscreen-derived inorganic UVFs in aquatic environments. We find that the associated primary nanoparticle characteristics and coating materials significantly affect the environmental behavior and fate of inorganic UVFs. The consequential ecotoxicological risks and underlying mechanisms are discussed at the individual and trophic transfer levels. Due to their persistence and bioaccumulation, more attention and efforts should be redirected to investigating the sources, fate, and trophic transfer of inorganic UVFs in ecosystems.

Keywords: cosmetics; nanoparticles; environmental behavior; ecosystem; toxic mechanism

Citation: Yuan, S.; Huang, J.; Jiang, X.; Huang, Y.; Zhu, X.; Cai, Z. Environmental Fate and Toxicity of Sunscreen-Derived Inorganic Ultraviolet Filters in Aquatic Environments: A Review. *Nanomaterials* **2022**, *12*, 699. https://doi.org/10.3390/nano12040699

Academic Editor: Laura Canesi

Received: 25 January 2022
Accepted: 14 February 2022
Published: 19 February 2022

1. Introduction

Sunscreen is one of the personal care products (PCPs) used to provide protection against ultraviolet radiation (UVR, 10–400 nm) damage [1–3]. Recently, with rising production and consumption, sunscreens have been increasingly released into aquatic environments, including oceans, rivers, lakes, and other water bodies, via several means of discharge (e.g., wastewater treatment plant effluents, runoff input, and recreational activities) [4–6]. The rapid growth in global tourism, especially coastal and marine tourism, where the number of international tourists worldwide grew from 463 million in 1992 to 763 million in 2004 and is estimated to have reached 1.56 billion in 2020 [7], has contributed to the increasing application of sunscreen [7,8]. Moreover, in these tropical countries, at least 25% of the sunscreens applied to skin are eventually released into the ocean during water recreational activities [9], which could pose potential risks to the aquatic environment.

Sunscreen is a multicomponent product that contains both active ingredients to shield or reflect UVR and commodity coatings to prevent bleaching and the loss of color [10]. The active ultraviolet filters (UVFs) in sunscreens can be organic or inorganic and can reflect and scatter UVR, which protects human skin from direct sunlight radiation [11,12]. Typically, organic UVFs are called chemical filters, as their mode of action (MoA) is related to the chemical changes in their molecules that prevent UVR from reaching the skin. The European Union regulates and authorizes 26 types of organic UVFs (summarized in our previous review) [13], which are widely used and globally recognized. In 2018, the Environmental Working Group (EWG) reported that two-thirds of the 1300 sunscreen products available contain chemicals that the EWG has deemed to be harmful to the environment, which are predominantly organic UVFs [14]. Inorganic UVFs are called physical filters or mineral filters, as their MoA is associated with physical phenomena, such as the scattering and reflection of UVR [15–20]. Titanium dioxide (TiO_2) and zinc oxide (ZnO) are the most widely used inorganic UVFs and are usually present in nanoparticle (NP) form, also known as nanosized TiO_2 ($nTiO_2$) and nanosized ZnO (nZnO), due to their greater dispersion and UV scattering superficial area [14]. Both $nTiO_2$ and nZnO are semiconductors with wide band gaps that can effectively shield UV light.

The adverse environmental effects of organic UVFs, including the bleaching effect on coral reefs and the negative hormonal effects on marine animals, were reviewed in a recent study [21]. The ecological risks of organic UVFs have resulted in warnings and restrictions on the application of chemical substances. The Hawaiian state legislature passed a bill on 1 May 2018 that bans the sale and distribution of sunscreens that contain certain organic UVFs (oxybenzone and octinoxate), which is anticipated to become effective in 2021 [22]. In addition, the EWG began to push the Food and Drug Administration in 2007 to update and improve cosmetic product regulations by urging the agency to set stricter standards to better protect public health [14].

Due to the ecotoxicological risks of organic UVFs, using inorganic UVFs for replacement has become a topic of interest for both producers and consumers. Although organic UVFs have dominated the market for PCPs in the past, inorganic UVFs as substitutions are increasing due to their broad UV spectrum protection and limited skin penetration and health risks [23,24]. It is believed that 60% of $nTiO_2$ and 80% of nZnO produced globally are used in cosmetic products [25,26]. With the increasing production and application, the discharge of inorganic UVFs into environments is inevitable. In the United States, hundreds of tons of TiO_2 and ZnO are disposed of in the environment every year [27]. To date, studies have shown that inorganic UVFs have been detected in marine waters, sediments, and organisms at increasing concentrations [1]. For example, Botta et al. [28] estimated that in reef areas, 36–56 tons of TiO_2 were released from sunscreens, where the concentration of TiO_2 could reach tens of milligram liters in surface microlayer [4]. Inorganic UVFs are prone to persisting in the environment due to continuous emissions and refractory degradation, which pose health threats to aquatic organisms at different trophic levels.

We comprehensively reviewed the current study status of the environmental sources, occurrences, behaviors, and impacts of sunscreen-derived inorganic UVFs in aquatic environments. The associated primary nanoparticle characteristics and coating materials significantly affect the environmental behavior and fate of inorganic UVFs. The consequential ecotoxicological risks and underlying mechanisms are discussed at the individual and trophic transfer levels. Accordingly, suggestions are given for future study and recommendations for the scientific attention and control of inorganic UVF-containing products.

2. Inorganic UVFs in Aquatic Environments

2.1. Sources and Occurrences

UVFs have been detected in surface waters [29], urban groundwater [30], sediments [31–33], marine water, and biota [1,34]. The environmental sources and distribution of organic UVFs have been well reviewed in recent years [1,34]. However, very little is known about the occurrences and distributions of the two increasingly used inorganic UVFs ($nTiO_2$ and nZnO). It has been shown that these substances are released into waters, either directly through human activities or indirectly through wastewater treatment plant (WWTP) drainage and atmospheric deposition (shown in Figure 1) [11,29,35]. Some studies have indicated that there is a direct relationship between the amounts of sunscreen components in waters and recreational activities, such as swimming, diving, surfing, etc [4,36,37]. In addition, the effluents of WWTPs and domestic sewage indirectly release UVFs, as sunscreen components cannot be completely removed [6,11]. Atmospheric aerosols containing UVFs may occur from different sources, including directly after spraying sunscreen on the skin, with effluents from WWTPs, and indirectly with the incineration of WWTP sludge.

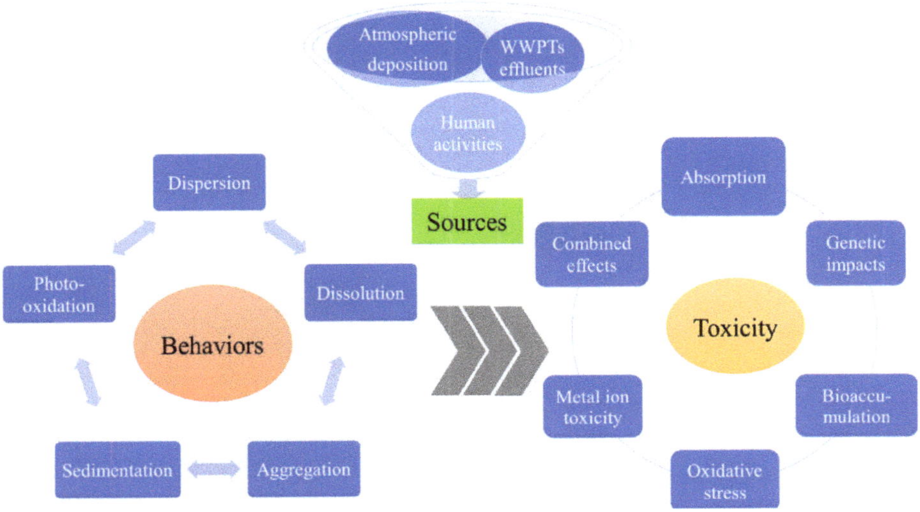

Figure 1. The sources, behaviors, and toxicity of sunscreen-derived inorganic UVFs in aquatic environments.

According to a survey study, there are approximately 16,000–25,000 tons per annual (t/a) of sunscreens that contain $nTiO_2$ in tropical countries, and at least 25% of sunscreen applied to the skin enters the ocean during water recreational activities [9]. It is estimated that the content of $nTiO_2$ in sunscreens is approximately 4%, and the amount of $nTiO_2$ released annually is approximately 160–250 t in these tropical countries [1,38]. Specifically, Sánchez-Quiles and Tovar-Sánchez [9] estimated that over 4 kg of $nTiO_2$ can be released from sunscreen into seawater during a summer day on a tropical touristic beach. Another study suggested that the recreational activities that take place at Old Danube Lake (Vienna, Austria) may involve the consumption of sunscreen of 8.1 t per year, and they estimated that 94.5 kg of TiO_2 per year may be released into lake waters [39]. A recent study has shown that inorganic UVFs present in the formulation of sunscreens are detected in nearshore water and are concentrated in the surface microlayer that ranges from 6.9 to 37.6 mg/L for TiO_2 and from 1.0 to 3.3 mg/L for ZnO [4].

2.2. Environmental Behaviors

The specific behavior of inorganic UVFs released from sunscreens into aquatic environments has not been well addressed. As sunscreen is a complex chemical mixture; once it is in water, the inorganic UVFs released from sunscreen are complex and can exist in the form of aggregates of various complex components [40,41], including surface-modified complexes or raw NPs. For raw NPs, their environmental fate generally includes dispersing, aggregating, and dissolving/releasing metal ions and settling onto sediments or being absorbed and bioaccumulated by organisms (shown in Figure 1) [28,39,42]. Many studies have confirmed that nZnO UVFs rapidly dissolve in water and form hydrated Zn^{2+} cations [43,44]. Other inorganic UVFs, e.g., $nTiO_2$, are regarded as relatively stable and rather insoluble in water [45]. Thus, these UVFs tend to aggregate into larger particles, which remain suspended or precipitate to the bottom of the aquatic environment. In general, the higher the content of UVFs, the higher the SPR the sunscreen obtained. For organic UVFs, they absorb UVR, thus their spectral characteristics determined the absorbance of

UVR as well as the sun protection factor (SPR); most of them are photo-instability effects related to UVR exposure [46]. For inorganic UVFs, they mean to scatter and reflect UVR; thus, they are more stable than organic UVFs, but their particle size would affect the SPR and transparency (aesthetics of the products), thus most inorganic UVFs are nanosized. The stability of physical sunscreens was influenced by the coating materials, with these organic materials in physical sunscreens tend to perform photodegradation and photo-instability effects related to UVR exposure, thus making inorganic UVFs easier to bear in the environment [46]. In addition, photooxidation and photodegradation are also proposed to occur when inorganic UVFs are exposed to sunlight. Inorganic UVFs, including $nTiO_2$ and nZnO, are often used as photocatalytic materials; once released into water, they can be photooxidized during irradiation by ultraviolet light and generate hole-electron pairs; reactive oxygen species (ROS) are produced when hole-electron pairs react with H_2O or O_2 on the surface of NPs, which also decreases the particle size and produces more ROS [47,48]. Studies have shown that inorganic UVFs are photooxidized, produce ROS, and cause photocatalytic toxicity to aquatic organisms [49]. In addition to these behaviors, inorganic UVFs easily settle into sediments due to gravitational force, thereby aggregating into larger NPs. UVFs, both the organic and inorganic varieties, are absorbed or captured by aquatic organisms during the above processes, which causes damage to organisms and even bioaccumulation in organisms or sediments in the water. We recently found that physical sunscreens and related inorganic UVFs exhibit bioattachment on the surfaces of button coral and cause significant growth inhibition and expulsion of zooxanthellae (*Symbiodinium* sp., unpublished data), which demonstrates the importance of further exploring the environmental fate of inorganic UVF-containing cosmetic products and the derived UVFs.

The $nTiO_2$ and nZnO were dispersed (partial dissolved) in physical sunscreens during the manufacturing process, which would be modified first sometimes. Thus inorganic UVFs in sunscreens often exist as surface-modified complexes. For surface-modified complexes, their potential environmental behavior presents some differences that need to be discussed. Primarily, coexisting surface coatings affect the fate of NPs to some extent. In addition to UVFs, sunscreens also contain other ingredients, such as preservatives (e.g., paraben derivates) [50], coloring agents (e.g., ammonium sulfate, ferric ammonium ferrocyanide, copper powder, and iron) [51], film-forming agents (e.g., acrylates and acrylamides) [52], surfactants, chelators, viscosity controllers (e.g., potassium cetyl phosphate and pentasodium ethylenediamine tetramethylene phosphonate), and fragrances [53]. Some of these ingredients have been detected in coastal waters [54–56]. Thus, $nTiO_2$ (and nZnO) may be present in the form of bare or coated NPs in the aquatic environment, and their dimension, shape, crystal phase, and surface area vary among different sunscreen products [27]. A recent study showed that sunscreen-derived $nTiO_2$ exhibits a larger particle size but a smaller hydrodynamic diameter and lower zeta potential than industrial uncoated $nTiO_2$, which exhibits significant aggregation [57]. In contrast, the presence of carboxymethyl cellulose (CMC) or polyvinylpyrrolidone (PVP) significantly enhances the stability of uncoated $nTiO_2$, as determined by the zeta potential values measured at pH 7, with substantial shape changes that result in spherical particles and relatively small $nTiO_2$ sizes [57]. Similar substantial shape transformations induced by stabilizers have been found in other studies [58,59]. Inorganic UVFs generally have a small particle size, strong hydrophobicity, and are insoluble in water; thus, Brownian motion, eddy motion, and runoff shear force result in some inorganic UVF particles remaining in suspension [60]. Engineered polymers or organic and inorganic substances that serve as coating materials or act as stabilizers have been found to modify the physicochemical properties of raw NPs, thereby affecting particle stability and mobility through electrostatic repulsion [61–63] and by maintaining the dispersion of nanosized inorganic UVFs. For example, $nTiO_2$ has been found to be fully dispersed and stabilized in natural water that contains organic materials [64]. Therefore, the stability of inorganic UVFs depends on their physicochemical properties and coating materials [27,57].

An early study indicated that eight of nine commercial sunscreen products are coated with nonvolatile inorganic residues, typically Al_2O_3 or SiO_2, to minimize the photochemical activity of TiO_2 [27]. Adsorbed or covalently bonded surfactants affect aggregation stability by increasing the surface charge and electrostatic repulsion or by reducing the interfacial energy between the particles and the solvent [65]. The interaction between steric repulsion and universal Coulomb attraction is caused by the surface coating layers, which may profoundly affect the aggregation kinetics. However, a recent study showed that sodium citrate provides higher stability for spherical $nTiO_2$ than PVP, sodium dodecyl sulfate, and polyethylene glycol, since sodium citrate results in lower critical coagulation concentrations [66]. Additionally, another study showed that the addition of coating materials such as CMC, PVP, and silica prevents significant TiO_2 aggregation by facilitating dispersion [60]. These stabilizers change the physicochemical properties (particle sizes and zeta potential) of $nTiO_2$ and produce a stable TiO_2 suspension with a cluster size smaller than that of uncoated $nTiO_2$ because they play the role of a dispersant that prevents nanoparticle aggregation [57]. A decrease in particle size results in a higher proportion of atoms on the particle surface, which alters the electronic structure, surface charge, and final degree of aggregation [67]. Small particles with high surface energy aggregate more readily than larger particles since aggregation reduces the free energy in the NP system.

It has been revealed that the dissolution of inorganic UVFs depends on the solubility of the materials themselves and on the concentration gradient in water [68,69]. For example, nZnO releases more Zn ions in seawater with a higher ionic strength than in fresh water [70]. Moreover, the dissolution of inorganic UVFs is clearly affected by the physicochemical properties of the material, such as the particle size, shape, and surface coating. Generally, the solubility of NPs is higher than that of the bulk phase because the decreased size increases the specific surface areas and the enthalpies of the formation of the ions [71]. Fairbairn et al. [72] also pointed out that nZnO is more easily dissolved in sea water than ZnO with ordinary particle sizes or Fe-doped nZnO. However, for nZnO, the impact of different sizes on dissolution is not as obvious for nanosized, bulk, or large particles due to the high solubility of ZnO, which can exhibit up to 80% dissolution [69,73,74]. Additionally, the shapes of NPs have been shown to affect both the rates of dissolution and the equilibrium concentrations [14]. The dissolution rate for spherical nCuO is faster than that of rod and spindle nCuO [75], while spherical nZnO induces lower toxicity than rod-shaped nZnO because the actual Zn ion concentration that results from the dissolution of rod-shaped nZnO is much higher than that of spherical nZnO [76].

Quite often, the dissolution rate of inorganic UVFs significantly decreases in the presence of surface coatings because the surface coating acts as a physical barrier or shield that prevents electrons or photons from reaching the NP surface [77]. In sunscreens, photoactivity problems may arise if particles are not treated with coatings, and manufacturers commonly employ inert surface coatings that dramatically reduce the potential for photoactivity; existing data suggests that these surface coatings reduce UV reactivity by as much as 99% [40,41]. However, organic coatings slow the dissolution process relative to that of uncoated ZnO but lead to an increased concentration of Zn^{2+} at equilibrium [78]. Otherwise, if the coatings are not stable or if manufacturers use forms of ZnO or TiO_2 that are not optimized for stability and sun protection, then sunscreens may not be protective [14]. These results suggest that inorganic UVFs might input substantial amounts of free metals into an aquatic environment and pose a toxicity risk to aquatic ecosystems.

In addition to the influence of internal NP properties, external environmental factors such as light, pH, and natural organic matter (NOM) can also make a difference. The interaction energy barrier decreases with a decreasing particle size according to the Derjaguin–Landau–Verwey–Overbeek (DLVO) theory, and it is affected by the properties of the primary NPs (e.g., size, shape, chemical composition, and surface coatings), solution chemistries (e.g., pH, ionic identity, electrolyte patterns, and reactions with NOM), and environmental conditions (e.g., temperature and dissolved oxygen level) [69,79]. For example, a large proportion of nZnO dissolves at a limit close to the solubility limit of ZnO(s) at a high pH of approximately 8.2, and both visible and UV light facilitate nZnO dissolution at lower pH values that range from 4.8 to 6.5 [80]. Light warms the water, enhances the release rates of inorganic UVFs, shortens the equilibrium time and even increases equilibrium concentrations [62]. Moreover, inorganic UVFs generate ROS under irradiation with visible and UV light; this results in the oxidation of metal ions and surface organic compounds, which increases the dissolution rates due to the decomposition of surface coatings and loss of the stabilizing effect of dissolved organic matter. The influence of solution properties on the dissolution of inorganic UVFs is dynamic and complex [62].

2.3. Substantial Environmental Impacts

The discharge of inorganic UVFs from sunscreens into waters is concomitant with the input of several other constituents, including nutrients (e.g., silicates, phosphates, and nitrates), metals (e.g., Al, Cd, Cu, Co, Mn, Mo, Ni, Pb, and Ti), and coating materials (e.g., preservatives, coloring agents, film-forming agents, surfactants, and stabilizers). Many of these coexisting substances are persistent; therefore, their effects might last beyond the most recent period of sunscreen use. These additional constituents influence the bioavailability and degradability of sunscreen ingredients since the biogeochemical routes into environmental media (water, sediment, and biota) and the hydrophobicity or hydrophilicity of the substances contained in sunscreens are diverse and complex [1,81]. Moreover, the effects of sunscreen contamination (especially from commercial formulations instead of individual compounds or ingredients) are sometimes difficult to perceive in laboratory studies because of their complex matrix [82,83] and unknown composition [84]. Additionally, because of the diverse formats of sunscreens (e.g., cream, gel, spray, and oil), their dilution and release of UVFs into water are different, as are their bioavailabilities and toxicities [4,85].

It is likely that environmental exposure to inorganic UVFs and the chemicals contained therein results from the production and consumption of sunscreens. Studies have indicated that UVFs and other ingredients from sunscreens have been detected in the tissues of marine organisms, such as clams, oysters, gastropods, and fish [86,87], and have shown toxicity in some aquatic species, such as the crustacean *Daphnia pulex* and the fish *Danio rerio* [88,89]. Rodríguez-Romero et al. [90] demonstrated with laboratory experiments and field measurements that sunscreens are an important source of nutrients, such as nitrogen compounds (NO_3^-, NO_2^-, and NH_4^+) and phosphate (PO_4^{3-}) in coastal marine environments, raising the possibility of algal blooms in oligotrophic waters. More specifically, some concentrations of the compounds (e.g., those of PO_4^{3-}, NH_4^+, NO_3^-, and Ti) released into water vary during the course of a day, which is known to be associated with variations in beach-goer activities and changes in solar radiation [4]. Sunscreens have also been identified as sources of high-risk metal substances [91], many of which (e.g., Al, Zn, Mg, Fe, Mn, Cu, Cr, and Pb) have been detected and quantified in aquatic environments [4,92]. Moreover, the organic components of sunscreens are readily removed from particle surfaces [93,94], which leaves the inorganic UVFs exposed to the surrounding environment. Although the ecological relevance of this input has not been well reviewed, Tovar-Sánchez et al. [4] suggested that it could enhance primary production in the oligotrophic waters of the Mediterranean Sea.

In addition to the direct output of soluble substances from sunscreens, some indirect metabolites are also produced in the water environment under sunlight. A study carried out on a touristic beach indicated that both temporal (daily) and vertical (water column) distributions of H_2O_2 concentrations generated by inorganic UVFs ($nTiO_2$ and nZnO) were present in marine waters [9]. According to the authors, the concentrations of H_2O_2 found within the top centimeter of the surface layer were up to 41.6% higher than those in the immediate subsurface waters [9]. Similarly, a large number of studies have indicated that $nTiO_2$ and nZnO produce ROS under sunlight exposure and induce oxidative stress in organisms [62,95–98]. Therefore, more reliable information is required on the role of sunlight in the release of the main ingredients and byproducts of sunscreens into water.

Accordingly, sunscreen-derived inorganic UVFs are very likely to be released into the main water bodies of lakes, rivers, and oceans but do not remain suspended for a long time, with the most likely fates being aggregation, dissolution, and settling onto the sediments due to the water chemistry conditions and the presence of natural colloids. However, their environmental behaviors will be affected by the surface coating and various physical and chemical factors, such as ocean currents, waves, and high salinity, and they will undergo complex aggregation and dissolution reactions; moreover, their structural form, distribution, and toxic effects will constantly change. Nevertheless, these behaviors and transformation processes for inorganic UVFs must influence their bioavailability and toxicity, which cause great impacts on natural aquatic ecosystems [80].

3. Toxicity of Inorganic UVFs on Aquatic Organisms

The adverse effects of organic UVFs on aquatic organisms have been reviewed in recent literature [21], but studies on the ecological risk of inorganic UVFs are limited. Although studies have found that inorganic UVFs do not cause more damage to humans than organic UVFs [34,99–101], notably, the potential environmental effects of UVFs on aquatic organisms are not taken into consideration during their production, and even worse, few specific recommendations for the environmentally friendly use of sunscreens have been offered by agencies or governments worldwide.

3.1. Interaction of Inorganic UVFs with Organisms in Aquatic Environments

Although inorganic UVFs are often coated with complex stabilizers, they are released in particle form when sunscreen enters the water. When they enter the water environment, inorganic UVFs tend to disperse, aggregate, dissolve metal ions, settle, absorb, and/or bioaccumulate within organisms. Studies have shown that inorganic UVFs interact with aquatic organisms in a variety of ways [83,102]. First, inorganic UVFs or their aggregates can adsorb or wrap themselves around the surface of phytoplankton or microorganisms and eventually be ingested by organisms. Second, filter-feeding or devouring animals, such as planktonic amphipods, benthic shellfish, and polychaetes, can filter or swallow inorganic UVFs directly. Third, organisms of high trophic levels can directly consume water that contains inorganic UVFs or algae and other low trophic level organisms, and thus cause the accumulation, transfer, and even magnification of inorganic UVFs along the food web and result in unpredictable environmental effects and ecological risks.

3.2. Toxicity of Inorganic UVFs on Organisms at the Individual Level

Sunscreen-derived inorganic UVFs are widely distributed in all levels of water, including the surface microlayer, water column, and sediment, which also results in interactions with various environmental factors; thus, they are deemed to cause adverse effects on various organisms in the aquatic environment. It is still difficult to conduct exposure experiments specifically for sunscreen-derived inorganic UVFs since sunscreens in water release not only inorganic UVFs but also many latent toxic chemicals, such as surfactants. Thus, there is little direct laboratory evidence of the damage caused by sunscreen-derived inorganic UVFs that primarily focuses on $nTiO_2$ UVFs and nZnO UVFs in aquatic organisms (Table 1).

Table 1. Toxicity and potentially toxic mode of action (MoA) of inorganic UV filters on aquatic organisms.

Inorganic UVFs	Organism	Exposure Conditions	Effects	MoA	References
TiO_2 (release from cosmetic products)	Algae (*Thalassiosira pseudonana*)	0–96 h; 0.13–100 mg/L	Growth inhibition	Potential ROS production	[103]
$nTiO_2$ from sunscreens	Chaetoceros gracilis (*Bacillariophyceae*); Amphidinium carterae (*Dinophyceae*); Pleurochrysis roscoffensis (*Primnesiophycae*); Nannochloropsis gaditana (*Eustigmatophyceae*)	75 h; sunscreens (1–200 mg/L) or $nTiO_2$ (1–10 mg/L)	Distribution of phytoplankton	H_2O_2 produced adsorption and absorption by the phytoplankton, membrane damage, ROS, and perhaps genotoxic damage	[104]
$nTiO_2$ from sunscreen	Sea urchin (*Paracentrotus lividus*)	3 h, 24 h; 10, 20, and 50 μL/L sunscreen	Sea urchin development impairment	decrease in AChE activity	[105]
nZnO (sunscreen-derived)	Algae (*Thalassiosira pseudonana*)	0–96 h, 10 and 50 mg/L	Growth inhibition	Time- and concentration-dependent bioaccumulation	[106]
ZnO from sunscreen	Stony corals (*Acropora* spp.)	48 h of in situ condition 6.3 mg/L	Coral bleaching; release of zooxanthellae	dissolved Zn^{2+} Zn^{2+} shading effects	[43]
zinc-containing sunscreens	Sea urchin (*Strongylocentrotus purpuratus*) embryos	96 h; 0.01–1 mg/L	Malformations (skeletal abnormality, stage arrest, and axis determination disruption)	Zn^{2+} internalized	[49]
$nTiO_2$ and nZnO from sunscreen	Shrimp (*Palaemon varians*)	4 h 0–300 mg/L sunscreen	Repellency and mortality effects		[85]

3.2.1. $nTiO_2$ UVFs

Only a few studies have focused on the toxicity of inorganic UVFs to marine algae. Early findings suggested that the $nTiO_2$ from sunscreens alters the species density and composition of the microalgae community due to the impairment of cell growth; sunscreen toxicity levels are significantly related to UVR, which is commonly neglected in some bioassays, but this could alter the results in important ways and should be considered when performing environmentally relevant bioassays [104]. Because of its photochemical properties, $nTiO_2$ produces high concentrations of H_2O_2 as a result of UVR [9], which causes toxic effects such as damage to cell membranes or cell walls [93], lipid peroxidation, growth inhibition, and a decline in the proportion of healthy cells in microalgae populations [107]. Furthermore, the adsorption of $nTiO_2$ particles on the surfaces of algae cells can cause physical damage, such as shading effects, which inhibit cell growth [108].

Direct toxicology data on the effects of sunscreen-derived inorganic $nTiO_2$ on zooplankton, fish, and benthos are rare [49,103–106]. A recent study indicated that $nTiO_2$ released from sunscreens causes repellency and mortality in shrimp (*Palaemon varians*) and speculated that the avoidance response might be the main factor responsible for the reduction in the shrimp population due to increasing sunscreen concentrations at the local scale [85]. In addition, the $nTiO_2$ released from sunscreens impairs sea urchin development or causes malformations due to a decrease in AChE activity [49,105]. In realistic environmental scenarios, the self-aggregation of inorganic UVFs into larger masses and their incorporation into aggregate materials might increase the bioavailability and toxicity for algae, phytoplankton, zooplankton, and benthos along food chains.

3.2.2. nZnO UVFs

nZnO can absorb ultraviolet A-rays (UVA) and ultraviolet B-rays (UVB), while $nTiO_2$ can only absorb UVB; therefore, nZnO provides better UV protection than $nTiO_2$, and its use in physical sunscreens may even exceed that of $nTiO_2$ in the future [109]. Few studies have assessed the potential release and toxicity of sunscreen-derived nZnO in aquatic environments [43,49,85]. For instance, studies conducted with zooplankton and benthic animals exposed to nZnO-containing sunscreen showed repellency and mortality effects in shrimp [85], irreversible coral bleaching, and widespread mortality of symbiotic zooxanthellae [43], which primarily resulted from Zn^{2+} toxicity. Moreover, studies have shown that the toxicity of nZnO UVFs appears to be related to solubility or the release of toxic metal ions (Zn^{2+}) instead of aggregation, which leads to the conclusion that higher Zn^{2+} solubility is accompanied by higher toxicity [110]. Similarly, the nZnO released from sunscreens has caused impairments or malformations in sea urchin development due to Zn^{2+} internalization [49,105]. These results indicate that the solubility of nZnO plays a critical role in the toxicity of physical sunscreens to marine organisms [11].

It has been reported that the surface properties of inorganic UVFs, including the pH and ionic strength of the solution, affect their solubility, which largely determines the extent of toxicity [111,112]. Attempts have been made to reduce solubility and, consequently, ZnO toxicity through iron doping. Although this strategy has been shown to reduce ZnO cytotoxicity in cell cultures [113], Fairbairn et al. [72] found that 10% iron-doped ZnO is just as toxic as non-doped ZnO to sensitive marine embryos. The solution pH and ionic strength may affect the adsorption of NPs onto cells due to changes in surface charges [114–116]. In addition, Peng et al. [117] reported different sensitivities to nZnO in three marine diatoms (*Thalassiosira pseudonana, Chaetoceros gracilis*, and *Phaeodactylum tricornutum*) and introduced the idea that the morphologies of nZnO samples also affect their toxicities. These results confirm that the toxic mechanisms of inorganic UVFs are related to various toxic factors; thus, more systematic studies are needed to elucidate their toxicity profiles.

3.3. *Impacts of Inorganic UVFs on Multiple Trophic Levels*

Given the persistence and stability of inorganic UVFs such as $nTiO_2$, organisms can accumulate and even transfer these substances along food chains [35,49,85,118,119]. Previous studies have shown that $nTiO_2$ and nZnO can be internalized into the cells of bacteria and algae and accumulate in aquatic organisms, including zooplankton, swimming organisms, and benthos [1,83,120,121]. Notably, it is highly possible that inorganic UVFs are transferred from lower trophic organisms to higher trophic organisms through predator-prey relations and biomagnification in the food web [122,123]. In fact, the bioaccumulation of chemicals released from sunscreens has been detected in fish and mussels [124–126], while the mechanisms by which inorganic UVFs transfer in a food web are still not clear. Studies have shown significant amounts of $nTiO_2$ in the dietary exposure groups, which indicates that dietary intake may constitute a major route of trophic transfer [123]. For nZnO, the transfer behaviors can be divided into particle and metal ion accumulation routes since nZnO easily dissolves to produce Zn^{2+}. Considering that some aquatic organisms, such as fish and clams, are human food sources and provide food for wildlife, the bioaccumulation and trophic transfer of inorganic UVFs along the food chain have raised increasing concerns.

3.4. *Potential Mechanisms for the Toxicity of Sunscreen-Derived Inorganic UVFs*

Since the two most commonly used inorganic UVFs, i.e., nZnO and $nTiO_2$, are NPs, they share similar behaviors in aquatic environments, as mentioned above. Therefore, it has been hypothesized that the toxicity of sunscreen-derived NPs might arise from mechanisms similar to those of raw $nTiO_2$ and nZnO. Although the aquatic toxicities of raw $nTiO_2$ and nZnO have been well-studied in previous reviews [22,60,127], the toxicological evaluation of the mechanism on sunscreen-derived inorganic UVFs with aquatic organisms has only recently begun, and few studies have assessed the toxic performance of sunscreen-derived NPs compared with those of engineered raw NPs [106].

As shown in Figure 2, adsorption or absorption is important and constitutes the first step in the interaction between NPs and aquatic organisms. Engineered raw NPs may attach to the surfaces of aquatic organisms and cause physical effects such as shade photosynthesis, direct mechanical damage to phytoplankton, or blocking vital movement in zooplankton [60]. Wang et al. [128] reported that $nTiO_2$ significantly inhibits *Phaeodactylum tricornutum* growth directly through physical effects such as cell wall damage that arises from algae entrapment. Although we recently found that sunscreen-derived inorganic UVF particles can be absorbed on the surfaces of button corals (unpublished data) and result in the contraction of tentacles, related reports are rare; thus, more studies are encouraged with other aquatic organisms to provide direct evidence.

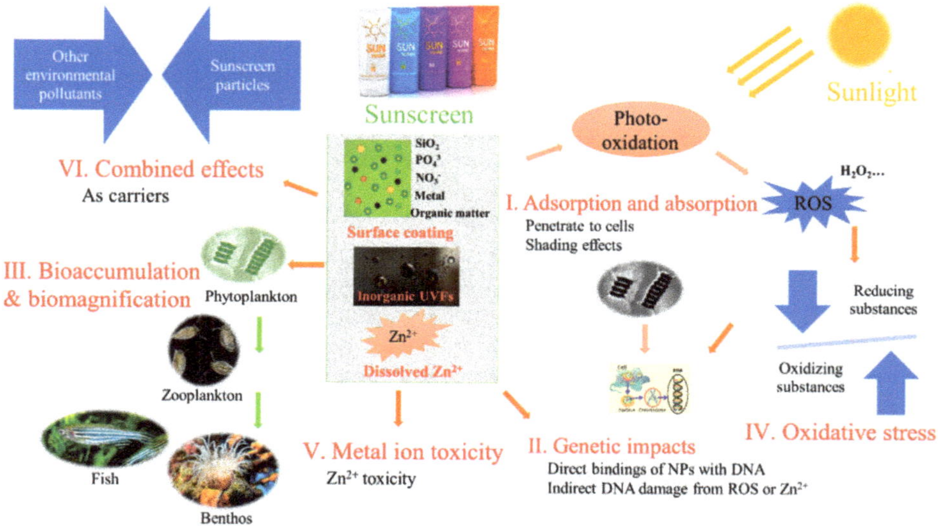

Figure 2. Potential mechanisms for sunscreen-derived inorganic UVF toxicity in aquatic organisms.

Internalization has been deemed a common pathway for the uptake of engineered NPs by algae [128,129]. Once they penetrate the cell barrier, NPs can undergo translocation into the intracellular environment via diffusion or endocytosis [130]. Here they can interact with DNA or attach to organelles in cells and block normal function or cause genetic impacts [60,127]. Genetic effects may be produced by the direct binding of NPs with DNA, by the indirect damage from the ROS generated by NPs, or by the toxic ions released from soluble NPs [60]. Although few studies have directly demonstrated the genetic damage induced by sunscreen-derived NPs, the ROS generation or Zn^{2+} dissolution from physical sunscreens can impact the DNA or RNA of aquatic organisms. In particular, small single NPs (<10 nm) can reach the nucleus through nuclear pores, while larger NPs may also have the opportunity to bind with DNA molecules when the nuclear membrane dissolves due to the division of cells during mitosis. The overall uptake of the NPs that reach the nucleus through diffusion across the nuclear membrane or that are transported through nuclear pore complexes presents the danger of subsequent direct interactions with cellular genetic material [60].

Following attachment, NPs may accumulate on cell surfaces or transfer to specific organs or tissues (e.g., stomach, gills, and liver) for storage [60]. Previous studies have shown that metal-based NPs can be ingested and can accumulate in single aquatic organisms [60,131] or undergo trophic transfer in the food chain [122,123], especially with higher trophic level organisms such as fish or filter-feeder organisms such as fleas and many benthic organisms (e.g., mussels, oysters, and clams), after waterborne or foodborne exposure [117,132–134]. The bioaccumulation of $nTiO_2$ and nZnO has been shown to inhibit the growth of aquatic organisms [117,132]. In fact, there is evidence that bioaccumulation is directly related to the toxicity of NPs [135,136]. These studies show that NPs mainly accumulate in specific organs or tissues in aquatic organisms and thus inhibit their biological intake and affect their biological metabolism and energy acquisition. Notably, NP absorption and bioaccumulation cause physical damage and then lead to adverse consequences for organisms, including oxidative stress, behavioral inhibition, and death. However, studies on the bioaccumulation of sunscreen particles are scarce. Although we recently found that the active components of physical sunscreen (Ti and Zn) are bioaccumulated in button corals (unpublished data), we still have not clarified how they enter coral individuals, the organs or tissues in which they prefer to accumulate, or the consequences that ultimately result. A remaining question is whether sunscreen-derived NPs exhibit toxic mechanisms similar to or different from those of raw NPs, since limited studies have only recently been published (shown in Table 1). However, the availability of studies on raw engineered NPs definitely shows that further studies to elucidate the toxicity profiles of physical sunscreens are urgently needed.

The specific toxicity of the MoA to metal-based NPs is related to ROS generation and subsequent ROS-induced oxidative stress. Oxidative stress and cellular toxicity are of concern because nZnO and $nTiO_2$ can penetrate the stratum corneum, enter the dermis, and ultimately reach the blood supply [120,137–140]. Previous results have suggested that the physical interactions of NPs induce significant oxidative stress, which provides direct evidence for the toxicological impact of engineered raw NPs in aquatic organisms [128,135]. In general, both sunscreen-derived and raw NPs can undergo photooxidation and generate ROS under sunlight irradiation, and ROS overgeneration is deemed to result in subsequent cell membrane damage, lipid peroxidation, growth inhibition, and other negative impacts [103,141,142]. Sunscreen-derived $nTiO_2$ has been indicated to induce the photocatalytic generation of ROS, such as H_2O_2, in vitro and cause growth inhibition and distribution changes in algae [9,103,104,143]. Moreover, inorganic UVFs enter aquatic organisms and induce ROS generation in vivo, which causes toxicological impacts on *Chlorella* spp. [144]. The production of ROS, either in vitro or in vivo, directly or indirectly, causes oxidative stress. According to a study on raw NPs, the in vitro aqueous production of ROS by raw NPs requires photosensitization; that is, the production of ROS is driven by light, especially UVR. Although the generation of ROS is instantaneous, ROS are usually quenched within seconds by reducing substances, and it has been reported that ROS produced in vitro exert harmful effects on organisms [9,60,104,145]. However, most related studies have been conducted to probe ROS generation in vivo and provide direct evidence of oxidative stress after NP exposure [128,135]. Unlike raw engineered NPs, sunscreen-derived NPs are often coated or modified when they are applied to cosmetic products. Once sunscreen is in water, inorganic UVFs released from sunscreen are complex and can exist in the form of aggregates of various complex components [40,41], including surface-modified complexes or raw NPs. To our knowledge, no study has been conducted to clarify whether coating modifications alter photosensitivity or affect the extent or duration of oxidative stress. However, for sunscreen-derived $nTiO_2$, studies have indicated that their toxicity is also affected by coexisting coatings, which might have determined their aggregated sizes or the levels of ROS generated. Studies have shown that UVFs coated with inert protective films (such as SiO_2, Al_2O_3, or organic matter) or coating materials such as organomodified silicon oxide exhibit significantly reduced production of ROS on the surfaces of NPs and alter the impacts of ROS on organisms, even during UVR [146,147]. That is, coating materials alleviate the impacts of ROS that result from sunscreen-derived UVFs on organisms. It is easy to understand that the coatings and modifications are meant not only to shield or reduce UV damage but also to prevent the adverse biological effects of UVFs [147]. Oxidative stress should be a common toxicity MoA for the two types of particles, but differences exist in the detailed MoA and sites affected (in vitro or in vivo). Raw engineered NPs are more often focused on the generation of ROS in vivo, while sunscreen-derived NPs are more often focused on the generation of ROS in vitro. Since these coating materials affect the behavior and toxicity of sunscreen-derived inorganic UVFs, the impacts of the different coating materials and their diverse characteristics on the toxicity of inorganic UVFs should be given more attention and considered during the development of safe sunscreens.

Furthermore, as with oxidative stress, metal ions can be released from both sunscreen-derived and raw NPs, which plays an important role in their toxicity to aquatic organisms. In contrast to the largely insoluble $nTiO_2$, nZnO can rapidly dissolve as Zn^{2+} in water [44], and Zn^{2+} is the major contributor to the toxicity of sunscreen-derived nZnO [49]. For example, nZnO toxicity to the marine diatom *Thalassiosira pseudonana* has been solely explained by the Zn^{2+} reaction [148]. Zn^{2+} toxicity constitutes another unique toxic MoA for nZnO UVFs. For raw engineered nZnO, the toxicity can be ascribed to Zn^{2+} concentrations; however, the coatings of sunscreen-derived nZnO often delay the dissolution equilibrium and lead to an increased concentration of Zn^{2+} cations at equilibrium [78]. Spisni et al. [106] reported that the toxicity of sunscreen-derived nZnO for the growth of algae (*Thalassiosira pseudonana*) appears to be lower than that of raw nZnO at relatively low concentrations, but the toxicity levels become similar when concentrations are increased to 50 mg/L. Recently, Corinaldesi et al. [43] found that sunscreen-derived nZnO induces the complete and latent irreversible bleaching of stony coral and rapid and widespread mortality of symbiotic zooxanthellae. Presumably, these effects are attributable to the toxicity of Zn^{2+}, which causes alterations in the composition of the cellular membrane lipids of hard corals and their symbiotic organisms [149]. This is of concern because an increasing number of manufacturers are using ZnO rather than TiO_2 in sunscreens.

Accordingly, the MoAs for the toxicity of sunscreen-derived inorganic UVFs are similar to those of raw engineered NPs, but they exhibit some differences due to the complex surface coatings and modifications. Sunscreen-derived NPs exist in forms that are more complicated than those of raw NPs, and nanoparticle monomer toxicity, agglomeration toxicity, or complex mixed toxicity may result when they enter water. In contrast to engineered NPs, inorganic UVFs are often coated with stabilizers in sunscreens to prevent aggregation [27,93,150,151]; thus, they result in altered interactions with organisms [152] and differences in the extent of toxicity [104,106]. Compared with raw $nTiO_2$, the presence of some stabilizers increases the toxicity of NPs and the inhibition of growth in *Escherichia coli* (*E. coli*) [57]. Moreover, the sizes of TiO_2 particles are relatively small and appear to contribute to *E. coli* cell damage [60], and $nTiO_2$ samples with small particle sizes, large surface areas, and strong electrostatic attractions easily act as carriers of other environmental pollutants [136,153,154], including the other components of sunscreens, which affect their toxicity to aquatic organisms. Accordingly, although sunscreen-derived inorganic UVF-engineered NPs exhibit some similarities in toxicity and MoA, the presence of surface coatings or modifications is known to cause differences and result in different toxicities; thus, further study is required to increase our understanding of these differences and their origins.

4. Conclusions and Future Perspectives

This study reviewed the fate and toxicity of inorganic UVFs in aquatic environments, with information on their sources, environmental behaviors, and toxicities to aquatic organisms from the individual to the trophic transfer levels. Inorganic UVFs derived from sunscreens are often dispersed, aggregated, dissolved into waters, and settled into sediments, and they tend to be absorbed and bioaccumulated by organisms; this results in adverse effects on various organisms in the aquatic environment, which are directly influenced by various environmental factors and the presence of coatings; resulting in different environmental fates and toxicities compared with raw engineered NPs.

Inorganic UVF-containing sunscreens are deemed to be a source of multiple environmental pollutants, and they pose new environmental risks to aquatic environments. As indicated by data on coastal-zone population growth and tourism activities, sunscreens exhibit the fastest growth in global sales. This fact, together with recent research that indicates the presence and accumulation of UVFs in environmental media, emphasizes the potential damage that could be caused in marine areas. Thus, future investigations are needed to understand the magnitude and real impacts of these emerging pollutants in marine systems, including studies on the distribution and partitioning in the water column, dissolution and speciation of their main components, evaluation of the ecological significance of nutrient input, and residence time, aging, persistence, accumulation, and toxicity in the trophic chain.

Most studies on the environmental behaviors of inorganic UVFs have been conducted under laboratory conditions, which may not represent realistic natural environments. Although some recent studies have investigated the aggregation, dissolution, and transformation of UVFs in natural water bodies by collecting lake water and seawater samples, knowledge of the environmental fate of inorganic UVFs in the real environment is still limited. In fact, UVFs can be greatly impacted by various factors in the natural environment, which complicates their behavior. Thus, further studies should be conducted under realistic environmental conditions to the fullest extent possible.

Moreover, a thorough understanding of the causal relationship between the properties of inorganic UVFs and toxicity remains largely elusive. Although many studies have been performed on the implications of these NPs for aquatic organisms, there is an insufficient characterization of the material properties and the relationship between the observed toxicity and specific features of inorganic UVFs, such as Zn^{2+} toxicity, bioaccumulation, shading effects, and ROS generation. Thus, establishing a quantitative correlation between environmental behaviors and toxicity would facilitate the future evaluation and prediction of the toxicity of related cosmetic products.

Finally, many previous studies have attributed the toxicity of inorganic UVFs to one or two major aspects of material properties or solution behaviors. Nevertheless, material properties are often interrelated and interdependent. Moreover, after undergoing the abovementioned processes, the coating materials, size distribution, and surface properties of the particles will be dramatically affected. Thus, tracking dynamic aggregation or disaggregation to determine the actual fractions of nanosized inorganic UVFs and aggregated or agglomerated particles at cellular interfaces remains the most important issue identified thus far.

Notably, regardless of the recommended usage level and the ways in which people use cosmetic products, the potential environmental effects of UVFs on nontarget organisms were not taken into consideration when governments and agencies developed their recommendations. In particular, although large quantities of sunscreen can be released directly into seawater during recreational activities carried out on hot days, there are very few specific recommendations for the use of sunscreens in coastal areas. Therefore, the ecotoxicological testing of whole products should be included in future assessments of environmental risks and in developing recommendations and regulations for the usage and formulation of commercial sunscreens.

Author Contributions: Conceptualization, S.Y. and X.Z.; methodology, S.Y.; validation, S.Y.; formal analysis, S.Y.; investigation, S.Y. and X.J.; writing—original draft preparation, S.Y.; writing—review and editing, J.H., X.J., Y.H., X.Z. and Z.C.; project administration, S.Y. and X.Z. All authors have read and agreed to the published version of the manuscript.

Funding: This study was funded by the Guangdong Basic and Applied Basic Research Fund (2019A1515110930), the National Natural Science Foundation of China (42077227) and Shenzhen Key Laboratory of Marine IntelliSense and Computation (ZDSYS20200811142605016). It was also supported in part by the Shenzhen Fundamental Research and Discipline Layout project (JCYJ20180507182227257).

Institutional Review Board Statement: Not applicable.

Informed Consent Statement: Not applicable.

Data Availability Statement: The data presented in this study are available on request from the corresponding author.

Conflicts of Interest: The authors declare no conflict of interest.

References

1. Sanchez-Quiles, D.; Tovar-Sanchez, A. Are sunscreens a new environmental risk associated with coastal tourism? *Environ. Int.* **2015**, *83*, 158–170. [CrossRef] [PubMed]
2. Lim, H.W.; Draelos, Z.D. *Clinical Guide to Sunscreens and Photoprotection*; Informa Health Care: New York, NY, USA, 2008.
3. Urbach, F. The historical aspects of sunscreens. *J. Photochem. Photobiol. B Biol.* **2001**, *64*, 99–104.
4. Tovar-Sanchez, A.; Sanchez-Quiles, D.; Basterretxea, G.; Benede, J.L.; Chisvert, A.; Salvador, A.; Moreno-Garrido, I.; Blasco, J. Sunscreen products as emerging pollutants to coastal waters. *PLoS ONE* **2013**, *8*, e65451. [CrossRef] [PubMed]
5. Gormsen, E. The impact of tourism on coastal areas. *GeoJournal* **1997**, *42*, 39–54. [CrossRef]
6. Li, W.; Ma, Y.; Guo, C.; Hu, W.; Liu, K.; Wang, Y.; Zhu, T. Occurrence and behavior of four of the most used sunscreen UV filters in a wastewater reclamation plant. *Water Res.* **2007**, *41*, 3506–3512. [CrossRef]
7. Honey, M.; Krantz, D. *Global Trends in Coastal Tourism*; Center on Ecotourism and Sustainable Development: Washington, DC, USA, 2007.
8. Hall, C.M. Trends in ocean and coastal tourism: The end of the last frontier? *Ocean. Coast. Manag.* **2001**, *44*, 601–618. [CrossRef]
9. Sanchez-Quiles, D.; Tovar-Sanchez, A. Sunscreens as a source of hydrogen peroxide production in coastal waters. *Environ. Sci. Technol.* **2014**, *48*, 9037–9042. [CrossRef]
10. Manová, E.; von Goetz, N.; Hauri, U.; Bogdal, C.; Hungerbühler, K. Organic UV filters in personal care products in Switzerland: A survey of occurrence and concentrations. *Int. J. Hyg. Environ. Health* **2013**, *216*, 508–514. [CrossRef]
11. Giokas, D.L.; Salvador, A.; Chisvert, A. UV filters: From sunscreens to human body and the environment. *TrAC Trends Anal. Chem.* **2007**, *26*, 360–374. [CrossRef]
12. Salvador, A.; Chisvert, A. *Analysis of Cosmetic Products*; Elsevier Science: Amsterdam, The Netherlands; London, UK, 2011.
13. Zhu, X.S.; Huang, J.Y.; Lü, X.H.; Du, Y.F.; Cai, Z.H. Fate and Toxicity of UV Filters in Marine Environments. *Huanjing Kexue/Environ. Sci.* **2018**, *39*, 2991–3002.
14. Environmental Working Group. EWG's Sunscreen Guide. 2019. Available online: https://www.ewg.org/sunscreen/report/executive-summary/ (accessed on 20 December 2020).
15. Antoniou, C.; Kosmadaki, M.G.; Stratigos, A.J.; Katsambas, A.D. Sunscreens–what's important to know. *J. Eur. Acad. Dermatol. Venereol.* **2008**, *22*, 1110–1119. [CrossRef] [PubMed]
16. Jin, C.-Y.; Zhu, B.S.; Wang, X.-F.; Lu, Q.H. Cytotoxicity of titanium dioxide nanoparticles in mouse fibroblast cells. *Chem. Res. Toxicol.* **2008**, *21*, 1871–1877. [CrossRef] [PubMed]
17. Seite, S.; Colige, A.; Piquemal-Vivenot, P.; Montastier, C.; Fourtanier, A.; Lapiere, C.; Nusgens, B. A full-UV spectrum absorbing daily use cream protects human skin against biological changes occurring in photoaging. *Photodermatol. Photoimmunol. Photomed.* **2000**, *16*, 147–155. [CrossRef]
18. van der Pols, J.C.; Williams, G.M.; Pandeya, N.; Logan, V.; Green, A.C. Prolonged prevention of squamous cell carcinoma of the skin by regular sunscreen use. *Cancer Epidemiol. Prev. Biomark.* **2006**, *15*, 2546–2548. [CrossRef] [PubMed]
19. Balogh, T.S.; Velasco, M.; Pedriali, C.A.; Kaneko, T.M.; Baby, A.R. Ultraviolet radiation protection: Current available resources in photoprotection. *BrasDermatol* **2011**, *86*, 732–742.
20. Wang, S.Q.; Balagula, Y.; Osterwalder, U. Photoprotection: A review of the current and future technologies. *Dermatol. Ther.* **2010**, *23*, 31–47. [CrossRef]
21. Schlumpf, M.; Cotton, B.; Conscience, M.; Haller, V.; Steinmann, B.; Lichtensteiger, W. In vitro and in vivo estrogenicity of UV screens. *Environ. Health Perspect.* **2001**, *109*, 239–244. [CrossRef]
22. Schneider, S.L.; Lim, H.W. A review of inorganic UV filters zinc oxide and titanium dioxide. *Photodermatol. Photoimmunol. Photomed.* **2019**, *35*, 442–446. [CrossRef]
23. Lu, P.J.; Huang, S.C.; Chen, Y.P.; Chiueh, L.C.; Shih, D.Y.C. Analysis of titanium dioxide and zinc oxide nanoparticles in cosmetics. *J. Food Drug Anal.* **2015**, *23*, 587–594. [CrossRef]
24. Watkinson, A.C.; Bunge, A.L.; Hadgraft, J.; Lane, M.E. Nanoparticles do not penetrate human skin-a theoretical perspective. *Pharm. Res.* **2013**, *30*, 1943–1946. [CrossRef]
25. Piccinno, F.; Gottschalk, F.; Seeger, S.; Nowack, B. Industrial production quantities and uses of ten engineered nanomaterials in Europe and the world. *J. Nanoparticle Res.* **2012**, *14*, 1109. [CrossRef]
26. Danovaro, R.; Bongiorni, L.; Corinaldesi, C.; Giovannelli, D.; Damiani, E.; Astolfi, P.; Greci, L.; Pusceddu, A. Sunscreens cause coral bleaching by promoting viral infections. *Environ. Health Perspect.* **2008**, *116*, 441–447. [CrossRef] [PubMed]
27. Lewicka, Z.A.; Benedetto, A.F.; Benoit, D.N.; William, W.Y.; Fortner, J.D.; Colvin, V.L. The structure, composition, and dimensions of TiO_2 and ZnO nanomaterials in commercial sunscreens. *J. Nanoparticle Res.* **2011**, *13*, 3607. [CrossRef]
28. Botta, C.; Labille, J.; Auffan, M.; Borschneck, D.; Miche, H.; Cabié, M.; Masion, A.; Rose, J.; Bottero, J.-Y. TiO_2-based nanoparticles released in water from commercialized sunscreens in a life-cycle perspective: Structures and quantities. *Environ. Pollut.* **2011**, *159*, 1543–1550. [CrossRef]

29. Tsui, M.M.; Leung, H.; Wai, T.-C.; Yamashita, N.; Taniyasu, S.; Liu, W.; Lam, P.K.; Murphy, M.B. Occurrence, distribution and ecological risk assessment of multiple classes of UV filters in surface waters from different countries. *Water Res.* **2014**, *67*, 55–65. [CrossRef]
30. Jurado, A.; Gago-Ferrero, P.; Vàzquez-Suñé, E.; Carrera, J.; Pujades, E.; Díaz-Cruz, M.S.; Barceló, D. Urban groundwater contamination by residues of UV filters. *J. Hazard. Mater.* **2014**, *271*, 141–149. [CrossRef]
31. Amine, H.; Gomez, E.; Halwani, J.; Casellas, C.; Fenet, H. UV filters, ethylhexyl methoxycinnamate, octocrylene and ethylhexyl dimethyl PABA from untreated wastewater in sediment from eastern Mediterranean river transition and coastal zones. *Mar. Pollut. Bull.* **2012**, *64*, 2435–2442. [CrossRef]
32. Barón, E.; Gago-Ferrero, P.; Gorga, M.; Rudolph, I.; Mendoza, G.; Zapata, A.M.; Díaz-Cruz, S.; Barra, R.; Ocampo-Duque, W.; Páez, M. Occurrence of hydrophobic organic pollutants (BFRs and UV-filters) in sediments from South America. *Chemosphere* **2013**, *92*, 309–316. [CrossRef]
33. Rodil, R.; Moeder, M. Development of a simultaneous pressurised-liquid extraction and clean-up procedure for the determination of UV filters in sediments. *Anal. Chim. Acta* **2008**, *612*, 152–159. [CrossRef]
34. Rainieri, S.; Barranco, A.; Primec, M.; Langerholc, T. Occurrence and toxicity of musks and UV filters in the marine environment. *Food Chem. Toxicol.* **2017**, *104*, 57–68.
35. Díaz-Cruz, M.S.; Barceló, D. Chemical analysis and ecotoxicological effects of organic UV-absorbing compounds in aquatic ecosystems. *TrAC Trends Anal. Chem.* **2009**, *28*, 708–717. [CrossRef]
36. Balmer, M.E.; Buser, H.R.; Muller, M.D.; Poiger, T. Occurrence of some organic UV filters in wastewater, in surface waters, and in fish from Swiss lakes. *Environ. Sci. Technol.* **2005**, *39*, 953–962. [CrossRef] [PubMed]
37. Poiger, T.; Buser, H.R.; Balmer, M.E.; Bergqvist, P.A.; Muller, M.D. Occurrence of UV filter compounds from sunscreens in surface waters: Regional mass balance in two Swiss lakes. *Chemosphere* **2004**, *55*, 951–963. [CrossRef] [PubMed]
38. Wong, S.W.; Leung, P.T.; Djurisic, A.B.; Leung, K.M. Toxicities of nano zinc oxide to five marine organisms: Influences of aggregate size and ion solubility. *Anal. Bioanal. Chem.* **2010**, *396*, 609–618. [CrossRef]
39. Gondikas, A.P.; Kammer, F.; Reed, R.B.; Wagner, S.; Ranville, J.F.; Hofmann, T. Release of TiO_2 nanoparticles from sunscreens into surface waters: A one-year survey at the old Danube recreational Lake. *Environ. Sci. Technol.* **2014**, *48*, 5415–5422. [CrossRef]
40. SCCNFP, Opinion Concerning Titanium Dioxide. Opinion: European Commission—The Scientific Committee on Cosmetic Products and Non-Food Products Intended for Consumers. *Sci. Comm. Cosmet. Prod. Non-Food Prod. Intend. Consum.* **2000**, *0005/98*, 2–43. Available online: https://ec.europa.eu/health/ph_risk/committees/sccp/documents/out135_en.pdf (accessed on 15 November 2021).
41. Pan, Z.; Lee, W.; Slutsky, L.; Clark, R.A.; Pernodet, N.; Rafailovich, M.H. Adverse effects of titanium dioxide nanoparticles on human dermal fibroblasts and how to protect cells. *Small* **2009**, *5*, 511–520. [CrossRef] [PubMed]
42. von der Kammer, F.; Ottofuelling, S.; Hofmann, T. Assessment of the physico-chemical behavior of titanium dioxide nanoparticles in aquatic environments using multi-dimensional parameter testing. *Environ. Pollut.* **2010**, *158*, 3472–3481. [CrossRef]
43. Corinaldesi, C.; Marcellini, F.; Nepote, E.; Damiani, E.; Danovaro, R. Impact of inorganic UV filters contained in sunscreen products on tropical stony corals (*Acropora* spp.). *Sci. Total Environ.* **2018**, *637*, 1279–1285. [CrossRef]
44. Brun, N.R.; Lenz, M.; Wehrli, B.; Fent, K. Comparative effects of zinc oxide nanoparticles and dissolved zinc on zebrafish embryos and eleuthero-embryos: Importance of zinc ions. *Sci. Total Environ.* **2014**, *476*, 657–666. [CrossRef]
45. Schmidt, J.; Vogelsberger, W. Dissolution kinetics of titanium dioxide nanoparticles: The observation of an unusual kinetic size effect. *J. Phys. Chem. B* **2006**, *110*, 3955–3963. [CrossRef]
46. Garoli, D.; Pelizzo, M.G.; Bernardini, B.; Nicolosi, P.; Alaibac, M. Sunscreen tests: Correspondence between in vitro data and values reported by the manufacturers. *J. Dermatol. Sci.* **2008**, *52*, 193–204. [CrossRef] [PubMed]
47. Uchino, T.; Tokunaga, H.; Ando, M.; Utsumi, H. Quantitative determination of OH radical generation and its cytotoxicity induced by TiO2–UVA treatment. *Toxicology* **2002**, *16*, 629–635. [CrossRef]
48. Dunford, R.; Salinaro, A.; Cai, L.; Serpone, N.; Horikoshi, S.; Hidaka, H.; Knowland, J. Chemical oxidation and DNA damage catalysed by inorganic sunscreen ingredients. *FEBS Lett.* **1997**, *418*, 87–90. [CrossRef]
49. Cunningham, B.; Torres-Duarte, C.; Cherr, G.; Adams, N. Effects of three zinc-containing sunscreens on development of purple sea urchin (*Strongylocentrotus purpuratus*) embryos. *Aquat. Toxicol.* **2020**, *218*, 105355. [CrossRef]
50. Peck, A.M. Analytical methods for the determination of persistent ingredients of personal care products in environmental matrices. *Anal. Bioanal. Chem.* **2006**, *386*, 907–939. [CrossRef]
51. Valet, B.; Mayor, M.; Fitoussi, F.; Capellier, R.; Dormoy, M.; Ginestar, J. Colouring agents in cosmetic products (excluding hair dyes): Types of decorative cosmetic products. In *Analysis of Cosmetic Products*; Elsevier: Amsterdam, The Netherlands, 2007; pp. 141–152.
52. Quartier, S.; Garmyn, M.; Becart, S.; Goossens, A. Allergic contact dermatitis to copolymers in cosmetics–case report and review of the literature. *Contact Dermat.* **2006**, *55*, 257–267. [CrossRef]
53. Tønning, K.; Jacobsen, E.; Pedersen, E.; Strange, M.; Poulsen, P.B.; Møller, L.; Boyd, H.B. *Survey and Health Assessment of the exposure of 2 year-olds to chemical substances in Consumer Products*; Danish Ministry of the Environment, Environmental Protection Agency: Odense, Denmark, 2009; Chapter 5; p. 91.
54. Vecchiato, M.; Gregoris, E.; Barbaro, E.; Barbante, C.; Piazza, R.; Gambaro, A. Fragrances in the seawater of Terra Nova Bay, Antarctica. *Sci. Total Environ.* **2017**, *593*, 375–379. [CrossRef]

55. Kung, T.A.; Lee, S.H.; Yang, T.C.; Wang, W.H. Survey of selected personal care products in surface water of coral reefs in Kenting National Park, Taiwan. *Sci. Total Environ.* **2018**, *635*, 1302–1307. [CrossRef]
56. Zhao, X.; Qiu, W.; Zheng, Y.; Xiong, J.; Gao, C.; Hu, S. Occurrence, distribution, bioaccumulation, and ecological risk of bisphenol analogues, parabens and their metabolites in the Pearl River Estuary, South China. *Ecotoxicol. Environ. Saf.* **2019**, *180*, 43–52. [CrossRef]
57. Baek, S.; Joo, S.H.; Blackwelder, P.; Toborek, M. Effects of coating materials on antibacterial properties of industrial and sunscreen-derived titanium-dioxide nanoparticles on *Escherichia coli*. *Chemosphere* **2018**, *208*, 196–206. [CrossRef]
58. Othman, S.H.; Abdul Rashid, S.; Mohd Ghazi, T.I.; Abdullah, N. Dispersion and stabilization of photocatalytic TiO_2 nanoparticles in aqueous suspension for coatings applications. *J. Nanomater.* **2012**, *2012*, 718214. [CrossRef]
59. Tejamaya, M.; Römer, I.; Merrifield, R.C.; Lead, J.R. Stability of citrate, PVP, and PEG coated silver nanoparticles in ecotoxicology media. *Environ. Sci. Technol.* **2012**, *46*, 7011–7017. [CrossRef] [PubMed]
60. Peng, C.; Zhang, W.; Gao, H.; Li, Y.; Tong, X.; Li, K.; Zhu, X.; Wang, Y.; Chen, Y. Behavior and potential impacts of metal-based engineered nanoparticles in aquatic environments. *Nanomaterials* **2017**, *7*, 21. [CrossRef] [PubMed]
61. Jafry, H.R.; Liga, M.V.; Li, Q.; Barron, A.R. Simple route to enhanced photocatalytic activity of P25 titanium dioxide nanoparticles by silica addition. *Environ. Sci. Technol.* **2010**, *45*, 1563–1568. [CrossRef]
62. Dalai, S.; Pakrashi, S.; Kumar, R.S.; Chandrasekaran, N.; Mukherjee, A. A comparative cytotoxicity study of TiO_2 nanoparticles under light and dark conditions at low exposure concentrations. *Toxicol. Res.* **2012**, *1*, 116–130. [CrossRef]
63. Wang, J.; Fan, Y. Lung injury induced by TiO_2 nanoparticles depends on their structural features: Size, shape, crystal phases, and surface coating. *Int. J. Mol. Sci.* **2014**, *15*, 22258–22278. [CrossRef]
64. Planchon, M.; Ferrari, R.; Guyot, F.; Gélabert, A.; Menguy, N.; Chanéac, C.; Thill, A.; Benedetti, M.F.; Spalla, O. Interaction between Escherichia coli and TiO_2 nanoparticles in natural and artificial waters. *Colloids Surf. B Biointerfaces* **2013**, *102*, 158–164. [CrossRef]
65. Rosen, M.; Kunjappu, J. Adsorption of surface-active agents at interfaces: The electrical double layer. *Surfactants Interfacial Phenom.* **2004**, *4*, 98.
66. Raza, G.; Amjad, M.; Kaur, I.; Wen, D. Stability and aggregation kinetics of titania nanomaterials under environmentally realistic conditions. *Environ. Sci. Technol.* **2016**, *50*, 8462–8472. [CrossRef]
67. Waychunas, G.A.; Kim, C.S.; Banfield, J.F. Nanoparticulate iron oxide minerals in soils and sediments: Unique properties and contaminant scavenging mechanisms. *J. Nanoparticle Res.* **2005**, *7*, 409–433. [CrossRef]
68. Borm, P.; Klaessig, F.C.; Landry, T.D.; Moudgil, B.; Pauluhn, J.; Thomas, K.; Trottier, R.; Wood, S. Research strategies for safety evaluation of nanomaterials, part V: Role of dissolution in biological fate and effects of nanoscale particles. *Toxicol. Sci.* **2006**, *90*, 23–32. [CrossRef] [PubMed]
69. Bian, S.; Mudunkotuwa, I.A.; Rupasinghe, T.; Grassian, V.H. Aggregation and dissolution of 4 nm ZnO nanoparticles in aqueous environments: Influence of pH, ionic strength, size, and adsorption of humic acid. *Langmuir* **2011**, *27*, 6059–6068. [CrossRef] [PubMed]
70. Zhang, L.; Li, J.; Yang, K.; Liu, J.; Lin, D. Physicochemical transformation and algal toxicity of engineered nanoparticles in surface water samples. *Environ. Pollut.* **2016**, *211*, 132–140. [CrossRef]
71. Gilbert, B.; Huang, F.; Zhang, H.; Waychunas, G.A.; Banfield, J.F. Nanoparticles: Strained and stiff. *Science* **2004**, *305*, 651–654. [CrossRef] [PubMed]
72. Fairbairn, E.A.; Keller, A.A.; Mädler, L.; Zhou, D.; Pokhrel, S.; Cherr, G.N. Metal oxide nanomaterials in seawater: Linking physicochemical characteristics with biological response in sea urchin development. *J. Hazard. Mater.* **2011**, *192*, 1565–1571. [CrossRef] [PubMed]
73. Franklin, N.M.; Rogers, N.J.; Apte, S.C.; Batley, G.E.; Gadd, G.E.; Casey, P.S. Comparative toxicity of nanoparticulate ZnO, bulk ZnO, and $ZnCl_2$ to a freshwater microalga (*Pseudokirchneriella subcapitata*): The importance of particle solubility. *Environ. Sci. Technol.* **2007**, *41*, 8484–8490. [CrossRef]
74. Odzak, N.; Kistler, D.; Behra, R.; Sigg, L. Dissolution of metal and metal oxide nanoparticles in aqueous media. *Environ. Pollut.* **2014**, *191*, 132–138. [CrossRef]
75. Misra, S.K.; Nuseibeh, S.; Dybowska, A.; Berhanu, D.; Tetley, T.D.; Valsami-Jones, E. Comparative study using spheres, rods and spindle-shaped nanoplatelets on dispersion stability, dissolution and toxicity of CuO nanomaterials. *Nanotoxicology* **2014**, *8*, 422–432. [CrossRef]
76. Lee, J.; Kang, B.; Hicks, B.; Chancellor Jr, T.F.; Chu, B.H.; Wang, H.-T.; Keselowsky, B.G.; Ren, F.; Lele, T.P. The control of cell adhesion and viability by zinc oxide nanorods. *Biomaterials* **2008**, *29*, 3743–3749. [CrossRef]
77. Li, X.; Lenhart, J.J.; Walker, H.W. Aggregation kinetics and dissolution of coated silver nanoparticles. *Langmuir* **2012**, *28*, 1095–1104. [CrossRef]
78. Gelabert, A.; Sivry, Y.; Ferrari, R.; Akrout, A.; Cordier, L.; Nowak, S.; Menguy, N.; Benedetti, M.F. Uncoated and coated ZnO nanoparticle life cycle in synthetic seawater. *Environ. Toxicol. Chem.* **2014**, *33*, 341–349. [CrossRef] [PubMed]
79. Lowry, G.V.; Gregory, K.B.; Apte, S.C.; Lead, J.R. *Transformations of Nanomaterials in the Environment*; ACS Publications: Washington, DC, USA, 2012.
80. Odzak, N.; Kistler, D.; Sigg, L. Influence of daylight on the fate of silver and zinc oxide nanoparticles in natural aquatic environments. *Environ. Pollut.* **2017**, *226*, 1–11. [CrossRef] [PubMed]

81. Fel, J.-P.; Lacherez, C.; Bensetra, A.; Mezzache, S.; Béraud, E.; Léonard, M.; Allemand, D.; Ferrier-Pagès, C. Photochemical response of the scleractinian coral *Stylophora pistillata* to some sunscreen ingredients. *Coral Reefs* **2019**, *38*, 109–122. [CrossRef]
82. Rodil, R.; Moeder, M.; Altenburger, R.; Schmitt-Jansen, M. Photostability and phytotoxicity of selected sunscreen agents and their degradation mixtures in water. *Anal. Bioanal. Chem.* **2009**, *395*, 1513–1524. [CrossRef] [PubMed]
83. Paredes, E.; Perez, S.; Rodil, R.; Quintana, J.B.; Beiras, R. Ecotoxicological evaluation of four UV filters using marine organisms from different trophic levels *Isochrysis galbana, Mytilus galloprovincialis, Paracentrotus lividus*, and *Siriella armata*. *Chemosphere* **2014**, *104*, 44–50. [CrossRef] [PubMed]
84. Sureda, A.; Capo, X.; Busquets-Cortes, C.; Tejada, S. Acute exposure to sunscreen containing titanium induces an adaptive response and oxidative stress in *Mytillus galloprovincialis*. *Ecotoxicol. Environ. Saf.* **2018**, *149*, 58–63. [CrossRef]
85. Araujo, C.V.M.; Rodriguez-Romero, A.; Fernandez, M.; Sparaventi, E.; Medina, M.M.; Tovar-Sanchez, A. Repellency and mortality effects of sunscreens on the shrimp *Palaemon varians*: Toxicity dependent on exposure method. *Chemosphere* **2020**, *257*, 127190. [CrossRef] [PubMed]
86. Kim, J.-W.; Isobe, T.; Ramaswamy, B.R.; Chang, K.-H.; Amano, A.; Miller, T.M.; Siringan, F.P.; Tanabe, S. Contamination and bioaccumulation of benzotriazole ultraviolet stabilizers in fish from Manila Bay, the Philippines using an ultra-fast liquid chromatography-tandem mass spectrometry. *Chemosphere* **2011**, *85*, 751–758. [CrossRef]
87. Nakata, H.; Murata, S.; Filatreau, J. Occurrence and concentrations of benzotriazole UV stabilizers in marine organisms and sediments from the Ariake Sea, Japan. *Environ. Sci. Technol.* **2009**, *43*, 6920–6926. [CrossRef]
88. Fent, K.; Chew, G.; Li, J.; Gomez, E. Benzotriazole UV-stabilizers and benzotriazole: Antiandrogenic activity in vitro and activation of aryl hydrocarbon receptor pathway in zebrafish eleuthero-embryos. *Sci. Total Environ.* **2014**, *482*, 125–136. [CrossRef]
89. Kim, J.-W.; Chang, K.-H.; Isobe, T.; Tanabe, S. Acute toxicity of benzotriazole ultraviolet stabilizers on freshwater crustacean (*Daphnia pulex*). *J. Toxicol. Sci.* **2011**, *36*, 247–251. [CrossRef] [PubMed]
90. Rodríguez-Romero, A.; Ruiz-Gutiérrez, G.; Viguri, J.R.; Tovar-Sánchez, A. Sunscreens as a New Source of Metals and Nutrients to Coastal Waters. *Environ. Sci. Technol.* **2019**, *53*, 10177–10187. [CrossRef] [PubMed]
91. Zmozinski, A.V.; Pretto, T.; Borges, A.R.; Duarte, A.T.; Vale, M.G.R. Determination of Pb and Cr in sunscreen samples by high-resolution continuum source graphite furnace atomic absorption spectrometry and direct analysis. *Microchem. J.* **2016**, *128*, 89–94. [CrossRef]
92. Zachariadis, G.; Sahanidou, E. Multi-element method for determination of trace elements in sunscreens by ICP-AES. *J. Pharm. Biomed. Anal.* **2009**, *50*, 342–348. [CrossRef]
93. Labille, J.; Feng, J.; Botta, C.; Borschneck, D.; Sammut, M.; Cabie, M.; Auffan, M.; Rose, J.; Bottero, J.-Y. Aging of TiO_2 nanocomposites used in sunscreen. Dispersion and fate of the degradation products in aqueous environment. *Environ. Pollut.* **2010**, *158*, 3482–3489. [CrossRef]
94. Auffan, M.; Pedeutour, M.; Rose, J.; Masion, A.; Ziarelli, F.; Borschneck, D.; Chaneac, C.; Botta, C.; Chaurand, P.; Labille, J. Structural degradation at the surface of a TiO_2-based nanomaterial used in cosmetics. *Environ. Sci. Technol.* **2010**, *44*, 2689–2694. [CrossRef]
95. Chen, T.H.; Lin, C.C.; Meng, P.J. Zinc oxide nanoparticles alter hatching and larval locomotor activity in zebrafish (*Danio rerio*). *J. Hazard. Mater.* **2014**, *277*, 134–140. [CrossRef]
96. Zhu, X.; Zhou, J.; Cai, Z. TiO_2 nanoparticles in the marine environment: Impact on the toxicity of tributyltin to abalone (*Haliotis diversicolor supertexta*) embryos. *Environ. Sci. Technol.* **2011**, *45*, 3753–3758. [CrossRef]
97. Sharma, V.; Anderson, D.; Dhawan, A. Zinc oxide nanoparticles induce oxidative DNA damage and ROS-triggered mitochondria mediated apoptosis in human liver cells (HepG2). *Apoptosis* **2012**, *17*, 852–870. [CrossRef]
98. Zhu, X.; Zhou, J.; Cai, Z. The toxicity and oxidative stress of TiO_2 nanoparticles in marine abalone (*Haliotis diversicolor supertexta*). *Mar. Pollut. Bull.* **2011**, *63*, 334–338. [CrossRef]
99. Krause, M.; Klit, A.; Blomberg Jensen, M.; Søeborg, T.; Frederiksen, H.; Schlumpf, M.; Lichtensteiger, W.; Skakkebaek, N.; Drzewiecki, K. Sunscreens: Are they beneficial for health? An overview of endocrine disrupting properties of UV-filters. *Int. J. Androl.* **2012**, *35*, 424–436. [CrossRef]
100. Zhang, Q.; Ma, X.; Dzakpasu, M.; Wang, X.C. Evaluation of ecotoxicological effects of benzophenone UV filters: Luminescent bacteria toxicity, genotoxicity and hormonal activity. *Ecotoxicol. Environ. Saf.* **2017**, *142*, 338–347. [CrossRef] [PubMed]
101. Kaiser, D.; Sieratowicz, A.; Zielke, H.; Oetken, M.; Hollert, H.; Oehlmann, J. Ecotoxicological effect characterisation of widely used organic UV filters. *Environ. Pollut.* **2012**, *163*, 84–90. [CrossRef] [PubMed]
102. Baker, T.J.; Tyler, C.R.; Galloway, T.S. Impacts of metal and metal oxide nanoparticles on marine organisms. *Environ. Pollut.* **2014**, *186*, 257–271. [CrossRef] [PubMed]
103. Spisni, E.; Seo, S.; Joo, S.H.; Su, C. Release and toxicity comparison between industrial- and sunscreen-derived nano-ZnO particles. *Int. J. Environ. Sci. Technol.* **2016**, *13*, 2485–2494. [CrossRef]
104. Sendra, M.; Sanchez-Quiles, D.; Blasco, J.; Moreno-Garrido, I.; Lubian, L.M.; Perez-Garcia, S.; Tovar-Sanchez, A. Effects of TiO_2 nanoparticles and sunscreens on coastal marine microalgae: Ultraviolet radiation is key variable for toxicity assessment. *Environ. Int.* **2017**, *98*, 62–68. [CrossRef]
105. Corinaldesi, C.; Damiani, E.; Marcellini, F.; Falugi, C.; Tiano, L.; Brugè, F.; Danovaro, R. Sunscreen products impair the early developmental stages of the sea urchin *Paracentrotus lividus*. *Sci. Rep.* **2017**, *7*, 7815. [CrossRef]
106. Contier, S.M.D. *Nanotechnology: Basic Calculations for Engineers and Scientists*; Wiley-Interscience: Hoboken, NJ, USA, 2006; p. 52.

107. Chaudhuri, R.K.; Majewski, G. Amphiphilic microfine titanium dioxide: Its properties and application in sunscreen formulations. *DCI* **1998**, *162*, 24–31.
108. Hundrinke, K.; Simon, M. Ecotoxic effect of photocatalytic active nanoparticles (TiO_2) on algae and daphnids. *Environ. Sci. Pollut. Res.* **2006**, *13*, 225–232. [CrossRef]
109. Navarro, E.; Baun, A.; Behra, R.; Hartmann, N.B.; Filser, J.; Miao, A.-J.; Quigg, A.; Santschi, P.H.; Sigg, L. Environmental behavior and ecotoxicity of engineered nanoparticles to algae, plants, and fungi. *Ecotoxicology* **2008**, *17*, 372–386. [CrossRef]
110. Perron, H.; Domain, C.; Roques, J.; Drot, R.; Simoni, E.; Catalette, H. Optimisation of accurate rutile TiO_2 (110),(100),(101) and (001) surface models from periodic DFT calculations. *Theor. Chem. Acc.* **2007**, *117*, 565–574. [CrossRef]
111. Hotze, E.M.; Phenrat, T.; Lowry, G.V. Nanoparticle aggregation: Challenges to understanding transport and reactivity in the environment. *J. Environ. Qual.* **2010**, *39*, 1909–1924. [CrossRef] [PubMed]
112. George, S.; Pokhrel, S.; Xia, T.; Gilbert, B.; Ji, Z.; Schowalter, M.; Rosenauer, A.; Damoiseaux, R.; Bradley, K.A.; Maedler, L.; et al. Use of a Rapid Cytotoxicity Screening Approach To Engineer a Safer Zinc Oxide Nanoparticle through Iron Doping. *Acs Nano* **2010**, *4*, 15–29. [CrossRef] [PubMed]
113. Jiang, W.; Mashayekhi, H.; Xing, B. Bacterial toxicity comparison between nano-and micro-scaled oxide particles. *Environ. Pollut.* **2009**, *157*, 1619–1625. [CrossRef] [PubMed]
114. Li, B.; Logan, B.E. Bacterial adhesion to glass and metal-oxide surfaces. *Colloids Surf. B Biointerfaces* **2004**, *36*, 81–90. [CrossRef]
115. Matranga, V.; Corsi, I. Toxic effects of engineered nanoparticles in the marine environment: Model organisms and molecular approaches. *Mar. Environ. Res.* **2012**, *76*, 32–40. [CrossRef]
116. Peng, X.; Palma, S.; Fisher, N.S.; Wong, S.S. Effect of morphology of ZnO nanostructures on their toxicity to marine algae. *Aquat. Toxicol.* **2011**, *102*, 186–196. [CrossRef]
117. Santos, A.J.M.; Miranda, M.S.; da Silva, J.C.E. The degradation products of UV filters in aqueous and chlorinated aqueous solutions. *Water Res.* **2012**, *46*, 3167–3176. [CrossRef]
118. Manzo, S.; Schiavo, S.; Oliviero, M.; Toscano, A.; Ciaravolo, M.; Cirino, P. Immune and reproductive system impairment in adult sea urchin exposed to nanosized ZnO via food. *Sci. Total Environ.* **2017**, *599–600*, 9–13. [CrossRef]
119. Osmond, M.J.; Mccall, M.J. Zinc oxide nanoparticles in modern sunscreens: An analysis of potential exposure and hazard. *Nanotoxicology* **2010**, *4*, 15–41. [CrossRef]
120. Miller, R.J.; Lenihan, H.S.; Muller, E.B.; Tseng, N.; Hanna, S.K.; Keller, A.A. Impacts of metal oxide nanoparticles on marine phytoplankton. *Environ. Sci. Technol.* **2010**, *44*, 7329–7334. [CrossRef] [PubMed]
121. Wang, Z.; Xia, B.; Chen, B.; Sun, X.; Zhu, L.; Zhao, J.; Du, P.; Xing, B. Trophic transfer of TiO 2 nanoparticles from marine microalga (*Nitzschia closterium*) to scallop (*Chlamys farreri*) and related toxicity. *Environ. Sci. Nano* **2017**, *4*, 415–424. [CrossRef]
122. Zhu, X.; Wang, J.; Zhang, X.; Chang, Y.; Chen, Y. Trophic transfer of TiO_2 nanoparticles from daphnia to zebrafish in a simplified freshwater food chain. *Chemosphere* **2010**, *79*, 928–933. [CrossRef] [PubMed]
123. Fent, K.; Kunz, P.Y.; Gomez, E. UV Filters in the Aquatic Environment Induce Hormonal Effects and Affect Fertility and Reproduction in Fish. *Chim. Int. J. Chem.* **2008**, *62*, 368–375. [CrossRef]
124. Gomez, E.; Bachelot, M.; Boillot, C.; Munaron, D.; Chiron, S.; Casellas, C.; Fenet, H. Bioconcentration of Two Pharmaceuticals (Benzodiazepines) and Two Personal Care Products (UV Filters) in Marine Mussels (*Mytilus galloprovincialis*) under Controlled Laboratory Conditions. *Environ. Sci. Pollut. Res.* **2012**, *19*, 2561–2569. [CrossRef]
125. Gonzalez, L.; Lison, D.; Kirschvolders, M. Genotoxicity of engineered nanomaterials: A critical review. *Nanotoxicology* **2008**, *2*, 252–273. [CrossRef]
126. Magdolenova, Z.; Collins, A.; Kumar, A.; Dhawan, A.; Stone, V.; Dusinska, M. Mechanisms of genotoxicity. A review of in vitro and in vivo studies with engineered nanoparticles. *Nanotoxicology* **2014**, *8*, 233–278. [CrossRef]
127. Wang, Y.; Zhu, X.; Lao, Y.; Lv, X.; Tao, Y.; Huang, B.; Wang, J.; Zhou, J.; Cai, Z. TiO_2 nanoparticles in the marine environment: Physical effects responsible for the toxicity on algae Phaeodactylum tricornutum. *Sci. Total Environ.* **2016**, *565*, 818–826. [CrossRef]
128. Miao, A.-J.; Luo, Z.; Chen, C.-S.; Chin, W.-C.; Santschi, P.H.; Quigg, A. Intracellular uptake: A possible mechanism for silver engineered nanoparticle toxicity to a freshwater alga *Ochromonas danica*. *PLoS ONE* **2010**, *5*, e15196. [CrossRef]
129. Schwegmann, H.; Feitz, A.J.; Frimmel, F.H. Influence of the zeta potential on the sorption and toxicity of iron oxide nanoparticles on *S. cerevisiae* and *E. coli*. *J. Colloid Interface Sci.* **2010**, *347*, 43–48. [CrossRef]
130. Zhu, X.; Chang, Y.; Chen, Y. Toxicity and bioaccumulation of TiO_2 nanoparticle aggregates in *Daphnia magna*. *Chemosphere* **2010**, *78*, 209–215. [CrossRef]
131. Chen, J.; Dong, X.; Xin, Y.; Zhao, M. Effects of titanium dioxide nano-particles on growth and some histological parameters of zebrafish (*Danio rerio*) after a long-term exposure. *Aquat. Toxicol.* **2011**, *101*, 493–499. [CrossRef] [PubMed]
132. Doyle, J.J.; Ward, J.E.; Mason, R. Exposure of bivalve shellfish to titania nanoparticles under an environmental-spill scenario: Encounter, ingestion and egestion. *J. Mar. Biol. Assoc.* **2016**, *96*, 137–149. [CrossRef]
133. Montes, M.O.; Hanna, S.K.; Lenihan, H.S.; Keller, A.A. Uptake, accumulation, and biotransformation of metal oxide nanoparticles by a marine suspension-feeder. *J. Hazard. Mater.* **2012**, *225*, 139–145. [CrossRef] [PubMed]
134. Lv, X.; Yang, Y.; Tao, Y.; Jiang, Y.; Chen, B.; Zhu, X.; Cai, Z.; Li, B. A mechanism study on toxicity of graphene oxide to *Daphnia magna*: Direct link between bioaccumulation and oxidative stress. *Environ. Pollut.* **2018**, *234*, 953–959. [CrossRef] [PubMed]

135. Tian, S.; Zhang, Y.; Song, C.; Zhu, X.; Xing, B. Bioaccumulation and biotransformation of polybrominated diphenyl ethers in the marine bivalve (*Scapharca subcrenata*): Influence of titanium dioxide nanoparticles. *Mar. Pollut. Bull.* **2015**, *90*, 48–53. [CrossRef] [PubMed]
136. Schilling, K.; Bradford, B.; Castelli, D.; Dufour, E.; Nash, J.F.; Pape, W.; Schulte, S.; Tooley, I.; van den Bosch, J.; Schellauf, F. Human safety review of "nano" titanium dioxide and zinc oxide. *Photochem. Photobiol. Sci.* **2010**, *9*, 495–509. [CrossRef] [PubMed]
137. Senzui, M.; Tamura, T.; Miura, K.; Ikarashi, Y.; Watanabe, Y.; Fujii, M. Study on penetration of titanium dioxide (TiO_2) nanoparticles into intact and damaged skin in vitro. *J. Toxicol. Sci.* **2010**, *35*, 107–113. [CrossRef]
138. Dussert, A.S.; Gooris, E.; Hemmerle, J. Characterization of the mineral content of a physical sunscreen emulsion and its distribution onto human stratum corneum. *Int. J. Cosmet. Sci.* **1997**, *19*, 119–129. [CrossRef]
139. Zvyagin, A.V.; Zhao, X.; Gierden, A.; Sanchez, W.; Ross, J.; Roberts, M.S. Imaging of zinc oxide nanoparticle penetration in human skin in vitro and in vivo. *J. Biomed. Opt.* **2008**, *13*, 064031. [CrossRef]
140. Clemente, Z.; Castro, V.; Feitosa, L.; Lima, R.; Jonsson, C.; Maia, A.; Fraceto, L. Fish exposure to nano-TiO_2 under different experimental conditions: Methodological aspects for nanoecotoxicology investigations. *Sci. Total Environ.* **2013**, *463*, 647–656. [CrossRef] [PubMed]
141. Clemente, Z.; Castro, V.; Moura, M.; Jonsson, C.; Fraceto, L. Toxicity assessment of TiO_2 nanoparticles in zebrafish embryos under different exposure conditions. *Aquat. Toxicol.* **2014**, *147*, 129–139. [CrossRef] [PubMed]
142. Lee, W.-M.; An, Y.-J. Effects of zinc oxide and titanium dioxide nanoparticles on green algae under visible, UVA, and UVB irradiations: No evidence of enhanced algal toxicity under UV pre-irradiation. *Chemosphere* **2013**, *91*, 536–544. [CrossRef]
143. Ji, J.; Long, Z.; Lin, D. Toxicity of oxide nanoparticles to the green algae *Chlorella* sp. *Chem. Eng. J.* **2011**, *170*, 525–530. [CrossRef]
144. Hanson, K.M.; Gratton, E.; Bardeen, C.J. Sunscreen enhancement of UV-induced reactive oxygen species in the skin. *Free. Radic. Biol. Med.* **2006**, *41*, 1205–1212. [CrossRef] [PubMed]
145. Tsuzuki, T.; He, R.; Wang, J.; Sun, L.; Wang, X.; Hocking, R. Reduction of the photocatalytic activity of ZnO nanoparticles for UV protection applications. *Int. J. Nanotechnol.* **2012**, *9*, 1017–1029. [CrossRef]
146. Livraghi, S.; Corazzari, I.; Paganini, M.C.; Ceccone, G.; Giamello, E.; Fubini, B.; Fenoglio, I. Decreasing the oxidative potential of TiO_2 nanoparticles through modification of the surface with carbon: A new strategy for the production of safe UV filters. *Chem. Commun.* **2010**, *46*, 8478–8480. [CrossRef] [PubMed]
147. Miao, A.J.; Zhang, X.Y.; Luo, Z.; Chen, C.S.; Chin, W.C.; Santschi, P.H.; Quigg, A. Zinc oxide–engineered nanoparticles: Dissolution and toxicity to marine phytoplankton. *Environ. Toxicol. Chem.* **2010**, *29*, 2814–2822. [CrossRef]
148. Tang, C.H.; Lin, C.Y.; Lee, S.H.; Wang, W.H. Membrane lipid profiles of coral responded to zinc oxide nanoparticle-induced perturbations on the cellular membrane. *Aquat. Toxicol.* **2017**, *187*, 72–81. [CrossRef]
149. Smijs, T.G.; Pavel, S. Titanium dioxide and zinc oxide nanoparticles in sunscreens: Focus on their safety and effectiveness. *Nanotechnol. Sci. Appl.* **2011**, *4*, 95. [CrossRef]
150. Jeon, S.-k.; Kim, E.-j.; Lee, J.; Lee, S. Potential risks of TiO_2 and ZnO nanoparticles released from sunscreens into outdoor swimming pools. *J. Hazard. Mater.* **2016**, *317*, 312–318. [CrossRef]
151. Batley, G.E.; Kirby, J.K.; McLaughlin, M.J. Fate and risks of nanomaterials in aquatic and terrestrial environments. *Acc. Chem. Res.* **2012**, *46*, 854–862. [CrossRef] [PubMed]
152. Tian, S.; Zhang, Y.; Song, C.; Zhu, X.; Xing, B. Titanium dioxide nanoparticles as carrier facilitate bioaccumulation of phenanthrene in marine bivalve, ark shell (*Scapharca subcrenata*). *Environ. Pollut.* **2014**, *192*, 59–64. [CrossRef] [PubMed]
153. Lu, J.; Tian, S.; Lv, X.; Chen, Z.; Chen, B.; Zhu, X.; Cai, Z. TiO_2 nanoparticles in the marine environment: Impact on the toxicity of phenanthrene and Cd2+ to marine zooplankton *Artemia salina*. *Sci. Total Environ.* **2018**, *615*, 375–380. [CrossRef] [PubMed]
154. Hu, X.; Zhou, Q. Health and ecosystem risks of graphene. *Chem. Rev.* **2013**, *113*, 3815–3835. [CrossRef] [PubMed]

nanomaterials

Article

Enhanced Bioaccumulation and Toxicity of Arsenic in Marine Mussel *Perna viridis* in the Presence of CuO/Fe_3O_4 Nanoparticles

Shuang Zhou [1,2], Wei Qian [1,3], Zigong Ning [4] and Xiaoshan Zhu [1,5,*]

1 Shenzhen International Graduate School, Tsinghua University, Shenzhen 518055, China; mzhou0327@163.com (S.Z.); jokloy@126.com (W.Q.)
2 Shenzhen Honglue Research Institute of Innovation Management, Shenzhen 518119, China
3 School of Environment, Tsinghua University, Beijing 100084, China
4 School of Civil and Environmental Engineering, Harbin Institute of Technology, Shenzhen 518055, China; ningzigong@hit.edu.cn
5 Southern Laboratory of Ocean Science and Engineering, Zhuhai 519000, China
* Correspondence: zhu.xiaoshan@sz.tsinghua.edu.cn

Abstract: Leakage of metal oxide nanoparticles (MNPs) into marine environments is inevitable with the increasing use of MNPs. However, little is known about the effects of these lately emerged MNPs on the bioaccumulation and toxicity of pre-existing contaminants in marine biota. The current study therefore investigated the effects of two common MNPs, CuO nanoparticles (nCuO) and Fe_3O_4 nanoparticles (nFe_3O_4), on bioaccumulation and toxicity of arsenic (As) in green mussel *Perna viridis*. Newly introduced MNPs remarkably promoted the accumulation of As and disrupted the As distribution in mussels because of the strong adsorption of As onto MNPs. Moreover, MNPs enhanced the toxicity of As by disturbing osmoregulation in mussels, which could be supported by decreased activity of Na^+-K^+-ATPase and average weight loss of mussels after MNPs exposure. In addition, the enhanced toxicity of As in mussels might be due to that MNPs reduced the biotransformation efficiency of more toxic inorganic As to less toxic organic As, showing an inhibitory effect on As detoxifying process of mussels. This could be further demonstrated by the overproduction of reactive oxygen species (ROS), as implied by the rise in quantities of superoxide dismutase (SOD) and lipid peroxidation (LPO), and subsequently restraining the glutathione-S-transferases (GST) activity and glutathione (GSH) content in mussels. Taken together, this study elucidated that MNPs may elevate As bioaccumulation and limit As biotransformation in mussels, which would result in an enhanced ecotoxicity of As towards marine organisms.

Keywords: As; metal oxide NPs; detoxification; biotransformation; *Perna virids*

Citation: Zhou, S.; Qian, W.; Ning, Z.; Zhu, X. Enhanced Bioaccumulation and Toxicity of Arsenic in Marine Mussel *Perna viridis* in the Presence of CuO/Fe_3O_4 Nanoparticles. *Nanomaterials* **2021**, *11*, 2769. https://doi.org/10.3390/nano11102769

Academic Editor: Alberto Villa

Received: 29 August 2021
Accepted: 12 October 2021
Published: 19 October 2021

1. Introduction

The commercial use of metal oxide nanoparticles (MNPs) has increased drastically over the decades in various industrial applications such as biochemical coatings, drug delivery, magnetic resonance imaging and catalysts, as well as in controlling environmental pollution [1–6]. Among these MNPs, CuO nanoparticles (nCuO) and Fe_3O_4 nanoparticles (nFe_3O_4) are importantly and commercially used nanomaterials, due to their unique properties such as large surface area, high surface reactivity and excellent affinity for heavy metals [7]. Nevertheless, the extensive use of nCuO and nFe_3O_4 in a variety of consumer products has resulted in their release into aquatic environment [8]. Unintentionally released nCuO and nFe_3O_4 may co-occur with pre-existing contaminants in aquatic environment, which would eventually alter the contaminants' environmental behaviors, fate and toxicity to the ecosystem and even human health [9].

Arsenic (As) has been ubiquitously found in coastal regions, with a background concentration of up to dozens of μg/L due to anthropogenic processes (such as industrial, agricultural and mining effluents) [10,11]. As can be accumulated by marine organisms [12] and cause adverse biochemical and physiological effects such as immune disorders, reduced reproduction and growth, cell and tissue damage, and cell death [13]. However, there has been little information about the effect of newly emerged nCuO and nFe_3O_4 on the pre-existing As bioaccumulation and toxicity in marine organisms. Previous studies have focused on the ecotoxicity of target MNPs (including bacteria, protozoa, water flea, fish, bivalves and so on) rather than their effect on co-existing contaminants [14–17]. Interestingly, recent research found that TiO_2 nanoparticles could inhibit biotransformation of inorganic As to organic As in the mussel *Perna viridis* [18], which confirmed that the MNPs could affect the toxicity of co-existing contaminants in aquatic organisms. However, it is still too little evidence to make comprehensive conclusions on the information about how the more recently introduced MNPs affect the behavior and toxicity of pre-existing contaminants in marine environments. More research is needed in this field to better assess the ecological risk of MNPs and As in marine environments.

Therefore, the purpose of this research is to investigate the effects of two typical MNPs (nCuO and nFe_3O_4) on the biotransformation and detoxification of As in the green mussel *Perna viridis* (*P. viridis*), following a series of long-term waterborne As exposures and As MNPs co-exposures. *P. viridis* is one of the most widely distributed bivalve species in marine environments. They are proficient at taking up suspended particles and accumulating contaminants under a wide range of environmental conditions [19,20] thus making them a good biomonitor and a key species for assessing the ecotoxicity of MNPs [21]. Here, we designed an aquaculture system to mimic the real exposure in marine environments where As exist chronically, and *P. viridis* already is acclimated to it. Then, the MNPs were considered as the new contaminant emerging in this system, hence providing opportunity to study their contributions to the ecotoxicity of As in *P. viridis*. We consequently analyzed As bioaccumulation and distribution, measured As speciation and As biotransformation-related biomarkers in order to facilitate the interpretation of the underlying mechanisms. We hypothesise that the lately introduced MNPs would enhance bioaccumulation and toxicity of As in marine mussel *P. viridis*. Overall, our findings would provide useful information for assaying the ecological risks of MNPs and As.

2. Materials and Methods

2.1. Chemicals and Nanoparticles Characterization

nCuO and nFe_3O_4 (<10 nm, purity ≥99.5%) stock suspensions were prepared in ultrapure water, sonicated (50 W, 40 kHz, KQ2200, Kun Shan Ultrasonic Instruments Co., Ltd, KunShan, China) for 30 min to reach a concentration of 1 mg/L, respectively. The MNPs were characterized prior to the toxicity tests. Specifically, the morphology of MNPs were analyzed at a dilution of 1 mg/L by transmission electron microscopy (TEM, Tecnai G2 Spirit, FEI, Hillsboro, OR, USA) and the particle sizes of MNPs during 24 h in seawater were determined at 20 °C by dynamic light scattering (DLS) with a zeta potential analyzer (ZetaPALS, Brookhaven, Holtsville, NY, USA), more details on the characterization methods are described in Gomes et al. [3]. $Na_2HAsO_4{\cdot}7H_2O$ was purchased from Sigma (Saint Louis, MO, USA), and the 1000 mg/L As(V) stock solution was made by dissolving it in artificial seawater.

2.2. Experimental Design

Green mussels *P. viridis* (n = 72, 6 ± 1.1 cm) were collected South of Guangdong Province (114°64′ E, 22°46′ N) and acclimated for seven days in artificial seawater at a constant temperature with aeration. After acclimation, half of the mussels were placed in 50 µg/L As(V) exposure media in a triplicate design, along with a control group kept in artificial seawater, for a period of 21 days. Water was completely changed every day with redosing after each change. Mussels were collected from control, 50 µg/L As(V) in the beginning of the experiment and after 1, 3, 7, 10, 14, and 21 days of exposure. After sampling, the mussels were washed, wet-weighed and stored at −80 °C for further use.

After 21 days of single As(V) exposure, co-exposure with MNPs were prepared by introducing 1 mg/L MNPs to control groups and 50 μg/L As(V) exposure groups (control groups: control + nCuO and control + nFe_3O_4, 50 μg/L As(V) exposure groups: As(V) + nCuO and As(V) + nFe_3O_4), for a period of 14 days, each co-exposure had three independent replicates. The concentration of MNPs selected was environmentally relevant. Three mussels were collected from control + nCuO, control + nFe_3O_4, As(V) + nCuO and As(V) + nFe_3O_4 in the beginning of the experiment and after 1, 3, 7, 10 and 14 days of exposure. After sampling, the mussels were washed, wet-weighed, dissected, and stored at −80 °C for further use. More details of the experiment design can be found in the Supplementary Materials.

2.3. Analysis of As in Mussels

The total As content in mussel was analyzed according to previously used methods [22] with minor modifications; microwave digestion was used to treat mussels in this study. The As concentration was measured by using an inductively coupled plasma mass spectrometer (Thermo Scientific™ ICAP-Q, ICP-MS, Thermo Fisher Scientific, Waltham, MA, USA). The As species were extracted by using a two-step sequential extraction as described previously [23]. The concentration of As in mussels with different species (inorganic As(III) and As(V), organic monomethylarsonic acid (MMA), dimethylarsinic acid (DMA), and arsenobetaine (AsB)) was determined by using a Thermo Scientific IC5000 ion chromatography system combined with a Thermo Scientific™ ICAP-Q, ICP-MS (IC-ICP-MS) [18].

2.4. As Bioconcentration Factor (BCF)

In the present study, we used BCF (which expresses the accumulation of a chemical substance directly from water through the gill apparatus and shells) to determine the accumulation of As in mussels based on the As concentration in the dry tissues of mussels. Specifically, BCF (L/kg, dw) was calculated as: BCF = As concentration in tissues (μg/g)/As concentration in culture water (mg/L).

2.5. As(V) Adsorption to MNPs in Seawater

Kinetic adsorption of As(V) onto MNPs were conducted in artificial seawater, using the same concentration as the exposure condition (50 μg/L As(V), 1 mg/L nCuO/nFe_3O_4). The details are provided in the Supplementary Materials.

2.6. Biomarkers Determination

Dried mussel tissues (50 mg) were homogenized with a 0.86% NaCl solution by using a tissue homogenizer. The supernatant was collected to assess the biomarkers after the homogenate of mussels' tissues was centrifuged at 3500 rpm at 4 °C for 10 min. In this study, Na^+-K^+-ATPase (NKA), Superoxide dismutase (SOD) activity, lipid peroxidation (LPO) levels, reduced glutathione content (GSH) and glutathione-S-transferases (GST) in mussels' tissues were chosen as the biomarkers, which were measured spectrophotometrically using the commercial kits (Nanjing Jiancheng Bioengineering Institute, Nanjing, China) on the basis of the manufacturers' protocols. Specifically, the SOD activity was measured with a spectrophotometer (Milton Roy Spec20, Milton Roy Co., Rochester, NY, USA) at 550 nm, the LPO levels of different groups were detected using a thiobarbituric acid reactive substances (TBARS) assay, by measuring the amount of MDA-thiobarbituric acid (TBA) complex at 535 nm. In addition, the GSH content was estimated by the 5,5′ dithiobis-(2-nitrobenzoic acid) (DTNB)-glutathione reductase coupled assay at 420 nm, the GST was determined spectrophotometrically using commercially available GST activity kits based upon the GST-catalyzed reaction between glutathione (GSH) and the GST substrate 1-chloro-2,4-dinitrobenzene (CDNB) at 412 nm. The NKA activity was measured spectrophotometrically using NKA assay kit at 636 nm.

2.7. Statistical Analysis

SPSS version 19.0 (IBM, Armonk, NY, USA) was used to perform statistical analysis on the obtained data. As the data was normally distributed and the variances were homogenous, the differences within treated groups were evaluated by one-way analysis of variance (ANOVA) with Tukey's posthoc tests. A probability level (p-value) of less than 0.05 was regarded as statistically significant.

3. Results and Discussion

3.1. MNPs Characterization

The morphology, size and distribution of nCuO and nFe_3O_4 were obtained by TEM analysis and DLS analysis. nCuO are spherical in shape with a mean size of 10 ± 3 nm. nFe_3O_4 are mainly spherical in shape and not strongly aggregated (Figure 1A,B). The size of both nCuO and nFe_3O_4 (<10 nm) reported by the manufacturer is broadly in agreement with the size obtained by TEM. In addition, we also determined the mean particle size by using DLS, both nCuO and nFe_3O_4 aggregated immediately when they were introduced to artificial seawater, and their sizes kept increasing during the 24 h exposure period (Figure 1C). Moreover, high polydispersity indexes were observed for nCuO (polydispersity index between 0.21 and 0.53) and nFe_3O_4 (polydispersity index between 0.22 and 0.56), suggesting that under the exposure conditions, both MNPs tendency to aggregate produces suspensions with the presence of both single particles and large aggregates, with size ranging from 10 to 414 nm for nCuO and 40 to 679 nm for nFe_3O_4. Several reports have shown the tendency of MNPs to form aggregates while in suspension by using the same particles [21,24,25].

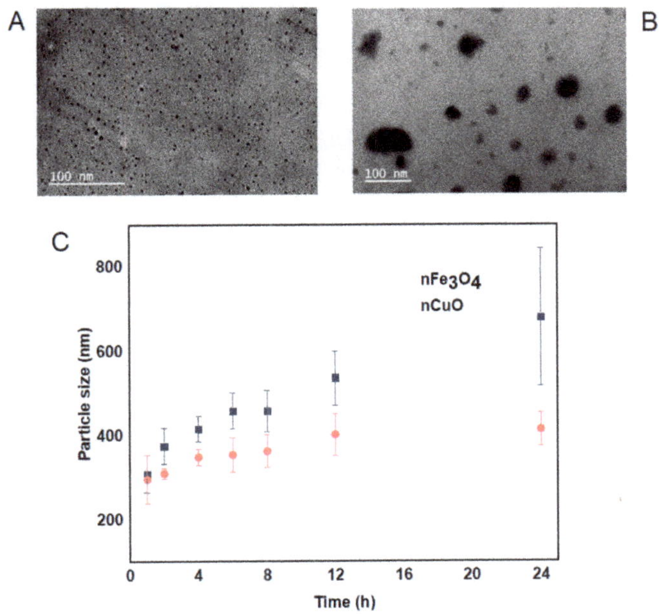

Figure 1. Transmission electron microscopic image of nCuO (**A**) and nFe_3O_4 (**B**). Particle size distribution (nm) during a 24 h time periods by dynamic light scattering for nFe_3O_4 and nCuO (**C**).

3.2. Exposure to MNPs Increased As Bioaccumulation and Altered As Distribution in Mussels

In the current study, *P. viridis* were firstly acclimated to the artificial seawater condition (as control mussels) and 50 µg/L As(V) exposure condition. Later, MNPs were introduced to the culture system in order to mimic the contaminated natural environment where As has pre-existed and MNPs were lately introduced as a new contaminant. The single As(V) exposure experiment showed that after 21 days exposure, no mortality was observed in both 50 μg/L As(V) exposure condition and control exposure condition, which confirmed that As could be detoxified in marine mussels [26]. Moreover, total body As concentrations in control mussels remained stable after 21 days exposure. However, the total body As concentration in mussels exposed to 50 μg/L As(V) after single As(V) exposure was significantly higher (10.14 ± 0.71 μg/g dw) in comparison to the ones in control mussels (6.26 ± 0.09 μg/g dw) (Figure 2A). These results validated that As can be accumulated and retained inside in mussels when mussels exposed to it through seawater, similar to the previous study in marine medaka (*Oryzias melastigma* and *Oryzias latipes*) [27,28].

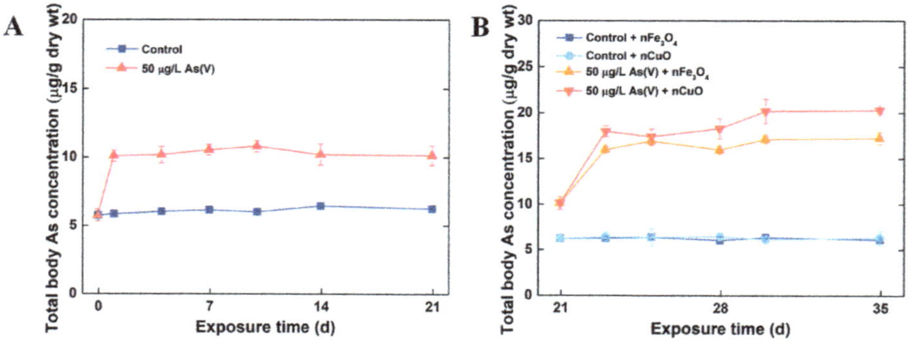

Figure 2. Dynamics of total body As concentrations in mussels without MNPs (**A**) and with MNPs (**B**). Data are mean ± SD (n = 4).

After 21 days single As(V) exposure, co-exposure with MNPs were prepared by introducing 1 mg/L MNPs to control groups and 50 μg/L As(V) exposure groups (control groups: control + nCuO and control + nFe_3O_4, 50 μg/L As(V) exposure groups: As(V) + nCuO and As(V) + nFe_3O_4) for a period of 14 days. It was apparent that both nCuO and nFe_3O_4 did not change total body As concentrations in control + nCuO and control + nFe_3O_4 mussels. By contrast, As accumulation in As(V) + MNPs co-exposure mussels was increased (Figure 2B). Specifically, after co-exposure to As(V) + nFe_3O_4 for 14 days, the total body of As concentration in mussel increased from a concentration of 10.14 ± 0.71 μg/g dw to that of 17.23 ± 0.67 μg/g dw. Likewise, total body As concentration in mussel reached a new equilibrium (20.3 ± 0.36 μg/g dw) compared to a previous one (10.14 ± 0.71 μg/g dw) after co-exposure to As(V) + nCuO for 14 days (Figure 2B). Moreover, as the bioaccumulation potential of As by mussel can be measured by BCF value, the BCF for As(V) + nCuO and As(V) + nFe_3O_4 were calculated to be 406 L/kg dw and 344.6 L/kg dw at new equilibrium, respectively, which were remarkably higher than that (202.8 L/kg dw) in single As(V) exposure. These results implied that both nCuO and nFe_3O_4 could elevate As bioaccumulation in mussels.

The elevated As bioaccumulation in mussels could be due to the effectiveness of MNPs for adsorption of As [29,30]. Regarding metalloids such as As, previous studies identified the 'Trojan horse effect' consisting of MNPs capacity to adsorb co-existing pollutants and thus enabling for their uptake by organisms [31–33], which may increase their toxicity impacts. In the present study, we also investigated the adsorption of As onto nCuO/nFe_3O_4 in artificial seawater. There was a rapid uptake of As(V) in the first 30 min, and then it reached equilibrium. The adsorption equilibrium of As(V) on nCuO and nFe_3O_4 were 86.37% and 80.21%, respectively (Figure S1). In addition, as introduced in the first section, *P. viridis* are proficient at taking up suspended particles and accumulating contaminants, thus, mussels might increase the total As concentrations in their body effectively by filter water and suspended particles containing As.

In addition, as previous researchers have pointed out the excellent affinity of nCuO and nFe_3O_4 towards As [7], it is likely that the distribution of As among mussels' tissues would be disrupted due to this unique property of MNPs. Thus, we analyzed As distribution in both single As(V) exposure and co-exposure mussels. The results showed that As was mainly retained in visceral mass under single As(V) exposure (Figure 3A). However, the distribution of As in mussels' tissues altered remarkably where gill was the main tissue for As retaining rather than visceral mass under co-exposure with As+MNPs. Furthermore, the visceral mass retained ever lower As concentrations (although not significantly) after co-exposure with MNPs compared to that in single As(V) exposure mussels (Figure 3A). This indicated that although As ingestion by mussels was elevated after co-exposure to either one of nCuO and nFe_3O_4, gill was the main tissue responsible for As retaining instead of visceral mass, resulting in higher stress levels in gill tissue [34,35].

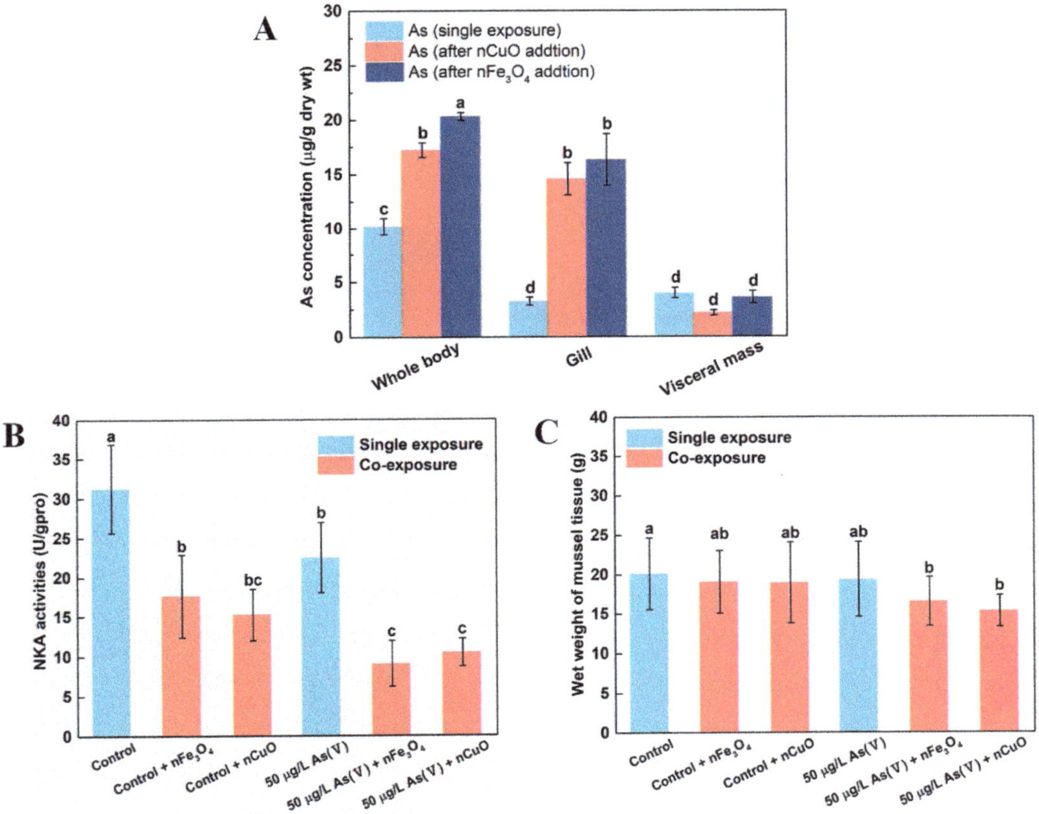

Figure 3. As Distribution before and after exposed to nFe_3O_4 and nCuO in mussels (**A**). Na^+-K^+-ATPase (NKA) activities in mussels exposed to different treatments at the end of the exposure (**B**). Body weight of mussels at the end of single As(V) exposure and co-exposure (**C**). Values are the mean ± SD (n = 6). Significant differences ($p < 0.05$) among exposure conditions were represented with different letters. Control: Mussels exposed to artificial seawater for 21 days (single exposure). Control + nCuO: Mussels exposed to 1 mg/L nCuO for 2 weeks after control exposure (co-exposure). Control + nFe_3O_4: Mussels exposed to 1 mg/L nFe_3O_4 for 2 weeks after control exposure (co-exposure). 50 μg/L As(V) only: Mussels were cultured in 50 μg/L As(V) solution for 21 days (single exposure). 50 μg/L As(V) + nCuO: Mussels exposed to 50 μg/L As(V) + nCuO for 2 weeks after 50 μg/L As(V) exposure (co-exposure). 50 μg/L As(V) + nFe_3O_4: Mussels exposed to 50 μg/L As(V) + nFe_3O_4 for 2 weeks after 50 μg/L As(V) exposure (co-exposure).

3.3. Exposure to MNPs Enhanced the Toxicity of As in Mussels

It is possible that the toxicity of As for mussels might be enhanced since the As bioaccumulation and distribution in mussels significantly changed after exposure to MNPs.

As mentioned above, the distribution of As in mussels' tissues altered remarkably under co-exposure with As + MNPs, where gill was the main tissue responsible for As retaining instead of visceral mass. Gill, as a unique organ for mussel osmoregulation, might be damaged by the elevated As concentrations [36]. To test our hypothesis, we assessed the osmoregulation capacity of gill by estimating the activities of Na^+-K^+-ATPase (NKA), since NKA is important not only for osmoregulation, but also for providing a driving force for many transporting systems in marine organisms [36]. Indeed, NKA activities decreased after exposure to nCuO/nFe_3O_4 (Figure 3B), which confirmed the osmoregulation disorder in mussels after MNPs exposure. Specifically, compared with the control exposure and As(V) single exposure, the NKA activity after exposure to nCuO/nFe_3O_4 was decreased by 51.1%/43.8% and by 53.3%/59.6%, respectively (Figure 3B). The energy metabolism of mussels was inhibited by the decreasing of osmoregulation capacity of mussels due to the decrease in NKA activity, causing a potentially physiological response, which resulted in affecting the growth of mussels. Although no mortality was observed in both single As(V) exposure condition and co-exposure condition, the average body weight of mussels after As(V) + MNPs exposure were significantly lower than that of the mussels in single As(V) exposure (Figure 3C), which confirmed the increased toxicity toward mussels after MNPs exposure.

Previous studies have reviewed that As could be detoxified in aquatic organisms through a series of detoxification strategies. One of the main strategies is As biotransformation. On the one hand, marine organisms could reduce less toxic As(V) to more toxic As(III) and subsequently excrete it, since As(III) is more easy to excrete compared to As(V) [37]. On the other hand, marine organisms can firstly reduce As(V) to As(III), afterwards, As methylation process occurred where As(III) was methylated to organic As species such as MMA, DMA and AsB [26,38]. Thus, the enhanced toxicity of As in mussels could also be attributed to the disrupting of As biotransformation in mussels after addition of MNPs. To test our hypothesis, we then analyzed the contents of both organic As species (i.e., MMA, DMA and AsB) and inorganic As species (i.e., As(III) and As(V)) in mussels after both single As(V) exposure and co-exposure.

As shown in Figure 4 and Table S1, the concentrations of organic As species and inorganic As species increased in mussels after co-exposure to As(V) + nCuO and As(V) + nFe_3O_4, which confirmed an elevated As bioaccumulation in mussels due to the lately introduced nCuO and nFe_3O_4. Surprisingly, after co-exposure to As(V) + nCuO and As(V) + nFe_3O_4, the percentage of inorganic As species in mussels increased from 23.5% to 44.0% and 44.7%, respectively (Figure 4D–F, Table S1). As a result, the percentage of organic As in mussels after co-exposure to As(V) + nCuO and As(V) + nFe_3O_4 decreased from 76.5% to 56.0% and 55.3%, respectively (Figure 4D–F, Table S1). Particularly, inorganic As to organic As ratios in mussels after co-exposure to As(V) + nCuO and As(V) + nFe_3O_4 were 0.81 and 0.79, respectively, which were significantly higher than that (0.30) before the introduction of nCuO and nFe_3O_4 (Table S1). These results therefore implied that nCuO and nFe_3O_4 may restrict As biotransformation by limiting the transformation of inorganic As to organic As.

More importantly, As(V) to As(III) ratio was 6.54 in mussels before introduction of nCuO/nFe_3O_4, it dramatically decreased to 1.97 and 2.31 in mussels after As(V) + nCuO and As(V) + nFe_3O_4 exposure, which indicated a more efficient As(III) bioaccumulation than that of As(V) bioaccumulation in mussels after introduction of nCuO/nFe_3O_4. On the other hand, the proportion of MMA after As(V) + nCuO and As(V) + nFe_3O_4 exposure were less than that of MMA in mussels before As(V) + nCuO and As(V) + nFe_3O_4 exposure. It has been reported that organic MMA is the main product in the biotransformation of inorganic As(III) during the As methylation process, which is the crucial stage for As detoxification in mussels [39], the higher efficient As(III) bioaccumulation and less MMA proportion in mussels both implied that As methylation process (i.e., transformation of inorganic As forms to organic As forms) was inhibited by MNPs. In addition, As methylation process inhibited by MNPs was further demonstrated by the experiment of mussel exposure to

$nCuO/nFe_3O_4$ only (Figure 4B,C and Table S1). Without another introduced As (i.e., no other As sources), there is a strong possibility that the variation of different pre-existing As species proportion in mussels was due to the intervention of $nCuO/nFe_3O_4$ on As biotransformation. As a whole, the results showed that the average body weight loss of mussels attributed to the enhanced toxicity of As, because the presence of MNPs led to the increasing inorganic As contents and decreasing organic As contents in mussels.

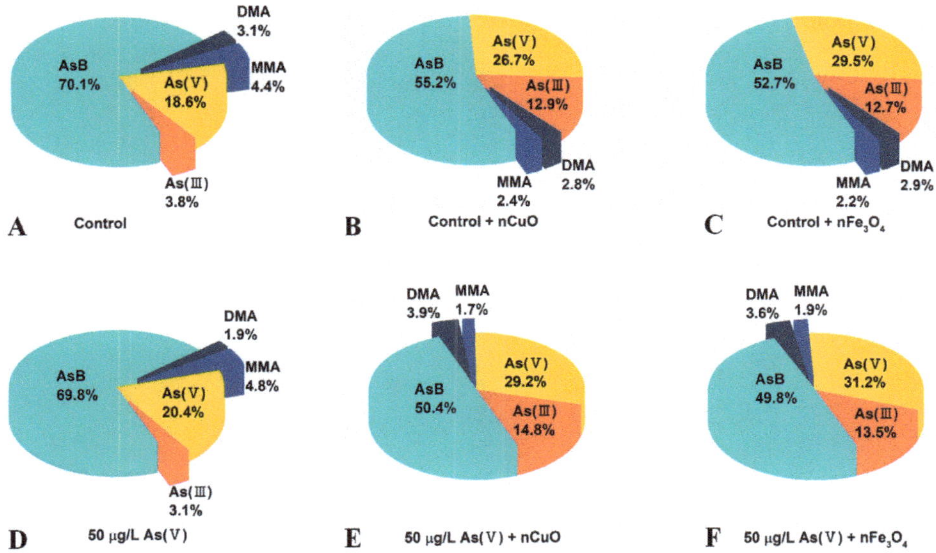

Figure 4. Content of As speciation in mussels after different exposure (organic monomethylarsonic acid (MMA), dimethylarsinic acid (DMA), and arsenobetaine (AsB)). (**A**) Control: Mussels exposed to artificial seawater for 21 days. (**B**) Control + nCuO: Mussels exposed to 1 mg/L nCuO for 2 weeks after control exposure. (**C**) Control + nFe_3O_4: Mussels exposed to 1 mg/L nFe_3O_4 for 2 weeks after control exposure. (**D**) 50 μg/L As(V) only: Mussels were cultured in 50 μg/L As(V) solution for 21 days. (**E**) 50 μg/L As(V) + nCuO: Mussels exposed to 50 μg/L As(V) + nCuO for 2 weeks after 50 μg/L As(V) exposure. (**F**) 50 μg/L As(V) + nFe_3O_4: Mussels exposed to 50 μg/L As(V) + nFe_3O_4 for 2 weeks after 50 μg/L As(V) exposure.

Existing research recognized that GST was a biotransformation rate limiting enzyme in mussels which have significant effect on As biotransformation and detoxification processes [22]. Moreover, GSH might also bind to As(V) which was a common and important mechanism during As metabolism [40]. For that reason, we measured GST activity and GSH content in mussels to elucidate their functions during As biotransformation and detoxification processes [22,41–43]. Indeed, our previous study about the $nTiO_2$ effect on biotransformation of As in mussels showed that the biotransformation of As was limited because of the down-regulated GST and GSH content in mussels [18]. In the current study, the decreasing of GST activity and GSH content in mussels were similar to those described in our previous study [18]. After exposing to As(V) + nCuO and As(V) + nFe_3O_4, compared with single 50 μg/L As(V) exposure, GST activity greatly decreased by 84.8% and 67.0% in 50 μg/L As(V) co-exposure, respectively. GST activity also decreased by 74.6% and 71.9% in control + nCuO and control + nFe_3O_4 mussels, respectively, compared with single control exposure (Figure 5A). Likewise, there was a similar reduction in the GSH contents, in As(V) + nCuO and As(V) + nFe_3O_4 exposure, which was lowered by 43.9% and 36.4% than in 50 μg/L As(V) exposure, respectively. It also decreased by 24.6% and 31.0% in control + nCuO and control + nFe_3O_4 mussels, compared with the control exposure, respectively (Figure 5B). All these results indicate that MNPs can mediate biotransformation of inorganic As to organic As in mussels by decreasing As metabolism enzymes such as GST and GSH, thereby enhancing toxicity of As towards mussels.

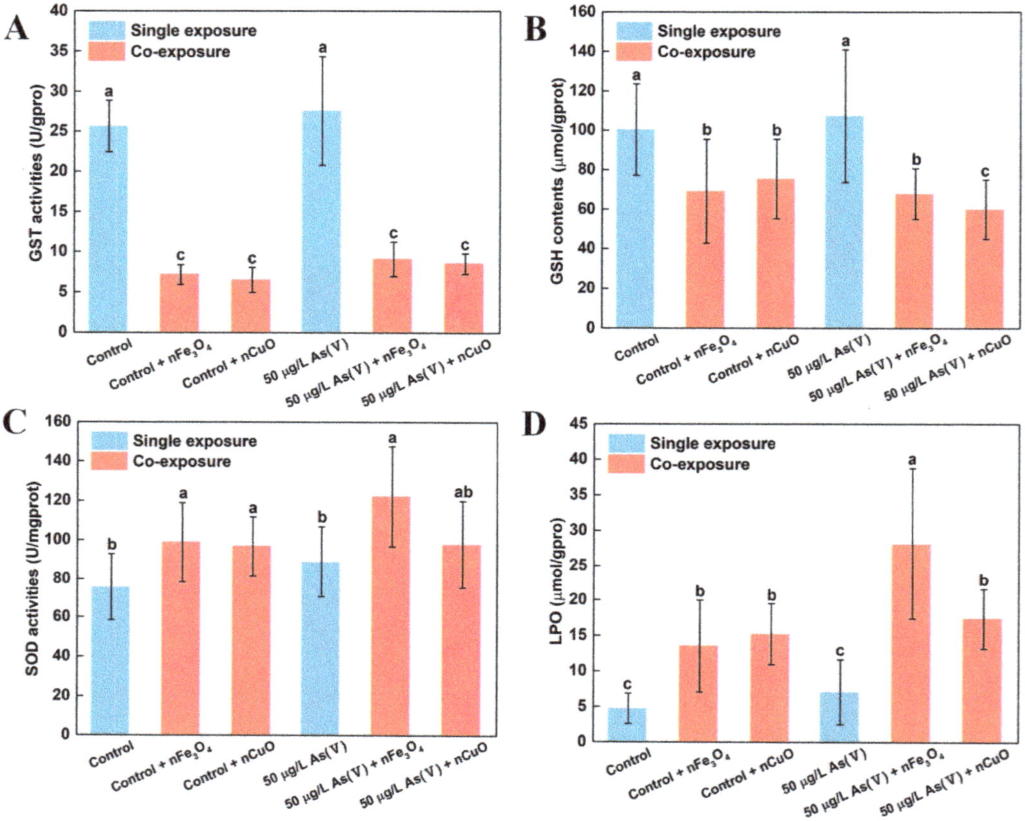

Figure 5. Biomarker values in mussels after different exposure. (**A**) GST activities, (**B**) GSH contents, (**C**) SOD activities and (**D**) LPO levels in mussel exposed to different treatments at the end of the exposure (mean ± SD, n = 6). Significant differences ($p < 0.05$) among exposure conditions were represented with different letters. Control: Mussels exposed to artificial seawater for 21 days (single exposure). Control + nCuO: Mussels exposed to 1 mg/L nCuO for 2 weeks after control exposure (co-exposure). Control + nFe_3O_4: Mussels exposed to 1 mg/L nFe_3O_4 for 2 weeks after control exposure (co-exposure). 50 μg/L As(V) only: Mussels were cultured in 50 μg/L As(V) solution for 21 days (single exposure). 50 μg/L As(V) + nCuO: Mussels exposed to 50 μg/L As(V) + nCuO for 2 weeks after 50 μg/L As(V) exposure (co-exposure). 50 μg/L As(V) + nFe_3O_4: Mussels exposed to 50 μg/L As(V) + nFe_3O_4 for 2 weeks after 50 μg/L As(V) exposure (co-exposure).

Another probable explanation for the inhibitory effects of MNPs on As biotransformation and detoxification in mussels might be that the specific surface characteristics as well as chemical properties enable NPs the capacity to generate ROS by interaction with subcellular structures [3]. In the case of MNPs, the physical contact between mussels and particles may cause overproduction of ROS, leading to an increase in activity of antioxidant enzymes such as SOD and LPO [44–46]. In the present study, the generation of ROS was confirmed after exposure to nCuO/nFe_3O_4. Both SOD activities and LPO levels increased in mussels exposed to nCuO/nFe_3O_4 compared with the control groups without nCuO/nFe_3O_4 exposure (Figure 5C,D). Accordingly, nCuO and nFe_3O_4 might inhibit the As biotransformation in mussels through overproduction of ROS. These inhibitory effects imply an overproduction of ROS that could have led to the degeneration of As metabolism related enzymes such as GST and GSH [47], which could further support the decreasing of

GST activity and GSH content in mussels and average weight loss of mussels after MNPs exposure. According to these results, it can be inferred that mussels could be subjected to a more toxic As environment after MNPs exposure.

4. Conclusions

The MNPs such as nCuO and nFe_3O_4 are increasingly applied in a variety of areas, and it is possible that they will end up in the environment in significant quantities, which makes it important to identify its effect on surrounding biota and environment. This study highlighted the importance of MNPs on biotransformation and toxicity of arsenic in green mussel *Perna viridis*. Exposure to MNPs elevated the bioaccumulation of As(V) and altered the As distribution in mussels, these alterations could be attributed to the adsorption of As on MNPs. What is more, newly introduced MNPs disturbed the osmoregulation system and enhanced the toxicity of As in mussels, which could be supported by decreased activities of Na^+-K^+-ATPase and average weight-loss of mussels after MNPs exposure. The present MNPs in mussels increase the content of inorganic As and motivate the ROS generation. The overproduction of ROS (SOD and LPO) restrains the activities of As metabolism enzymes (GST and GSH) in mussels, and then reduces As methylation and detoxification, subsequently, resulting in an increase in the toxicity of As to the mussels. The current work validates that MNPs enhance the bioaccumulation and toxicity of As in marine biota, resulting in an enhanced ecotoxicity of As towards marine ecosystems, and which improve our understanding about the ecological risks of MNPs and As.

Supplementary Materials: The following are available online at https://www.mdpi.com/article/10.3390/nano11102769/s1. Figure S1: As(V) adsorption onto 1 mg/L nCuO and nFe_3O_4, Table S1: As species concentrations (μg/g) and proportion (%). Supplementary data to this research article can be found in Supplementary Materials.

Author Contributions: S.Z.: Conceptualization, Methodology, Writing—original draft, Writing—review & editing, Investigation. W.Q.: Methodology, Investigation. Z.N.: Writing—review & editing. X.Z.: Conceptualization, Methodology, Writing—review & editing, Supervision, Project administration. All authors have read and agreed to the published version of the manuscript.

Funding: The National Natural Science Foundation of China (Nos. 41877352 and 42077227) supported this work. It was also supported in part by the Natural Science Foundation of Guangdong Province of China (2021A1515010158), China Postdoctoral Science Foundation (2020M680517) and Shenzhen Fundamental Research and Discipline Layout project (JCYJ20180507182227257).

Institutional Review Board Statement: Ethical review and approval were waived for this study, due to mussels are one kind of seafood.

Informed Consent Statement: Not applicable.

Conflicts of Interest: The authors declare that they have no known competing financial interests or personal relationships that could have appeared to influence the work reported in this paper.

References

1. Yoon, K.Y.; Hoon Byeon, J.; Park, J.H.; Hwang, J. Susceptibility constants of Escherichia coli and Bacillus subtilis to silver and copper nanoparticles. *Sci. Total Environ.* **2007**, *373*, 572–575. [CrossRef]
2. Fahmy, B.; Cormier, S.A. Copper oxide nanoparticles induce oxidative stress and cytotoxicity in airway epithelial cells. *Toxicol. Vitr.* **2009**, *23*, 1365–1371. [CrossRef]
3. Gomes, T.; Pinheiro, J.P.; Cancio, I.; Pereira, C.G.; Cardoso, C.; Bebianno, M.J. Effects of copper nanoparticles exposure in the mussel Mytilus galloprovincialis. *Environ. Sci. Technol.* **2011**, *45*, 9356–9362. [CrossRef] [PubMed]
4. Qiu, H.; Cui, B.; Li, G.; Yang, J.; Peng, H.; Wang, Y.; Li, N.; Gao, R.; Chang, Z.; Wang, Y. Novel Fe_3O_4@ZnO@$mSiO_2$ Nanocarrier for targeted drug delivery and controllable release with microwave irradiation. *J. Phys. Chem. C* **2014**, *118*, 14929–14937. [CrossRef]
5. Sadri, F.; Ramazani, A.; Massoudi, A.; Khoobi, M.; Tarasi, R.; Shafiee, A.; Azizkhani, V.; Dolatyari, L.; Joo, S.W. Green oxidation of alcohols by using hydrogen peroxide in water in the presence of magnetic Fe_3O_4 nanoparticles as recoverable catalyst. *Green Chem. Lett. Rev.* **2014**, *7*, 257–264. [CrossRef]
6. Zhang, Y.R.; Su, P.; Huang, J.; Wang, Q.R.; Zhao, B.X. A magnetic nanomaterial modified with poly-lysine for efficient removal of anionic dyes from water. *Chem. Eng. J.* **2015**, *262*, 313–318. [CrossRef]

7. Abbas, Q.; Yousaf, B.; Ullah, H.; Ali, M.U.; Ok, Y.S.; Rinklebe, J. Environmental transformation and nano-toxicity of engineered nano-particles (ENPs) in aquatic and terrestrial organisms. *Crit. Rev. Environ. Sci. Technol.* **2020**, *50*, 2523–2581. [CrossRef]
8. Vance, M.E.; Kuiken, T.; Vejerano, E.P.; McGinnis, S.P.; Hochella, M.F.; Hull, D.R. Nanotechnology in the real world: Redeveloping the nanomaterial consumer products inventory. *Beilstein J. Nanotechnol.* **2015**, *6*, 1769–1780. [CrossRef]
9. Deng, R.; Lin, D.; Zhu, L.; Majumdar, S.; White, J.C.; Gardea-Torresdey, J.L.; Xing, B. Nanoparticle interactions with co-existing contaminants: Joint toxicity, bioaccumulation and risk. *Nanotoxicology* **2017**, *11*, 591–612. [CrossRef]
10. Smedley, P.L.; Kinniburgh, D.G. A review of the source, behaviour and distribution of arsenic in natural waters. *Appl. Geochem.* **2002**, *17*, 517–568. [CrossRef]
11. Li, H.; Gao, X.; Gu, Y.; Wang, R.; Xie, P.; Liang, M.; Ming, H.; Su, J. Comprehensive large-scale investigation and assessment of trace metal in the coastal sediments of Bohai Sea. *Mar. Pollut. Bull.* **2018**, *129*, 126–134. [CrossRef]
12. Zhang, W.; Guo, Z.; Song, D.; Du, S.; Zhang, L. Arsenic speciation in wild marine organisms and a health risk assessment in a subtropical bay of China. *Sci. Total Environ.* **2018**, *626*, 621–629. [CrossRef] [PubMed]
13. Byeon, E.; Kang, H.; Yoon, C.; Lee, J. Toxicity mechanisms of arsenic compounds in aquatic organisms. *Aquat. Toxicol.* **2021**, *237*, 105901. [CrossRef]
14. Baumann, J.; Köser, J.; Arndt, D.; Filser, J. The coating makes the difference: Acute effects of iron oxide nanoparticles on Daphnia magna. *Sci. Total Environ.* **2014**, *484*, 176–184. [CrossRef]
15. Rotini, A.; Tornambè, A.; Cossi, R.; Iamunno, F.; Benvenuto, G.; Berducci, M.T.; Maggi, C.; Thaller, M.C.; Cicero, A.M.; Manfra, L.; et al. Salinity-based toxicity of CuO nanoparticles, CuO-bulk and Cu ion to vibrio anguillarum. *Front. Microbiol.* **2017**, *8*, 2076. [CrossRef]
16. Maurya, R.; Pandey, A.K. Importance of protozoa Tetrahymena in toxicological studies: A review. *Sci. Total Environ.* **2020**, *741*, 140058. [CrossRef] [PubMed]
17. Braz-Mota, S.; Campos, D.F.; MacCormack, T.J.; Duarte, R.M.; Val, A.L.; Almeida-Val, V.M.F. Mechanisms of toxic action of copper and copper nanoparticles in two Amazon fish species: Dwarf cichlid (Apistogramma agassizii) and cardinal tetra (Paracheirodon axelrodi). *Sci. Total Environ.* **2018**, *630*, 1168–1180. [CrossRef]
18. Qian, W.; Chen, C.C.; Zhou, S.; Huang, Y.; Zhu, X.; Wang, Z.; Cai, Z. TiO_2 Nanoparticles in the marine environment: Enhancing bioconcentration, while limiting biotransformation of arsenic in the Mussel Perna viridis. *Environ. Sci. Technol.* **2020**, *54*, 12254–12261. [CrossRef]
19. Zhelyazkov, G.; Yankovska-Stefanova, T.; Mineva, E.; Stratev, D.; Vashin, I.; Dospatliev, L.; Valkova, E.; Popova, T. Risk assessment of some heavy metals in mussels (Mytilus galloprovincialis) and veined rapa whelks (Rapana venosa) for human health. *Mar. Pollut. Bull.* **2018**, *128*, 197–201. [CrossRef] [PubMed]
20. Coppola, F.; Almeida, Â.; Henriques, B.; Soares, A.M.V.M.; Figueira, E.; Pereira, E.; Freitas, R. Biochemical responses and accumulation patterns of Mytilus galloprovincialis exposed to thermal stress and Arsenic contamination. *Ecotoxicol. Environ. Saf.* **2018**, *147*, 954–962. [CrossRef]
21. Hanna, S.; Miller, R.; Lenihan, H. Accumulation and toxicity of copper oxide engineered nanoparticles in a marine mussel. *Nanomaterials* **2014**, *4*, 535–547. [CrossRef]
22. Zhang, W.; Guo, Z.; Zhou, Y.; Liu, H.; Zhang, L. Biotransformation and detoxification of inorganic arsenic in Bombay oyster Saccostrea cucullata. *Aquat. Toxicol.* **2015**, *158*, 33–40. [CrossRef] [PubMed]
23. Whaley-Martin, K.J.; Koch, I.; Moriarty, M.; Reimer, K.J. Arsenic speciation in blue mussels (Mytilus edulis) along a highly contaminated arsenic gradient. *Environ. Sci. Technol.* **2012**, *46*, 3110–3118. [CrossRef] [PubMed]
24. Gomes, T.; Araújo, O.; Pereira, R.; Almeida, A.C.; Cravo, A.; Bebianno, M.J. Genotoxicity of copper oxide and silver nanoparticles in the mussel Mytilus galloprovincialis. *Mar. Environ. Res.* **2013**, *84*, 51–59. [CrossRef] [PubMed]
25. Zhang, Y.Q.; Dringen, R.; Petters, C.; Rastedt, W.; Köser, J.; Filser, J.; Stolte, S. Toxicity of dimercaptosuccinate-coated and un-functionalized magnetic iron oxide nanoparticles towards aquatic organisms. *Environ. Sci. Nano.* **2016**, *3*, 754–767. [CrossRef]
26. Zhang, W.; Huang, L.; Wang, W.X. Biotransformation and detoxification of inorganic arsenic in a marine juvenile fish Terapon jarbua after waterborne and dietborne exposure. *J. Hazard. Mater.* **2012**, *221–222*, 162–169. [CrossRef] [PubMed]
27. Chen, L.; Song, D.; Zhang, W.; Zhang, C.; Zhang, L. The dynamic changes of arsenic bioaccumulation and antioxidant responses in the marine medaka Oryzias melastigma during chronic exposure. *Aquat. Toxicol.* **2019**, *212*, 110–119. [CrossRef]
28. Suhendrayatna; Ohki, A.; Nakajima, T.; Maeda, S. Studies on the accumulation and transformation of arsenic in freshwater organisms I. Accumulation, transformation and toxicity of arsenic compounds on the Japanese Medaka, Oryzias latipes. *Chemosphere* **2002**, *46*, 319–324. [CrossRef]
29. Hua, M.; Zhang, S.; Pan, B.; Zhang, W.; Lv, L.; Zhang, Q. Heavy metal removal from water/wastewater by nanosized metal oxides: A review. *J. Hazard. Mater.* **2012**, *211–212*, 317–331. [CrossRef] [PubMed]
30. Amde, M.; Liu, J.; Tan, Z.Q.; Bekana, D. Transformation and bioavailability of metal oxide nanoparticles in aquatic and terrestrial environments: A review. *Environ. Pollut.* **2017**, *230*, 250–267. [CrossRef]
31. Fan, W.; Cui, M.; Liu, H.; Wang, C.; Shi, Z.; Tan, C.; Yang, X. Nano-TiO_2 enhances the toxicity of copper in natural water to Daphnia magna. *Environ. Pollut.* **2011**, *159*, 729–734. [CrossRef]
32. Zhu, X.; Zhou, J.; Cai, Z. TiO_2 nanoparticles in the marine environment: Impact on the toxicity of tributyltin to abalone (Haliotis diversicolor supertexta) embryos. *Environ. Sci. Technol.* **2011**, *45*, 3753–3758. [CrossRef]

33. Lu, J.; Tian, S.; Lv, X.; Chen, Z.; Chen, B.; Zhu, X.; Cai, Z. TiO_2 nanoparticles in the marine environment: Impact on the toxicity of phenanthrene and Cd^{2+} to marine zooplankton Artemia salina. *Sci. Total Environ.* **2018**, *615*, 375–380. [CrossRef] [PubMed]
34. Chandra Sekhar, K.; Chary, N.S.; Kamala, C.T.; Suman Raj, D.S.; Sreenivasa Rao, A. Fractionation studies and bioaccumulation of sediment-bound heavy metals in Kolleru lake by edible fish. *Environ. Int.* **2004**, *29*, 1001–1008. [CrossRef]
35. Zhao, S.; Meng, F.; Fu, H.; Xiao, J.; Gao, Y. Metallothionein levels in gills and visceral mass of Ruditapes philippinarum exposed to sublethal doses of cadmium and copper. In Proceedings of the International Conference on Challenges in Environmental Science and Computer Engineering, CESCE, Wuhan, China, 6–7 March 2010; Volume 2. [CrossRef]
36. Hwang, P.P.; Lee, T.H. New insights into fish ion regulation and mitochondrion-rich cells. *Comp. Biochem. Physiol. A Mol. Integr. Physiol.* **2007**, *148*, 479–497. [CrossRef] [PubMed]
37. Rosen, B.P. Families of arsenic transporters. *Trends Microbiol.* **1999**, *7*, 207–212. [CrossRef]
38. Sörös, C.; Bodó, E.T.; Fodor, P.; Morabito, R. The potential of arsenic speciation in molluscs for environmental monitoring. *Anal. Bioanal. Chem.* **2003**, *377*, 25–31. [CrossRef] [PubMed]
39. Vahter, M. Mechanisms of arsenic biotransformation. *Toxicology* **2002**, *181–182*, 211–217. [CrossRef]
40. Bagnyukova, T.V.; Luzhna, L.I.; Pogribny, I.P.; Lushchak, V.I. Oxidative stress and antioxidant defenses in goldfish liver in response to short-term exposure to arsenite. *Environ. Mol. Mutagen.* **2007**, *48*, 658–659. [CrossRef] [PubMed]
41. Bhattacharya, A.; Bhattacharya, S. Induction of oxidative stress by arsenic in Clarias batrachus: Involvement of peroxisomes. *Ecotoxicol. Environ. Saf.* **2007**, *66*, 178–187. [CrossRef]
42. Sampayo-Reyes, A.; Zakharyan, R.A. Inhibition of human glutathione s-transferase omega by tocopherol succinate. *Biomed. Pharmacother.* **2006**, *60*, 238–244. [CrossRef]
43. Vasken Aposhian, H.; Zakharyan, R.A.; Avram, M.D.; Sampayo-Reyes, A.; Wollenberg, M.L. A review of the enzymology of arsenic metabolism and a new potential role of hydrogen peroxide in the detoxication of the trivalent arsenic species. *Toxicol. Appl. Pharmacol.* **2004**, *198*, 327–335. [CrossRef]
44. Griffitt, R.J.; Weil, R.; Hyndman, K.A.; Denslow, N.D.; Powers, K.; Taylor, D.; Barber, D.S. Exposure to copper nanoparticles causes gill injury and acute lethality in zebrafish (Danio rerio). *Environ. Sci. Technol.* **2007**, *41*, 8178–8186. [CrossRef] [PubMed]
45. Heinlaan, M.; Ivask, A.; Blinova, I.; Dubourguier, H.C.; Kahru, A. Toxicity of nanosized and bulk ZnO, CuO and TiO_2 to bacteria Vibrio fischeri and crustaceans Daphnia magna and Thamnocephalus platyurus. *Chemosphere* **2008**, *71*, 1308–1316. [CrossRef] [PubMed]
46. Unfried, K.; Albrecht, C.; Klotz, L.O.; Von Mikecz, A.; Grether-Beck, S.; Schins, R.P.F. Cellular responses to nanoparticles: Target structures and mechanisms. *Nanotoxicology* **2007**, *1*, 52–71. [CrossRef]
47. Han, J.; Won, E.J.; Hwang, D.S.; Rhee, J.S.; Kim, I.C.; Lee, J.S. Effect of copper exposure on GST activity and on the expression of four GSTs under oxidative stress condition in the monogonont rotifer, Brachionus koreanus. *Comp. Biochem. Physiol. C Toxicol. Pharmacol.* **2013**, *158*, 91–100. [CrossRef] [PubMed]

nanomaterials

MDPI

Article

Effects of Nanoplastics on the Dinoflagellate *Amphidinium carterae* Hulburt from the Perspectives of Algal Growth, Oxidative Stress and Hemolysin Production

Su-Chun Wang [1], Fei-Fei Liu [1,*], Tian-Yuan Huang [1], Jin-Lin Fan [2], Zhi-Yin Gao [1] and Guang-Zhou Liu [1,*]

1 Institute of Marine Science and Technology, Shandong University, Qingdao 266237, China; suchunw@163.com (S.-C.W.); huangty1994@163.com (T.-Y.H.); gzy18864805967@163.com (Z.-Y.G.)
2 Department of Science and Technology Management, Shandong University, Jinan 250100, China; fanjinlin@sdu.edu.cn
* Correspondence: liufeifei@sdu.edu.cn (F.-F.L.); liuguangzhou@sdu.edu.cn (G.-Z.L.); Tel.: +86-532-5863-3262 (F.-F.L.)

Abstract: Recently, the effects of nanoplastics (NPs) on aquatic organisms have attracted much attention; however, research on the toxicity of NPs to microalgae has been insufficient. In the present study, the effects of polystyrene nanoplastics (nano-PS, 50 nm) on growth inhibition, chlorophyll content, oxidative stress, and algal toxin production of the marine toxigenic dinoflagellate *Amphidinium carterae* Hulburt were investigated. Chlorophyll synthesis was promoted by nano-PS on day 2 but was inhibited on day 4; high concentrations of nano-PS ($\geq$50 mg/L) significantly inhibited the growth of *A. carterae*. Moreover, despite the combined effect of superoxide dismutase (SOD) and glutathione (GSH), high reactive oxygen species (ROS) level and malondialdehyde (MDA) content were still induced by nano-PS ($\geq$50 mg/L), indicating severe lipid peroxidation. In addition, the contents of extracellular and intracellular hemolytic toxins in nano-PS groups were significantly higher than those in control groups on days 2 and 8, except that those of extracellular hemolytic toxins in the 100 mg/L nano-PS group decreased on day 8 because of severe adsorption of hemolytic toxins to the nano-PS. Hence, the effects of nano-PS on *A. carterae* are closely linked to nano-PS concentration and surface properties and exposure time. These findings provide a deep understanding of the complex effects of NPs on toxigenic microalgae and present valuable data for assessing their environmental risks.

Keywords: polystyrene nanoplastics; growth inhibition; oxidative stress; hemolytic toxin

Citation: Wang, S.-C.; Liu, F.-F.; Huang, T.-Y.; Fan, J.-L.; Gao, Z.-Y.; Liu, G.-Z. Effects of Nanoplastics on the Dinoflagellate *Amphidinium carterae* Hulburt from the Perspectives of Algal Growth, Oxidative Stress and Hemolysin Production. *Nanomaterials* **2021**, *11*, 2471. https://doi.org/10.3390/nano11102471

Academic Editors: Xiaoshan Zhu, Jian Zhao and David M Brown

Received: 22 August 2021
Accepted: 17 September 2021
Published: 22 September 2021

1. Introduction

Plastic pollution in aquatic environments has recently gained attention worldwide. With the in-depth study of microplastics (MPs), nanoplastics (NPs) have also drawn public awareness [1]. Nanoplastics, defined as plastic materials with at least one dimension on the nanoscale, have been found in natural waters and their concentrations are expected to continuously increase because of the degradation of primary micro(nano)plastics [2–5]. Owing to their small size and large surface area, the ecological effects of NPs on aquatic organisms differ from those of large plastic particles [6]. Polystyrene (PS; 100 nm, <10 mg/L) has been found to accumulate in crustaceans without affecting their mortality rate; however, the swimming speed and enzyme activity in individuals with high accumulated PS levels are significantly altered [7]. When exposed to PS-COOH (100 nm), a dose-dependent relationship between reactive oxygen species (ROS) production and PS-COOH concentration was observed in the sperm cells of Pacific oysters [8]. In addition, PS-NH_2 NPs interfere with the development of sea urchin embryos by modulating protein and gene profiles [9]. Overall, NPs pollution in aquatic environments has become a main challenge that requires further investigation.

As the basis of the food web, algae are crucial for the stability of marine ecosystems [10]. Owing to their short growth period and high sensitivity to toxic substances, microalgae have been considered a good choice for detecting environmental threats caused by MPs pollution [11]. Size-dependent negative effects of PS with the particle sizes of 0.05, 0.5, and 6 μm on the marine flagellate *Dunaliella tertiolecta* have been reported [12]. In addition to growth inhibition effects, polystyrene nanoplastics (nano-PS) can also reduce chlorophyll content and accelerate ROS production in algal cells [13]. For example, after 2 d of exposure to PS-NH_2 (200 nm), the chlorophyll content and photosynthetic efficiency of *Chaetoceros neogracile* decreased by 24% and 13%, respectively, and esterase activity also significantly decreased while the intracellular ROS level increased [14]. In addition, MPs and NPs can also affect the secretion of hemolytic toxins produced by harmful algal bloom species and absorb some substances of hemolytic toxins released by algal cells [15–19]. Considering the coexistence of NPs and harmful algal bloom species in aquatic environments, interactions between NPs and hemolytic toxins may negatively affect aquatic ecology and pose a potential risk to animals and humans. Therefore, more attention should be paid to the comprehensive evaluation of the effects of NPs on harmful algal bloom species.

Amphidinium carterae Hulburt (*A. carterae*), a harmful algal bloom species, is mainly distributed in tropical and temperate seas worldwide and can produce hemolytic toxins [20–22]. The synthesis of hemolytic toxins has been reported to be closely related to the salinity, pH, temperature, and light intensity of the algal growth environment [23–26]. *A. carterae* cells in the logarithmic phase have been shown to increase their hemolytic activity as light intensity increased, while hemolytic activity was greatly inhibited at low temperature (10 °C) and salinity (15) [26]. In addition, current research on the effects of NPs on the toxin production of harmful algal bloom species is still insufficient and unclear; for example, PS (100 nm) has no significant effect on the growth and photosynthetic activity of *Microcystis aeruginosa*, and promotes microcystin production only after 48 h. PS-NH_2 (50 nm) induces photosynthesis inhibition and oxidative stress, and enhances the synthesis of microcystin; while larger particle PS-NH_2 (200 nm, 5 mg/L) had no significant effect on microcystin production of *M. aeruginosa* [18,19,27]. Thus, a deeper understanding of the effects of NPs on harmful algal bloom species needs to be investigated.

In order to evaluate the effects of NPs on marine harmful algal bloom species, we chose nano-PS (50 nm) without functional groups as the test chemical and *A. carterae* as the test species. We investigated the growth inhibition, chlorophyll content, ROS level, antioxidant enzyme activity, and hemolytic toxin content of *A. carterae*. In addition, scanning electron microscopy (SEM) was performed to observe the interaction between nano-PS and algal cells. This research forms the basis for a more comprehensive evaluation of the toxicity of NPs to marine harmful algal bloom species and for the assessment of their environmental risks.

2. Materials and Methods

2.1. Algal Cultivation

A. carterae was provided by Shanghai Guangyu Biotechnology Co., Ltd. (Shanghai, China). Microalgae were cultivated in f/2 medium made with sterile artificial seawater (filtered through a 0.45 μm acetate filter membrane). Erlenmeyer flasks containing microalgal cells were cultivated at 20 ± 1 °C under cool white fluorescent lights (4000 lux) with a 12 h-light-dark cycle and were kept at a constant temperature oscillation incubator (ZQZY-CGF8, Zhichu Instrument Co., Ltd., Shanghai, China) at a speed of 50 rpm. According to the growth curves of *A. carterae* (Figure S1), the incubation lasted approximately 4 days until logarithmic phase growth prevailed. Cell density was calculated using an optical microscope (BX53, Olympus, Tokyo, Japan).

2.2. Nanoplastics Treatment

Nano-PS powder with a size of 50 nm was purchased from Changchun Lianyu Chemical Technology Co., Ltd. (Changchun, China) (Figure S2). The hydrodynamic diameter and zeta potential of nano-PS were measured using a particle size analyzer (Zetasizer Nano ZS90, Malvern Panalytical Ltd., Malvern, UK), which were 424 nm and −23.7 mV, respectively. Before treatment, the nano-PS and culture media (10 mL) were added to Erlenmeyer flasks and ultrasonicated to obtain a uniformly dispersed suspension. Then, 40 mL of algal cells in logarithmic phase growth were transferred into the flasks. The initial algal density was maintained at 9×10^5 cells/mL, while the concentrations of nano-PS were set at 10, 20, 50, 80, and 100 mg/L. Each concentration treatment was replicated three times, and all operations were performed under sterile conditions to avoid contamination by bacteria. All flasks were placed in an incubator for 8 days under pre-culture conditions.

2.3. Measurement of Chlorophyll Content

Chlorophyll content was determined using the acetone extraction method [28]. First, 20 mL of algal culture was centrifuged (5000 rpm, 10 min) to remove the supernatant. The pellets were mixed with 5 mL of 90% acetone to extract chlorophyll for 24 h at 4 °C in the dark. The samples were then centrifuged at 6500 rpm for 15 min, and the absorbance (OD) of the supernatant at 645 nm and 663 nm was measured. Total chlorophyll (T_{chl}) content was calculated using the following equation: $T_{chl} = 20.2OD_{645} + 8.02OD_{663}$.

2.4. Assays for ROS Level, MDA Content, SOD Activity, and GSH Content

In this study, ROS levels were detected using 2′, 7′-dichlorodihydrofluorescein diacetate (DCFH-DA; Sigma-Aldrich, St. Louis, MO, USA), based on previous studies [29,30]. The stock solution of DCFH-DA was prepared using N, N-dimethylformamide (DFM, Sinopharm Chemical Reagent Co., Ltd., Shanghai, China) and its final concentration used in the experiment was 10 μmol/L. The microalgal cells immersed in DCFH-DA solution were incubated in the dark at 37 °C for 20 min. Finally, these algal cells were resuspended in PBS and detected using a multi-mode microplate reader (Synergy H1, BioTek, Winooski, VT, USA) to obtain the fluorescence intensity values, which indicated the ROS level.

The SOD activity, MDA content, and GSH content were measured using detection kits purchased from the Jiancheng Bioengineering Institute (Nanjing, China). First, algal cells were crushed using an ultrasonic cell disruptor. After adding the relevant reagents, samples for SOD detection were incubated at 37 °C for 20 min. The samples for MDA were kept in a water bath at 95 °C for 40 min, and for GSH, at room temperature for 5 min before measurement. Then, SOD, MDA, and GSH were determined using the Synergy H1 system at 450, 532, and 420 nm, respectively.

2.5. Extraction and Determination of Hemolytic Toxin

In this experiment, the hemolytic toxin content was measured using an erythrocyte lysis assay (ELA) [31,32]. First, 0.4 mL of different concentrations of digitalis saponin and 1.6 mL of 0.5% rabbit red blood cells were incubated in a 37 °C water bath for 30 min. The reaction liquid was then centrifuged, and the absorbance of the supernatant was measured at 540 nm to obtain the hemolysis standard curve of digitalis saponin (Figure S3). The extracellular toxin of *A. carterae* was determined from the supernatant after centrifugation (5000 rpm, 10 min), and the remaining algal cells were used to determine the content of intracellular toxins. For the intracellular toxin extraction, algal cells were first crushed using an ultrasonic cell disruptor; then, an extraction solution prepared using chloroform, methanol, and water (13:7:5, *V:V:V*) was added for liquid phase extraction. After evaporation with a rotary evaporator, intracellular toxins were collected using 1 mL of methanol. The above ELA test was repeated to obtain the corresponding hemolytic toxin content based on the hemolysis standard curve [33,34].

2.6. Sample Preparation for SEM Assay

The morphology of the algal cells was observed using a scanning electron microscope (FEI Quanta 250 FEG; Thermo-Fisher Scientific, Waltham, MA, USA). Microalgal cells were collected by centrifugation (5000 rpm, 10 min) and fixed with 2.5% glutaraldehyde at

4 °C overnight. Then, the samples were washed three times with phosphate buffer (PBS, pH 7.4) and dehydrated using 30, 50, 70, 80, 90, 95 and 100% ethanol solutions for 15 min, respectively [27,35]. Finally, the samples were freeze-dried for SEM observation.

2.7. Statistical Analysis

All tests were performed in triplicate, and the results were expressed as the mean ± standard deviation. The enzyme analysis results were analyzed using one-way ANOVA and LSDs, with the analysis performed using the SPSS statistical software (IBM, Chicago, IL, USA). A value of $p < 0.05$ was used to denote a significant difference.

3. Results and Discussion

3.1. Cell Density and Chlorophyll Content

As shown in Figure 1A, the effects of nano-PS on the cell density of *A. carterae* were related to its concentration and experimental time. The nano-PS at 10 mg/L had no significant effect on the growth of algal cells, whereas 20 mg/L of nano-PS inhibited the growth of algal cells only on days 2 and 8 compared with the control group. The nano-PS at 50, 80, and 100 mg/L significantly inhibited the growth of the test algal cells with the *IR* of 17.5, 22.1, and 38.7% (Table S1), respectively. There was a clear negative dose-dependent relationship between cell density and nano-PS concentration on day 8. In addition, 100 mg/L of nano-PS induced the algal cells to enter the decline phase earlier than the other treatments. Moreover, we also observed that the *IR* of nano-PS groups, except that of the nano-PS at 10 mg/L, decreased initially and then increased (Table S1). This may have resulted from the resistance and adaptation of algae to nano-PS [36].

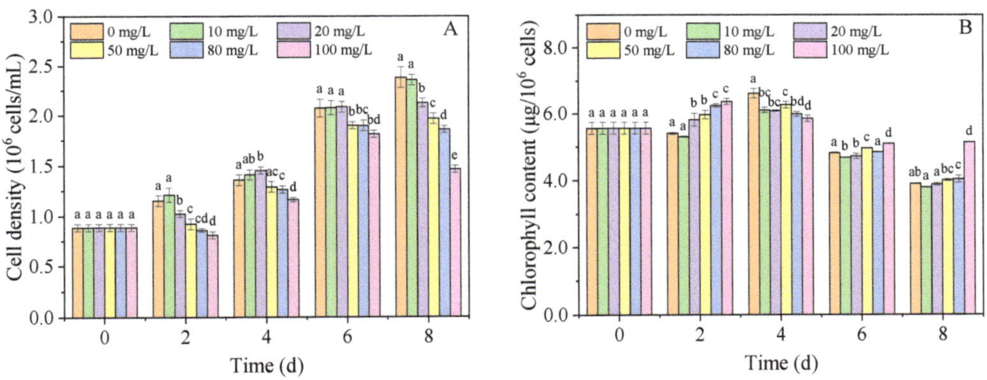

Figure 1. Cell density (**A**) and chlorophyll content (**B**) of *A. carterae* at different concentrations of nano-PS. Different letters represent significant differences ($p < 0.05$).

The chlorophyll content of *A. carterae* increased first and then decreased during the experiment (Figure 1B), owing to the limited nutrients with increasing cell density. On day 2, nano-PS at 20–100 mg/L promoted the synthesis of chlorophyll; the chlorophyll content in the 100 mg/L nano-PS group was 1.18-fold higher than that in the control group (Table S2). This may be a stress response to the decrease in light intensity caused by nano-PS [37,38]. As the cell density decreased in the nano-PS groups, pigment accumulated in the algal cells on 6–8 d, resulting in much higher chlorophyll content in the 100 mg/L nano-PS group than in the control group. However, nano-PS decreased the chlorophyll content on day 4, probably because of ROS accumulation, which could inhibit the synthesis of chlorophyll [39].

3.2. *Lipid Peroxidation in A. carterae*

Malonaldehyde is the main peroxidation product of cytomembrane lipids and is often caused by excessive ROS [40,41]. In the present study, the MDA content and ROS levels were measured, and the results are shown in Figure 2A,B. Compared with the control group, the nano-PS (≥20 mg/L) significantly increased the MDA content on days 2 and 4, indicating severe lipid peroxidation caused by nano-PS. In addition, low concentrations (≤20 mg/L) of nano-PS had no significant effect on MDA content of *A. carterae* on 6 and 8 d. However, the MDA content in the 50, 80, and 100 mg/L nano-PS groups was also significantly higher than that in the control group, but their difference gradually decreased, probably because of the adaptability of algal cells to the stress of nano-PS [42]. These results were confirmed by the relative ROS levels shown in Figure 2B. The relative ROS levels in the nano-PS groups were much higher than those in the control group on 2 and 4 d, and nano-PS at high concentrations (≥50 mg/L) induced high ROS levels over the experimental period. Similarly, Hazeem et al. reported that nano-PS (20 and 50 nm) could cause a significant increase in ROS levels in *Chlorella vulgaris* [43]. Overall, the high ROS level and MDA content indicate the occurrence of lipid peroxidation, which probably induces membrane damage.

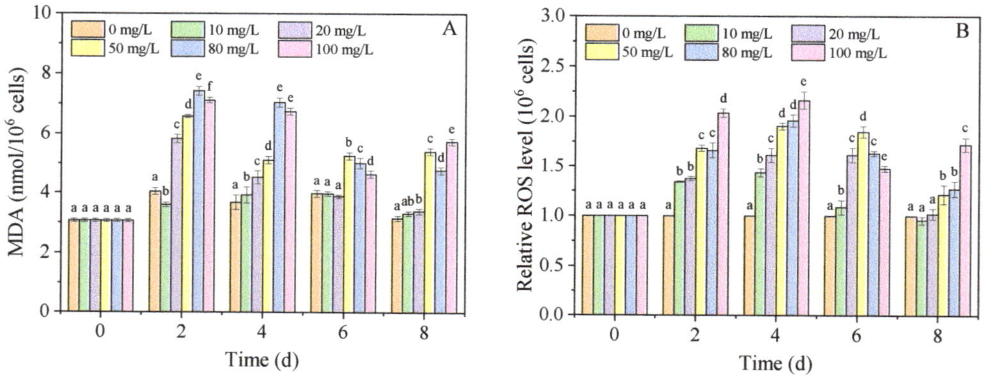

Figure 2. The MDA content (**A**) and relative ROS level (**B**) of *A. carterae* in different concentrations of nano-PS groups. Different letters represent significant differences ($p < 0.05$).

3.3. *SOD Activity and GSH Content*

As important antioxidants, both SOD and GSH can remove ROS in cells to protect cells against oxidative damage [41,44]. To explore the oxidative stress response caused by nano-PS in the algal cells, the SOD activity and GSH content were detected, and the results are displayed in Figures 3A and 3B, respectively. The SOD activity in the nano-PS groups, except the 10 mg/L nano-PS group, was significantly higher than that in the control group on day 2. The SOD activity in all groups gradually decreased, especially from 2 d to 4 d. The rate of decrease was 18.97% in the control group, and 40.6, 43.1, and 29.7 in 50, 80 and 100 mg/L of nano-PS groups, respectively. These changes are most likely because of the inhibition of SOD synthesis caused by the high ROS levels in the test algae cells [45]. In addition, the gap in SOD activity between the nano-PS groups and the control was obviously reduced, thus indicating that the antioxidant capacity of algal cells decreased [46]. Despite the decrease in SOD activity, 100 mg/L of nano-PS stimulated SOD activity at all times.

Similarly, the GSH content of *A. carterae* increased initially and then decreased over the experimental period (Figure 3B). The GSH content in nano-PS groups was significantly higher than that in the control group on days 2–4. Moreover, there was a dose-dependent relationship between the content of GSH and the concentrations of nano-PS; nano-PS at 100 mg/L induced the highest GSH content, which was up to 3.1- and 2.2-fold higher than that of the control group, respectively. In contrast to SOD activity, the GSH content in the high concentration nano-PS groups was still significantly higher than that in the control group from 4 d to 8 d. This can compensate for the decrease in SOD activity caused by nano-PS at high concentrations to reduce ROS levels. The complex variation trend of the SOD activity and GSH content may be caused by the content of antioxidant enzymes induced by nano-PS and the complementary effect between these enzymes [44,47,48]. Overall, although nano-PS induced high ROS levels, the algal cells still had an antioxidant capacity to resist lipid peroxidation throughout the experimental period [42,49].

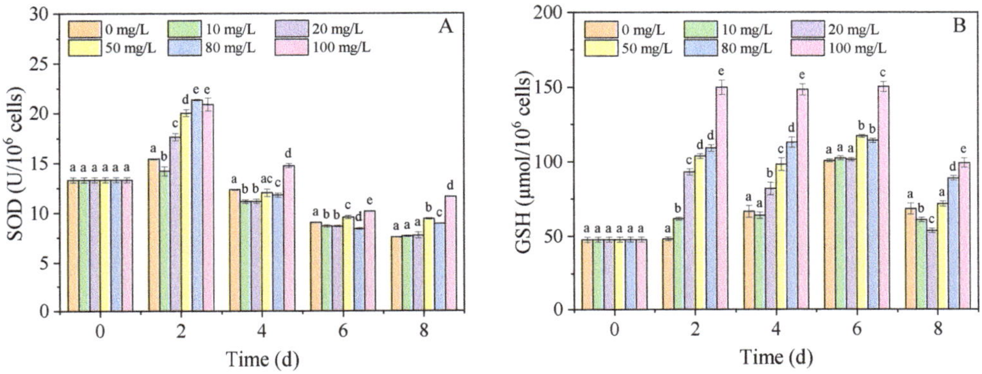

Figure 3. The SOD activity (**A**) and GSH content (**B**) of *A. carterae* in different concentrations of nano-PS groups. Different letters represent significant differences ($p < 0.05$).

3.4. SEM Analysis

As shown in Figure 4, although nano-PS induced membrane oxidative damage, it had no visible effect on size and morphology of the test algal cells compared with those in the control group because of the protection by the cell wall. However, the nano-PS at high concentrations aggregated easily and adsorbed on the surface of the algal cells (Figure 4B,C,E,F). Additionally, nano-PS and algal cells can even form large heterogeneous aggregates (Figure 4D) which sink easily; this aggregation is mainly caused by extracellular polymeric substances [50–52]. In all, the adsorption and aggregation of nano-PS with microalgae could limit the transfer of energy and nutrients, and the motility of the algal cells, thus inhibiting microalgal growth [53].

3.5. Hemolytic Toxins Content

In the present study, the hemolytic activities of extracellular and intracellular toxins were assessed to evaluate the effect of nano-PS on toxin production. Based on the results shown in Figure 5, the contents of extracellular and intracellular hemolytic toxins were higher than those of the control group under the stimulation of nano-PS on day 2. This can be attributed to the stress reaction of algae in adverse environments, in which the oxidative stress of nano-PS enhances the expression of toxic genes [18,23,54,55]. From 4 d to 6 d, the content of extracellular toxin in the nano-PS groups was not significantly different from that in the control group, while the content of intracellular hemolytic toxin in the nano-PS groups decreased, most likely because of the growth inhibition caused by nano-PS. On the last day, the content of intracellular hemolytic toxin in the nano-PS groups increased again. In addition, owing to the cell membrane damage caused by the high ROS level, the content of extracellular hemolytic toxin in the nano-PS groups (10–80 mg/L) also increased [19]. However, the content of extracellular hemolytic toxin in the 100 mg/L nano-PS group was significantly lower than that in the control group because some substances in hemolytic toxins, such as amphidinols, could be adsorbed by nano-PS [15,16]. Based on the above results, a high concentration of nano-PS can affect not only the synthesis of toxins but also the concentration of toxins in algal cells.

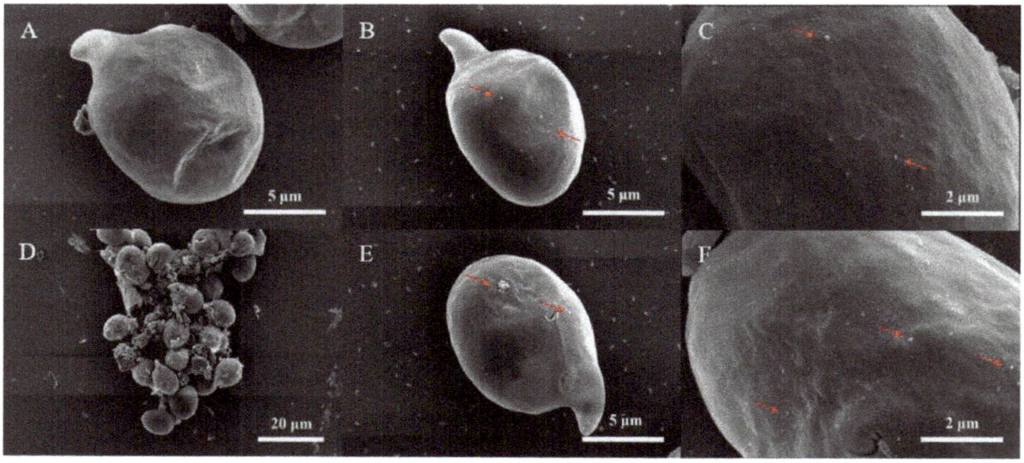

Figure 4. SEM images of *A. carterae* in the control (**A**), 100 mg/L of nano-PS (**B–D**) on day 2, and 100 mg/L of nano-PS (**E**,**F**) on day 8. The arrows point to the nano-PS adhered to the algae cell.

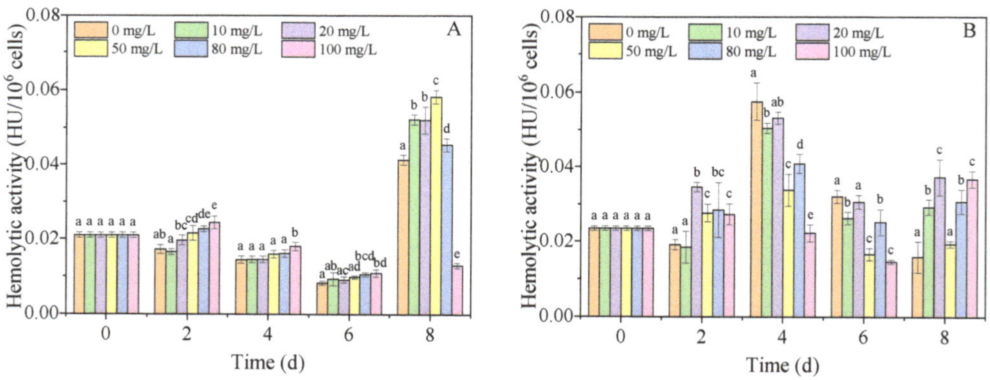

Figure 5. Hemolytic activity of extracellular toxins (**A**) and intracellular toxins (**B**) in different concentrations of nano-PS groups. Different letters represent significant differences ($p < 0.05$).

4. Conclusions

In the present study, the effects of nano-PS (50 nm) on the toxigenic dinoflagellate *A. carterae* were investigated in terms of algal growth, oxidative stress, and hemolysin production. The results showed that the effects of nano-PS on algal cells depended greatly on its concentration and exposure time. Nano-PS significantly inhibited chlorophyll synthesis only on day 4. Moreover, nano-PS at high concentrations inhibited cell growth at all time, while it stimulated first then limited cell growth at low concentrations. Although the antioxidant capacity of algal cells was higher than that in control group, nano-PS especially at high concentrations could still induce high ROS levels and lipid peroxidation, which were the main cause of cell growth inhibition. In addition, nano-PS can affect not only the synthesis of toxins, but also the toxin distribution in and out of algal cells. Considering the coexistence of NPs and harmful algal bloom species, their interaction may have negative consequences for aquatic ecology, thus further affecting the aquaculture industry and posing a potential risk to animals and humans. These findings of this research are valuable for us to understand the effects of NPs on harmful algal bloom species and provide insights into assessing their actual risks to the environment.

Supplementary Materials: The following are available online at https://www.mdpi.com/article/10.3390/nano11102471/s1, Figure S1: Growth curve of *A. carterae*, Figure S2: TEM image of nano-PS, Figure S3: The hemolysis standard curve of digitalis saponin. The value of EC_{50} is 1.53 μg/mL, which means 1.53 μg/mL digitalis saponins is equal to 1 HU, Table S1: Growth inhibition rate (*IR*) of nano-PS on *A. carterae*, Table S2: Inhibition rate of nano-PS on chlorophyll content of *A. carterae*.

Author Contributions: Conceptualization, F.-F.L. and G.-Z.L.; methodology, S.-C.W. and T.-Y.H.; software, S.-C.W. and T.-Y.H.; validation, S.-C.W., F.-F.L., T.-Y.H., Z.-Y.G. and G.-Z.L.; formal analysis, S.-C.W., F.-F.L., Z.-Y.G. and J.-L.F.; investigation, S.-C.W. and T.-Y.H.; resources, S.-C.W. and Z.-Y.G.; data curation, S.-C.W.; writing—original draft preparation, S.-C.W., T.-Y.H. and F.-F.L.; writing—review and editing, F.-F.L., J.-L.F. and G.-Z.L.; visualization, S.-C.W. and F.-F.L.; supervision, F.-F.L. and G.-Z.L.; project administration, F.-F.L. and G.-Z.L.; funding acquisition, F.-F.L., J.-L.F. and G.-Z.L. All authors have read and agreed to the published version of the manuscript.

Funding: This research was funded by the National Natural Science Foundation of China-Shandong Joint Fund (U1906224), the Natural Science Foundation of Shandong Province (ZR2020MD114, ZR2019MEE026), the Key Research and Development Program of Shandong Province (2019JZZY010333), the Young Scholars Program of Shandong University, and Shandong University Interdisciplinary Research and Innovation Team of Young Scholars (2020QNQT20).

Institutional Review Board Statement: Not Applicable.

Informed Consent Statement: Not Applicable.

Data Availability Statement: The datasets generated during and/or analyzed during the current study are available from the corresponding author on reasonable request.

Acknowledgments: We would like to thank Sen Wang, Haiyan Yu, and Xiaomin Zhao from the State Key Laboratory of Microbial Technology, Shandong University, for their assistance with the SEM analysis.

Conflicts of Interest: The authors declare no conflict of interest.

References

1. Da Costa, J.P.; Reis, V.; Paço, A.; Costa, M.; Duarte, A.C.; Rocha-Santos, T. Micro(nano)plastics—Analytical challenges towards risk evaluation. *TrAC Trends Anal. Chem.* **2019**, *111*, 173–184. [CrossRef]
2. Ter Halle, A.; Jeanneau, L.; Martignac, M.; Jardé, E.; Pedrono, B.; Brach, L.; Gigault, J. Nanoplastic in the North Atlantic Subtropical Gyre. *Environ. Sci. Technol.* **2017**, *51*, 13689–13697. [CrossRef]
3. Dawson, A.L.; Kawaguchi, S.; King, C.; Townsend, K.; King, R.; Huston, W.; Nash, S.B. Turning microplastics into nanoplastics through digestive fragmentation by Antarctic krill. *Nat. Commun.* **2018**, *9*, 1–8. [CrossRef]
4. Ferreira, I.; Venâncio, C.; Lopes, I.; Oliveira, M. Nanoplastics and marine organisms: What has been studied? *Environ. Toxicol. Pharmacol.* **2019**, *67*, 1–7. [CrossRef] [PubMed]
5. Guo, X.; Wang, J. The chemical behaviors of microplastics in marine environment: A review. *Mar. Pollut. Bull.* **2019**, *142*, 1–14. [CrossRef]
6. Mattsson, K.; Hansson, L.-A.; Cedervall, T. Nano-plastics in the aquatic environment. *Environ. Sci. Process. Impacts* **2015**, *17*, 1712–1721. [CrossRef] [PubMed]
7. Gambardella, C.; Morgana, S.; Ferrando, S.; Bramini, M.; Piazza, V.; Costa, E.; Garaventa, F.; Faimali, M. Effects of polystyrene microbeads in marine planktonic crustaceans. *Ecotoxicol. Environ. Saf.* **2017**, *145*, 250–257. [CrossRef] [PubMed]
8. González-Fernández, C.; Tallec, K.; Le Goïc, N.; Lambert, C.; Soudant, P.; Huvet, A.; Suquet, M.; Berchel, M.; Paul-Pont, I. Cellular responses of Pacific oyster (*Crassostrea gigas*) gametes exposed in vitro to polystyrene nanoparticles. *Chemosphere* **2018**, *208*, 764–772. [CrossRef]
9. Pinsino, A.; Bergami, E.; Della Torre, C.; Vannuccini, M.L.; Addis, P.; Secci, M.; Dawson, K.A.; Matranga, V.; Corsi, I. Amino-modified polystyrene nanoparticles affect signalling pathways of the sea urchin (*Paracentrotus lividus*) embryos. *Nanotoxicology* **2017**, *11*, 201–209. [CrossRef]
10. Mao, Y.; Ai, H.; Chen, Y.; Zhang, Z.; Zeng, P.; Kang, L.; Li, W.; Gu, W.; He, Q.; Li, H. Phytoplankton response to polystyrene microplastics: Perspective from an entire growth period. *Chemosphere* **2018**, *208*, 59–68. [CrossRef]
11. Zhang, C.; Chen, X.; Wang, J.; Tan, L. Toxic effects of microplastic on marine microalgae Skeletonema costatum: Interactions between microplastic and algae. *Environ. Pollut.* **2017**, *220*, 1282–1288. [CrossRef]

12. Sjollema, S.B.; Redondo-Hasselerharm, P.; Leslie, H.A.; Kraak, M.H.S.; Vethaak, A.D. Do plastic particles affect microalgal photosynthesis and growth? *Aquat. Toxicol.* **2016**, *170*, 259–261. [CrossRef]
13. Besseling, E.; Wang, B.; Lürling, M.; Koelmans, A. Nanoplastic Affects Growth of S. obliquus and Reproduction of D. magna. *Environ. Sci. Technol.* **2014**, *48*, 12336–12343. [CrossRef] [PubMed]
14. González-Fernández, C.; Toullec, J.; Lambert, C.; Le Goïc, N.; Seoane, M.; Moriceau, B.; Huvet, A.; Berchel, M.; Vincent, D.; Courcot, L.; et al. Do transparent exopolymeric particles (TEP) affect the toxicity of nanoplastics on Chaetoceros neogracile? *Environ. Pollut.* **2019**, *250*, 873–882. [CrossRef] [PubMed]
15. Martínez, K.A.; Lauritano, C.; Druka, D.; Romano, G.; Grohmann, T.; Jaspars, M.; Martín, J.; Díaz, C.; Cautain, B.; De La Cruz, M.; et al. Amphidinol 22, a New Cytotoxic and Antifungal Amphidinol from the Dinoflagellate Amphidinium carterae. *Mar. Drugs* **2019**, *17*, 385. [CrossRef] [PubMed]
16. Li, J.; Zhang, K.; Zhang, H. Adsorption of antibiotics on microplastics. *Environ. Pollut.* **2018**, *237*, 460–467. [CrossRef]
17. Zheng, X.; Zhang, W.; Yuan, Y.; Li, Y.; Liu, X.; Wang, X.; Fan, Z. Growth inhibition, toxin production and oxidative stress caused by three microplastics in Microcystis aeruginosa. *Ecotoxicol. Environ. Saf.* **2021**, *208*, 111575. [CrossRef] [PubMed]
18. Wu, D.; Wang, T.; Wang, J.; Jiang, L.; Yin, Y.; Guo, H. Size-dependent toxic effects of polystyrene microplastic exposure on Microcystis aeruginosa growth and microcystin production. *Sci. Total Environ.* **2021**, *761*, 143265. [CrossRef]
19. Feng, L.-J.; Sun, X.-D.; Zhu, F.-P.; Feng, Y.; Duan, J.-L.; Xiao, F.; Li, X.-Y.; Shi, Y.; Wang, Q.; Sun, J.-W.; et al. Nanoplastics Promote Microcystin Synthesis and Release from Cyanobacterial Microcystis aeruginosa. *Environ. Sci. Technol.* **2020**, *54*, 3386–3394. [CrossRef]
20. Baig, H.; Saifullah, S.; Dar, A. Occurrence and toxicity of Amphidinium carterae Hulburt in the North Arabian Sea. *Harmful Algae* **2006**, *5*, 133–140. [CrossRef]
21. Wang, L.; Chen, G.-F.; Zhang, C.-Y.; Wang, Y.-Y.; Sun, R. Rapid and sensitive detection of Amphidinium carterae by loop-mediated isothermal amplification combined with a chromatographic lateral-flow dipstick. *Mol. Cell. Probes* **2019**, *43*, 72–79. [CrossRef]
22. Yasumoto, T. Fish Poisoning Due to Toxins of Microalgal Origins in the Pacific. *Toxicon* **1998**, *36*, 1515–1518. [CrossRef]
23. Yasumoto, T. The chemistry and biological function of natural marine toxins. *Chem. Rec.* **2001**, *1*, 228–242. [CrossRef]
24. Guerrini, F.; Ciminiello, P.; Dell'Aversano, C.; Tartaglione, L.; Fattorusso, E.; Boni, L.; Pistocchi, R. Influence of temperature, salinity and nutrient limitation on yessotoxin production and release by the dinoflagellate Protoceratium reticulatum in batch-cultures. *Harmful Algae* **2007**, *6*, 707–717. [CrossRef]
25. Valenti, T.W., Jr.; James, S.V.; Lahousse, M.J.; Schug, K.A.; Roelke, D.L.; Grover, J.P.; Brooks, B.W. A mechanistic explanation for pH-dependent ambient aquatic toxicity of Prymnesium parvum carter. *Toxicon* **2010**, *55*, 990–998. [CrossRef] [PubMed]
26. Zimmermann, L.A. *Environmental Regulation of Toxin Producting: Comparison of Homlytic Activity of Amphidinium Carterae and Amphidinium Klebsii*; University of North Carolina Wilmington: Wilmington, NC, USA, 2006.
27. Zhang, Q.; Qu, Q.; Lu, T.; Ke, M.; Zhu, Y.; Zhang, M.; Zhang, Z.; Du, B.; Pan, X.; Sun, L.; et al. The combined toxicity effect of nanoplastics and glyphosate on Microcystis aeruginosa growth. *Environ. Pollut.* **2018**, *243*, 1106–1112. [CrossRef]
28. Arnon, D.I. Copper enzymes in isolated chloroplasts. Polyphenoloxidase in Beta vulgaris. *Plant Physiol.* **1949**, *24*, 1–15. [CrossRef]
29. Gu, S.; Zheng, H.; Xu, Q.; Sun, C.; Shi, M.; Wang, Z.; Li, F. Comparative toxicity of the plasticizer dibutyl phthalate to two freshwater algae. *Aquat. Toxicol.* **2017**, *191*, 122–130. [CrossRef]
30. Liu, Q.; Tang, X.; Wang, Y.; Yang, Y.; Zhang, W.; Zhao, Y.; Zhang, X. ROS changes are responsible for tributyl phosphate (TBP)-induced toxicity in the alga Phaeodactylum tricornutum. *Aquat. Toxicol.* **2019**, *208*, 168–178. [CrossRef] [PubMed]
31. Nayak, B.B.; Karunasagar, I. Influence of bacteria on growth and hemolysin production by the marine dinoflagellate Amphidinium carterae. *Mar. Biol.* **1997**, *130*, 35–39. [CrossRef]
32. Pagliara, P.; Caroppo, C. Toxicity assessment of Amphidinium carterae, Coolia cfr. monotis and Ostreopsis cfr. ovata (Dinophyta) isolated from the northern Ionian Sea (Mediterranean Sea). *Toxicon* **2012**, *60*, 1203–1214. [CrossRef]
33. Eschbach, E.; John, U.; Medlin, L.K. Improved erythrocyte lysis assay in microtitre plates for sensitive detection and efficient measurement of haemolytic compounds from ichthyotoxic algae. *J. Appl. Toxicol.* **2001**, *21*, 513–519. [CrossRef] [PubMed]
34. Ling, C.; Trick, C.G. Expression and standardized measurement of hemolytic activity in Heterosigma akashiwo. *Harmful Algae* **2010**, *9*, 522–529. [CrossRef]
35. Zhu, Z.-L.; Wang, S.-C.; Zhao, F.-F.; Wang, S.-G.; Liu, F.-F.; Liu, G.-Z. Joint toxicity of microplastics with triclosan to marine microalgae Skeletonema costatum. *Environ. Pollut.* **2019**, *246*, 509–517. [CrossRef] [PubMed]
36. Li, S.; Wang, P.; Zhang, C.; Zhou, X.; Yin, Z.; Hu, T.; Hu, D.; Liu, C.; Zhu, L. Influence of polystyrene microplastics on the growth, photosynthetic efficiency and aggregation of freshwater microalgae Chlamydomonas reinhardtii. *Sci. Total Environ.* **2020**, *714*, 136767. [CrossRef]
37. Falkowski, P.G.; Owens, T.G. Light-shade adaptation: Two strategies in marine phytoplankton. *Plant Physiol.* **1980**, *66*, 592–595. [CrossRef] [PubMed]
38. Falkowski, P.G.; Owens, T.G.; Ley, A.C.; Mauzerall, D.C. Effects of Growth Irradiance Levels on the Ratio of Reaction Centers in Two Species of Marine Phytoplankton. *Plant Physiol.* **1981**, *68*, 969–973. [CrossRef]
39. Wu, Y.; Guo, P.; Zhang, X.; Zhang, Y.; Xie, S.; Deng, J. Effect of microplastics exposure on the photosynthesis system of freshwater algae. *J. Hazard. Mater.* **2019**, *374*, 219–227. [CrossRef]
40. Li, F.M.; Liang, Z.; Zheng, X.; Zhao, W.; Wu, M.; Wang, Z.Y. Toxicity of nano-TiO2 on algae and the site of reactive oxygen species production. *Aquat. Toxicol.* **2015**, *158*, 1–13. [CrossRef]

41. Xia, B.; Chen, B.; Sun, X.; Qu, K.; Ma, F.; Du, M. Interaction of TiO_2 nanoparticles with the marine microalga Nitzschia closterium: Growth inhibition, oxidative stress and internalization. *Sci. Total Environ.* **2015**, *508*, 525–533. [CrossRef]
42. Sabatini, S.E.; Juárez, Á.B.; Eppis, M.R.; Bianchi, L.; Luquet, C.M.; Molina, M.D.C. Oxidative stress and antioxidant defenses in two green microalgae exposed to copper. *Ecotoxicol. Environ. Saf.* **2009**, *72*, 1200–1206. [CrossRef]
43. Hazeem, L.J.; Yesilay, G.; Bououdina, M.; Perna, S.; Cetin, D.; Suludere, Z.; Barras, A.; Boukherroub, R. Investigation of the toxic effects of different polystyrene micro-and nanoplastics on microalgae Chlorella vulgaris by analysis of cell viability, pigment content, oxidative stress and ultrastructural changes. *Mar. Pollut. Bull.* **2020**, *156*, 111278. [CrossRef]
44. Cirulis, J.T.; Scott, J.A.; Ross, G.M. Management of oxidative stress by microalgae. *Can. J. Physiol. Pharmacol.* **2013**, *91*, 15–21. [CrossRef] [PubMed]
45. Hong, Y.; Hu, H.-Y.; Xie, X.; Li, F.-M. Responses of enzymatic antioxidants and non-enzymatic antioxidants in the cyanobacterium Microcystis aeruginosa to the allelochemical ethyl 2-methyl acetoacetate (EMA) isolated from reed (*Phragmites communis*). *J. Plant Physiol.* **2008**, *165*, 1264–1273. [CrossRef] [PubMed]
46. Liu, F.; Pang, S.J. Stress tolerance and antioxidant enzymatic activities in the metabolisms of the reactive oxygen species in two intertidal red algae Grateloupia turuturu and Palmaria palmata. *J. Exp. Mar. Biol. Ecol.* **2010**, *382*, 82–87. [CrossRef]
47. Li, M.; Hu, C.; Zhu, Q.; Chen, L.; Kong, Z.; Liu, Z. Copper and zinc induction of lipid peroxidation and effects on antioxidant enzyme activities in the microalga Pavlova viridis (Prymnesiophyceae). *Chemosphere* **2006**, *62*, 565–572. [CrossRef]
48. Deng, X.-Y.; Cheng, J.; Hu, X.-L.; Wang, L.; Li, D.; Gao, K. Biological effects of TiO2 and CeO2 nanoparticles on the growth, photosynthetic activity, and cellular components of a marine diatom Phaeodactylum tricornutum. *Sci. Total Environ.* **2017**, *575*, 87–96. [CrossRef] [PubMed]
49. Li, X.; Yang, W.L.; He, H.; Wu, S.; Zhou, Q.; Yang, C.; Zeng, G.; Luo, L.; Lou, W. Responses of microalgae Coelastrella sp. to stress of cupric ions in treatment of anaerobically digested swine wastewater. *Bioresour. Technol.* **2018**, *251*, 274–279. [CrossRef]
50. Bergami, E.; Pugnalini, S.; Vannuccini, M.; Manfra, L.; Faleri, C.; Savorelli, F.; Dawson, K.A.; Corsi, I. Long-term toxicity of surface-charged polystyrene nanoplastics to marine planktonic species Dunaliella tertiolecta and Artemia franciscana. *Aquat. Toxicol.* **2017**, *189*, 159–169. [CrossRef]
51. Song, C.; Liu, Z.; Wang, C.; Li, S.; Kitamura, Y. Different interaction performance between microplastics and microalgae: The bio-elimination potential of Chlorella sp. L38 and Phaeodactylum tricornutum MASCC-0025. *Sci. Total Environ.* **2020**, *723*, 138146. [CrossRef] [PubMed]
52. Li, Z.; Yi, X.; Zhou, H.; Chi, T.; Li, W.; Yang, K. Combined effect of polystyrene microplastics and dibutyl phthalate on the microalgae Chlorella pyrenoidosa. *Environ. Pollut.* **2020**, *257*, 113604. [CrossRef] [PubMed]
53. Bhattacharya, P.; Lin, S.; Turner, J.P.; Ke, P.C. Physical Adsorption of Charged Plastic Nanoparticles Affects Algal Photosynthesis. *J. Phys. Chem. C* **2010**, *114*, 16556–16561. [CrossRef]
54. Shao, J.; Wu, Z.; Yu, G.; Peng, X.; Li, R. Allelopathic mechanism of pyrogallol to Microcystis aeruginosa PCC7806 (Cyanobacteria): From views of gene expression and antioxidant system. *Chemosphere* **2009**, *75*, 924–928. [CrossRef] [PubMed]
55. Qian, H.; Yu, S.; Sun, Z.; Xie, X.; Liu, W.; Fu, Z. Effects of copper sulfate, hydrogen peroxide and N-phenyl-2-naphthylamine on oxidative stress and the expression of genes involved photosynthesis and microcystin disposition in Microcystis aeruginosa. *Aquat. Toxicol.* **2010**, *99*, 405–412. [CrossRef]

MDPI
St. Alban-Anlage 66
4052 Basel
Switzerland
www.mdpi.com

Nanomaterials Editorial Office
E-mail: nanomaterials@mdpi.com
www.mdpi.com/journal/nanomaterials

www.ingramcontent.com/pod-product-compliance
Lightning Source LLC
Chambersburg PA
CBHW061602210926
51850CB00009B/2183
* 9 7 8 3 0 3 6 5 9 9 7 8 6 *